TRANSPORTATION!

DK SMITHSONIAN ✹

TRANSPORTATION!

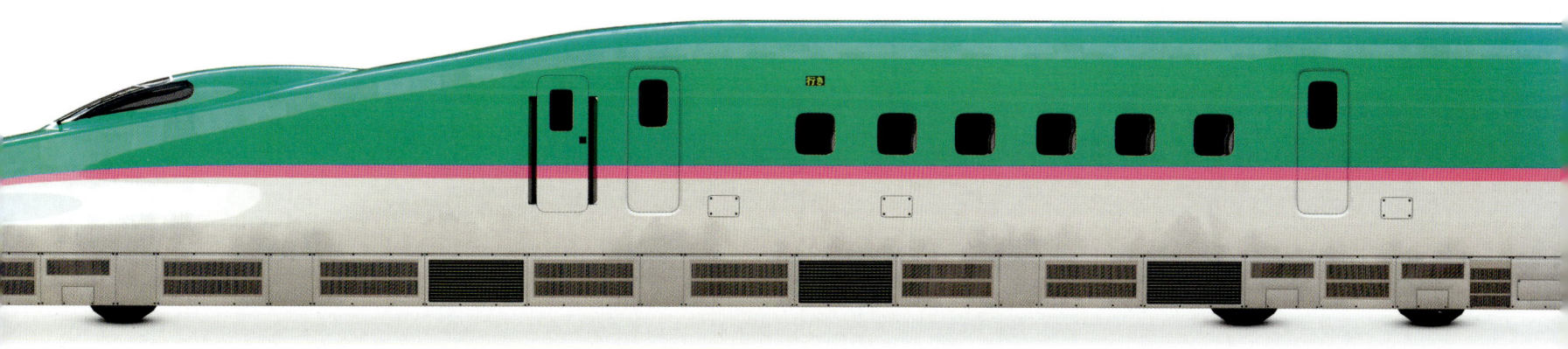

Senior Editors Sam Kennedy, Anna Streiffert Limerick
Senior Art Editor Stefan Podhorodecki
Project Editors Edward Pearce, Vicky Richards
Editor Binta Jallow
Senior US Editor Kayla Dugger
Executive US Editor Lori Cates Hand
Designers Kelly Adams, Mik Gates, Beth Johnston, Kit Lane
Jacket Design Development Manager Sophia MTT
Jacket Designer Tanya Mehrotra
DTP Designer Rakesh Kumar
Senior Jackets Coordinator Priyanka Sharma Saddi
DK Media Archive Romaine Werblow
Picture Research Geetam Biswas, Myriam Megharbi
Production Editor Gillian Reid
Production Controller Poppy David
Managing Editor Francesca Baines
Managing Art Editor Philip Letsu
Publisher Andrew Macintyre
Associate Publishing Director Liz Wheeler
Art Director Karen Self
Publishing Director Jonathan Metcalf

Written by Ian Fitzgerald, Clive Gifford,
Giles Sparrow, Giles Chapman

Consultant Roger Bridgman

Illustrators and Render Artists
Adam Benton, Peter Bull, Jason Harding, Richard Chasemore,
Jon@KJA, Chris@KJA, Stuart Jackson-Carter–SJC Illustration,
Simon Mumford, Tony Randazzo, Simon Tegg, Jack Williams

First American Edition, 2024
Published in the United States by DK Publishing
1745 Broadway, 20th Floor, New York, NY 10019

Copyright © 2024 Dorling Kindersley Limited
DK, a Division of Penguin Random House LLC
24 25 26 27 28 10 9 8 7 6 5 4 3 2
008-331873-May/2024

A catalog record for this book
is available from the Library of Congress.
ISBN 978-0-7440-9874-7

Printed and bound in China
www.dk.com

THE SMITHSONIAN
This trademark is owned by the Smithsonian Institution
and is registered in the U.S. Patent and Trademark Office.

MIX
Paper | Supporting
responsible forestry
FSC™ C018179

This book was made with Forest
Stewardship Council™ certified
paper—one small step in DK's
commitment to a sustainable future.
Learn more at
www.dk.com/uk/information/sustainability

CONTENTS

LAND

AIR AND SPACE

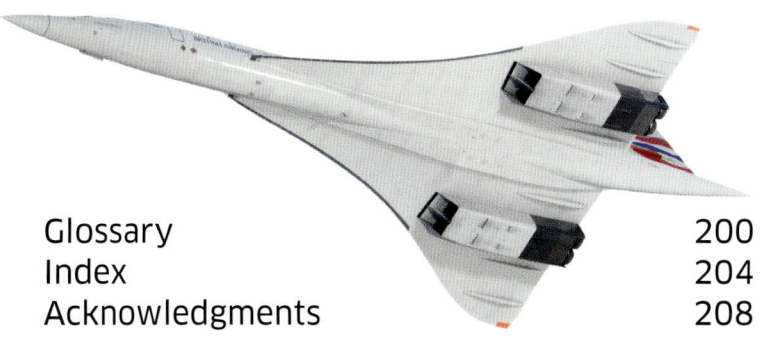

Scales and sizes

The data box for each vehicle or craft includes a scale drawing to indicate its size. These size comparisons are based on the height of an average adult human male, a school bus, or a Boeing 747 passenger airplane.

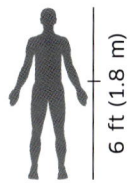

6 ft (1.8 m)

3½ ft (11 m)

251 ft (76.5 m)

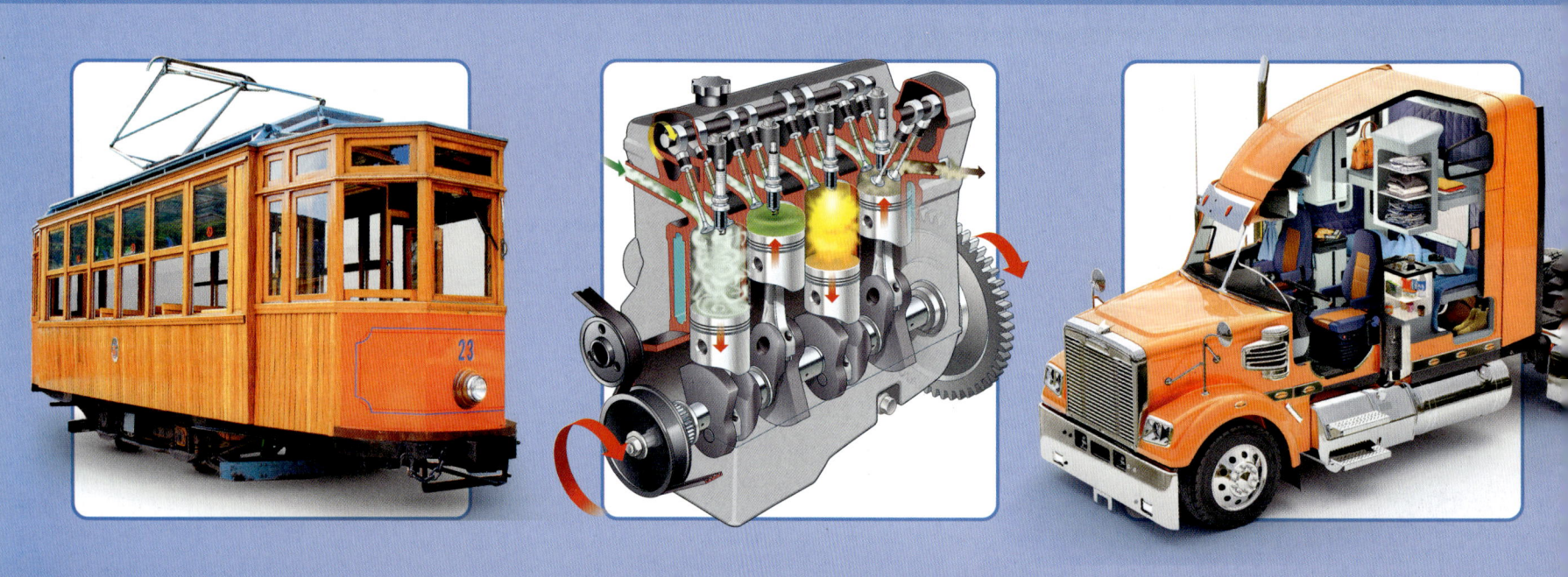

LAND

Today, there are more ways to travel across land than ever before. A range of vehicles exist to move people and their goods quickly and safely, from public transportation—such as buses and high-speed trains—to powerful construction vehicles, driverless cars, and nifty motorcycles.

8   land ∘ **LAND TRANSPORTATION THROUGH TIME**

85 million The **number of skate-boarders** in the world.

Cars become bigger
In the late 20th century, larger cars came into fashion, such as trucks and SUVs (sport utility vehicles).

TOYOTA 4RUNNER, 1998

MODERN ERA

Modern era
In recent decades, vehicles have smashed records and pushed new boundaries. Methods of travel that use AI or green technology are on the brink of being realized.

THRUST SSC

Land speed record
British-designed car Thrust SSC set a new record in 1997, with a speed of 763 mph (211.9 kph).

SHINKANSEN E5, 2011

Bullet trains
The first high-speed rail service using "bullet trains" opened in Japan in 1964, traveling between Tokyo and Osaka.

Smallest car
With three wheels and space for just one person, the Peel P50 became the smallest road-going car.

PEEL P50, 1962

First Formula One
The first Formula One World Championship, in 1950, was won by a team driving the Alfa Romeo 158, a race car originally built in 1938.

ALFA ROMEO 158

Land transportation through time

Ever since the discovery of the wheel, humans have invented new vehicles that would allow them to travel across vast stretches of land at faster and faster speeds.

People have always needed to move both themselves and goods between settlements—first by using simple horse-drawn carts and wagons, then moving to powered vehicles. Many advances in vehicle design happened in the 18th and 19th centuries, when sleek, powerful cars and trains were developed—the forerunners of those we use today. In recent years, manufacturers have focused on making vehicles not only technologically smarter, but with a less harmful impact on the environment.

Timeline of land transportation
This timeline shows some of the key advances in land travel throughout human history, featuring cars, trains, and other wheeled vehicles that changed the world.

BENZ PATENT MOTOR CAR, 1885

20TH CENTURY

Early 20th century
By the turn of the century, a range of powered vehicles had been developed. Engineers soon found ways to improve their efficiency and adapt them for more uses.

Early car
German engineer Karl Benz designed one of the first cars that ran using an internal combustion engine.

METROPOLITAN 1

First underground line
Powered by steam engines, the Metropolitan line in London, UK, became the first underground railway to run through a major city, in 1863.

EARLY TRAVEL

Early travel
Humans' first attempts at finding better ways to travel began more than 5,000 years ago. The earliest simple wheeled vehicles gradually became more complex.

Invention of the wheel
The wheel, invented in around 3500 BCE in Mesopotamia, was originally used in making pottery but was soon adapted for use in transportation.

Four-legged transportation
Riding on horses, first domesticated in Central Asia, allowed humans to get places much faster than walking.

ROMAN CHARIOT

Chariots dominate
In the ancient Mediterranean empires, horse-drawn two-wheeled chariots were the dominant military vehicle.

AROUND 3500 BCE   AROUND 1700–500 BCE

34 miles (54.7 km)–the length of the **Prague pneumatic post system,** which used air to transport letters through long tubes.

2001 The year the Segway was first launched–a motorized, two-wheeled personal transporter.

9

NISSAN LEAF

Rise of electric cars
The first mass-market electric vehicle, the Nissan Leaf (2010) was powered by a lithium-ion battery.

ELECTRIC UNICYCLE

Going electric
More transportation began to use electricity, such as ebikes, electric scooters, and even electric unicycles.

Self-driving cars
In 2012, a Google driverless car passes a driving test on its own.

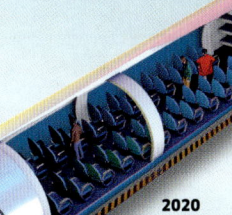

Testing future travel
The first passenger test takes place on the proposed Virgin hyperloop–a transport method that propels pods through a vacuum tube.

2020

WLA, 1942

Motorcycles gain popularity
Originally made during the war, the Harley-Davidson WLA was produced until the 1950s and became popular with civilians.

After World War II
After the end of the war, vehicle designs changed rapidly. Cars and motorcycles became more stylish and more affordable, and trains were soon able to travel at incredible speeds.

POST WWII

BIRMINGHAM AIRPORT MAGLEV SHUTTLE, UK (1984–1995)

Maglev trains
German Hermann Kemper patented his idea in 1937, but it took time before magnetic levitation trains took off.

Longest railway tunnel
At 12¼ miles (19.8 km) long, the Simplon Tunnel linking Switzerland and Italy was the longest in the world when it opened in 1906.

FORD MODEL T, 1913

Mass production begins
American industrialist Henry Ford was the first to build cars using a moving assembly line.

Early tanks
The British Mark I tank was the first tank to see combat when deployed on the battlefields of World War I in 1916.

First golf cart
The earliest electric golf carts were developed in the US in the 1930s, but only became popular later in the century.

Early bike
Designed by German inventor Karl von Drais in 1817, the *laufmaschine* (running machine), also called the "dandy horse," was one of the earliest bicycles.

First trolley
In 1807, the Swansea and Mumbles Railway in Wales, UK, was the first to open a trolley service, with trolley cars pulled by horses. They soon sprang up elsewhere.

Roman roads
Although some built-up roads already existed, the Roman Empire was the first to build vast networks of paved roads across Europe, West Asia, and Northern Africa.

Compass
The magnetic compass, first invented in China, was adapted to help people navigate across both land and sea.

INDUSTRIAL ERA

Industrial era
From the middle of the 18th century, industry began to advance rapidly, leading to new technologies that would power the earliest trains and cars.

1804 steam locomotive
Designed by British engineer Richard Trevithick, the first steam locomotive was powered by burning coal and ran on iron rails.

10   land • MOVING ACROSS LAND

1,270 miles (2,044 km)—the length of the **Beringia-92**, the longest dog sled race ever, held in Russia in 1992.

Moving across land

To travel anything more than the shortest distance over land, people have used wheeled vehicles and roads to make their journeys faster and easier.

The earliest transportation methods involved dragging people or goods from place to place, a slow and inefficient process. After the invention of the wheel, the friction that made traveling over land so difficult was less of a problem, but smooth surfaces had become a necessity. Today, large road networks crossing chasms and bridging rivers and advanced wheel technology make traveling over land easier and than ever before.

ANIMAL POWER

For thousands of years, humans have used the powerful muscles and steady endurance of animals for land transport tasks such as carrying heavy loads, pulling wagons and other vehicles, and hauling farming tools. Horses are the main means of fast travel in temperate climates, but many other types of animals carry out important tasks across the globe.

Plow puller
Before mechanized farming machines, oxen hauled tools. They still do in many places, such as Cambodia, where this farmer is using them to pull a plow.

Llama loads
High up in the South American Andes Mountains, sturdy llamas are the most efficient cargo carriers. Their thick coats keep them warm as they move quickly across the rugged terrain.

Dog power
Sleds pulled by dog teams provide fast, smooth rides across snow, where heavier animals would sink. Today, huskies like these help with reindeer herding and compete in races.

FIGHTING FRICTION

No matter how smooth an object might seem, it is actually covered with thousands of tiny dents and bumps. When two objects rub together, these bumps and dents snag on each other and slow the objects down. This slowing force is called friction.

Sliding friction
Dragging an object creates a lot of friction because two large surfaces are rubbing against each other at the same time. Lots of energy is needed to overcome the friction and start the object in motion.

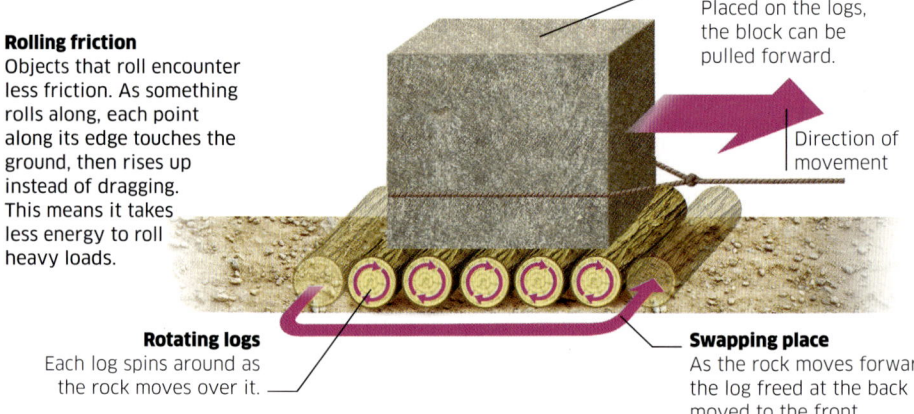

Stone block
The weight of the stone creates a downward force.

Direction of movement

Rope for pulling

Rolling friction
Objects that roll encounter less friction. As something rolls along, each point along its edge touches the ground, then rises up instead of dragging. This means it takes less energy to roll heavy loads.

Rolling rock
Placed on the logs, the block can be pulled forward.

Direction of movement

Rotating logs
Each log spins around as the rock moves over it.

Swapping place
As the rock moves forward, the log freed at the back is moved to the front.

INVENTING THE WHEEL

The first wheel was not invented to make a vehicle roll. Humans started to use wheels in Mesopotamia around 5,500 years ago to make pots from clay. But at some point, it was turned from a horizontal orientation to a vertical one. People found that when two wheels arranged vertically were attached to each other by an axle, they could be used to carry heavy loads.

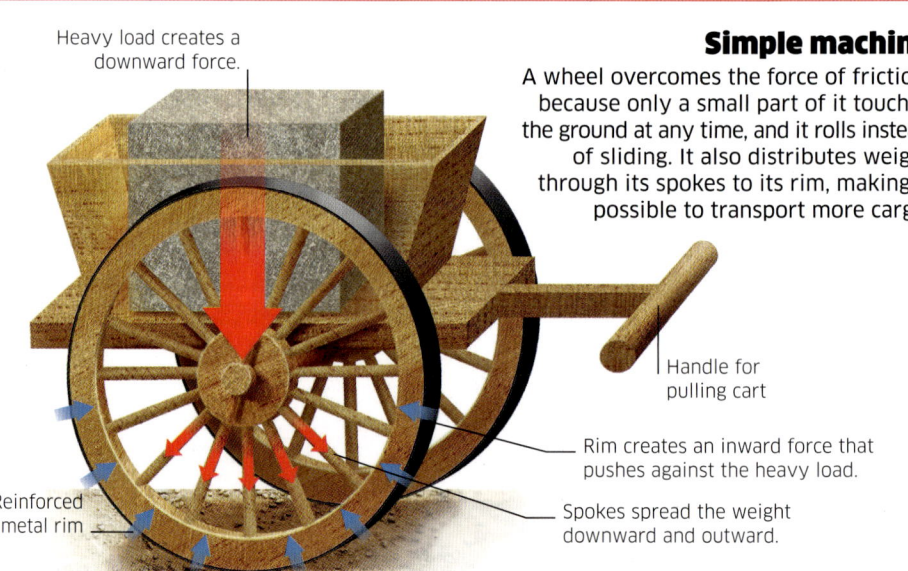

Heavy load creates a downward force.

Simple machine
A wheel overcomes the force of friction, because only a small part of it touches the ground at any time, and it rolls instead of sliding. It also distributes weight through its spokes to its rim, making it possible to transport more cargo.

Handle for pulling cart

Rim creates an inward force that pushes against the heavy load.

Spokes spread the weight downward and outward.

Reinforced metal rim

1839 The year the **first pneumatic tire** was **invented** for **horse-drawn carriages**, by Scotsman **Robert Thompson**.

102.4 miles (164.8 km)—the **length** of the **Danyang-Kunshan Grand Bridge** in China, the **world's longest** bridge.

11

ROLLING IMPROVEMENT

The first wheels were solid discs, initially made of stone, then of circular slices of wood. But they were heavy and people soon started working out ways to make them lighter, stronger, and less bumpy.

1 Solid Wheel
The first wheels, made around 3500 BCE, were solid discs pierced by an axle. Later wheels were made from thick wooden planks held together with sturdy pegs.

2 Wooden spokes
Ancient Egyptians cut out parts of their wheels to make them lighter and faster. The Greeks continued this trend, removing more wood and adding spokes for strength.

3 Metal tires
From about 500 BCE, wooden wheels were reinforced by metal tires around the rim that helped them last longer and travel over rougher terrain.

4 Rubber tires
First used from the 1870s on bicycles, air-filled rubber tires help absorb shocks from bumpy roads and make the wheel lighter.

4 Car wheel
Today's cars have wheels made from alloys—a mix of metals such as aluminum and magnesium—that make wheels both lighter and stronger.

Axle evolution

Axles do more than just hold two wheels together. When an axle connected to a wheel rotates, the rim of the wheel also turns but covers a far greater distance. A bigger wheel pushes the vehicle farther for each turn of the axle but also requires a bigger force to turn it.

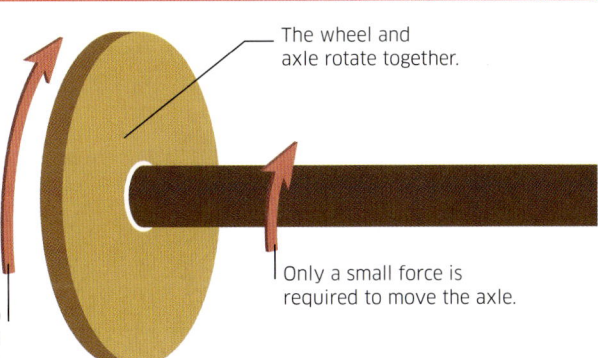

The wheel and axle rotate together.

Only a small force is required to move the axle.

The force applied to the axle is magnified at the wheel's rim.

Powered axles

The axles of a car are turned by the engine through gears (see page 19). The gears are changed as necessary to ensure that the engine always runs at a speed that makes it efficient. In two-wheel-drive cars, only one of the axles is powered, but in a four-wheel-drive car, both front and rear axles receive power from the engine, boosting traction on slippery surfaces.

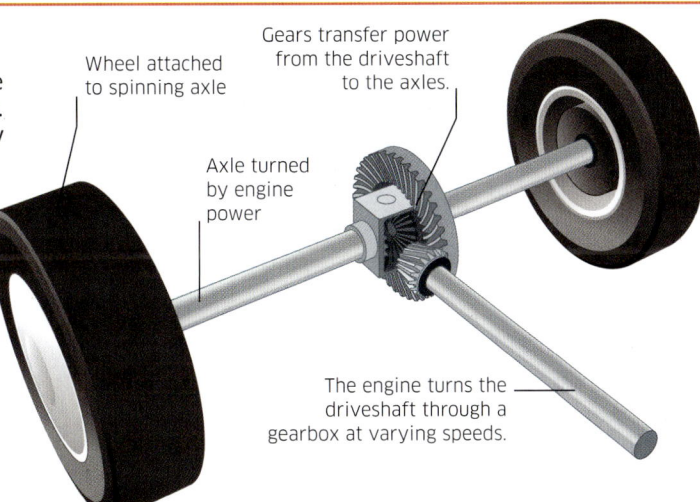

Gears transfer power from the driveshaft to the axles.

Wheel attached to spinning axle

Axle turned by engine power

The engine turns the driveshaft through a gearbox at varying speeds.

CROSSING OBSTACLES

As well as roads, a range of additional structures are needed to connect far-flung places. Bridges span gaps between areas of solid ground, while tunnels avoid steep climbs up mountainsides. Large junctions connect roads going in different directions.

Bridges

Bridges usually connect two banks of a river or land separated by larger waterways, such as a strait. Some bridges carry vehicles across gorges or uneven terrain, such as the Millau Viaduct in France.

The Millau Viaduct, France

Tunnels

When mountains are too high to drive over or too large to drive around, tunnels are drilled through the rock. Some tunnels run beneath water, and others under cities to reduce street-level traffic.

Zion-Mount Carmel Tunnel, Utah

Junctions

Simple crosswalks work for smaller roads, but modern multilane highways are designed with complex junctions with many entries and exits so that cars coming from different directions can join.

Bridge junction in Shanghai, China

Early land transport

For most of human history, people have relied on muscle power—their own or that of animals—to get from one place to another.

For thousands of years, from the first civilizations until the 20th century, few people ever traveled more than a short distance from their homes. When someone did go on a long journey, they walked, rode an animal, or traveled in a cart or coach pulled by a draft animal such as a horse. The invention of steam engines in the 19th century finally made it possible to travel in vehicles that could move without the aid of muscle power.

Charioteer
This man steers the horses and tells them how fast to go.

WHEELBARROW
Rolling transport
Origin: China
Date: 221 BCE–256 CE

The wheelbarrow was first invented in China as a way to transport a heavy load of goods or even people. Its Chinese name means "wooden ox."

Wooden handles
The longer the handles of a wheelbarrow, the less strength is needed to lift and push it.

CHARIOT
Lightweight two-wheeler
Origin: Mesopotamia
Date: 3000 BCE

With the invention of the spoked wheel, it became possible to make fast, light, two-wheeled vehicles that could be pulled along by horses.

Harness keeps horses attached to the chariot

CAMEL
Desert strider
Origin: Arabia
Date: 2200 BCE

Sometimes called "the ships of the desert," camels were favored by long-distance traders because they were capable of carrying loads great distances over harsh landscapes.

Broad toes keep the camel from sinking into the sand

HORSE
Mounted movement
Origin: Central Asia
Date: 2000 BCE

Riding on horseback allows people to travel far greater distances at a much faster speed than traveling on foot. This horse and rider come from Mongolia, where horses are small and strong.

Bracket

SEDAN CHAIR
People carrier
Origin: Croatia
Size: 18th century

For those wealthy enough to afford it, the sedan chair made it possible to travel without taking a step. Two poles were slipped through four brackets on either side of the box, which were then lifted off the ground by strong men.

30 minutes—the time it took a **steam car** to get going in cold **weather**.

5,000 The number of chariots that clashed at the **Battle of Kadesh**, Syria, in **1274 BCE**. It may have been the **largest chariot battle** ever.

13

ROYAL CARRIAGE
Status symbol
Origin: Brazil
Date: 19th century

Covered with intricate carvings and painted with a coat of arms, this coach belonged to the Brazilian emperors. The elaborate decoration reflected the imperial family's wealth and magnificence.

SLED
Horse-drawn winter ride
Origin: Netherlands
Date: 1880s

In wintery conditions, the sleek runners of a sled often provide a faster, smoother ride than wheeled vehicles, which can get stuck in snow or slip on ice. This sled was lined with fur for a cozier ride.

Driver's seat

GRENVILLE STEAM CARRIAGE
Powered road transport
Origin: UK
Date: 1880

This tricycle was powered by a steam engine and steered using the front wheel. Steam cars were initially more popular than gas ones, but as the latter improved, steam could not compete.

Boiler

STROLLER
Transport for children
Origin: UK
Date: 1773

The world's first stroller was built for the British Duke of Devonshire's children. The gilded carriage could be pulled along by a pony, dog, or goat. The baby's seat rested on suspended springs for a smoother ride.

Flourishes
The stroller curled back on itself like a wave.

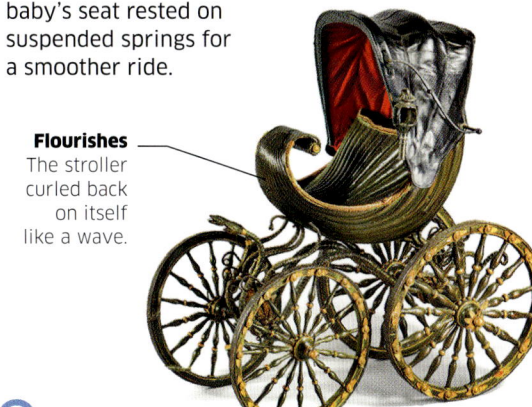

READING CARAVAN
Home on wheels
Origin: UK
Date: 1880s

Named for the British town in which many were built, these horse-drawn homes contained space for a bed and kitchen. They were popular among Romany Gypsies, who traveled between farm shows.

TROLLEY
Electric light railcar
Origin: Spain
Date: 1912

Pantograph takes electricity from overhead power lines

Trolleys follow set routes and run on rails installed along roads. The earliest trolleys were pulled by horses, but the many trolleys that still run today are powered by electricity.

CHUCK WAGON
Food cart
Origin: US
Date: 19th century

Settlers in the west of North America traveled in convoys—trains of wagons following each other. The chuck wagon carried the food and cooking equipment. A canvas cover could be fitted over the metal hoops.

PHAETON
Owner-driven carriage
Origin: US
Date: 19th century

These lightweight carriages were used by wealthy people to make short journeys and show off their horses. The driver's seat was protected from the elements by a collapsible hood.

Large back wheels

660,000 lb (300,000 kg)—the amount of Parisian horse poop produced each day in the 1820s.

Small horse trolleys without a rear deck were called **bobtails**.

Horse bus

In 1828, Paris became the first city in the world to introduce a new form of public transportation—horse-drawn "omnibuses," the ancestor of the modern bus.

The service ran citywide and was popular and efficient. But omnibuses were not built for comfort. The carriage's metal-rimmed wheels clattered jerkily over Paris's cobbled streets while passengers bounced up and down on hard wooden benches.

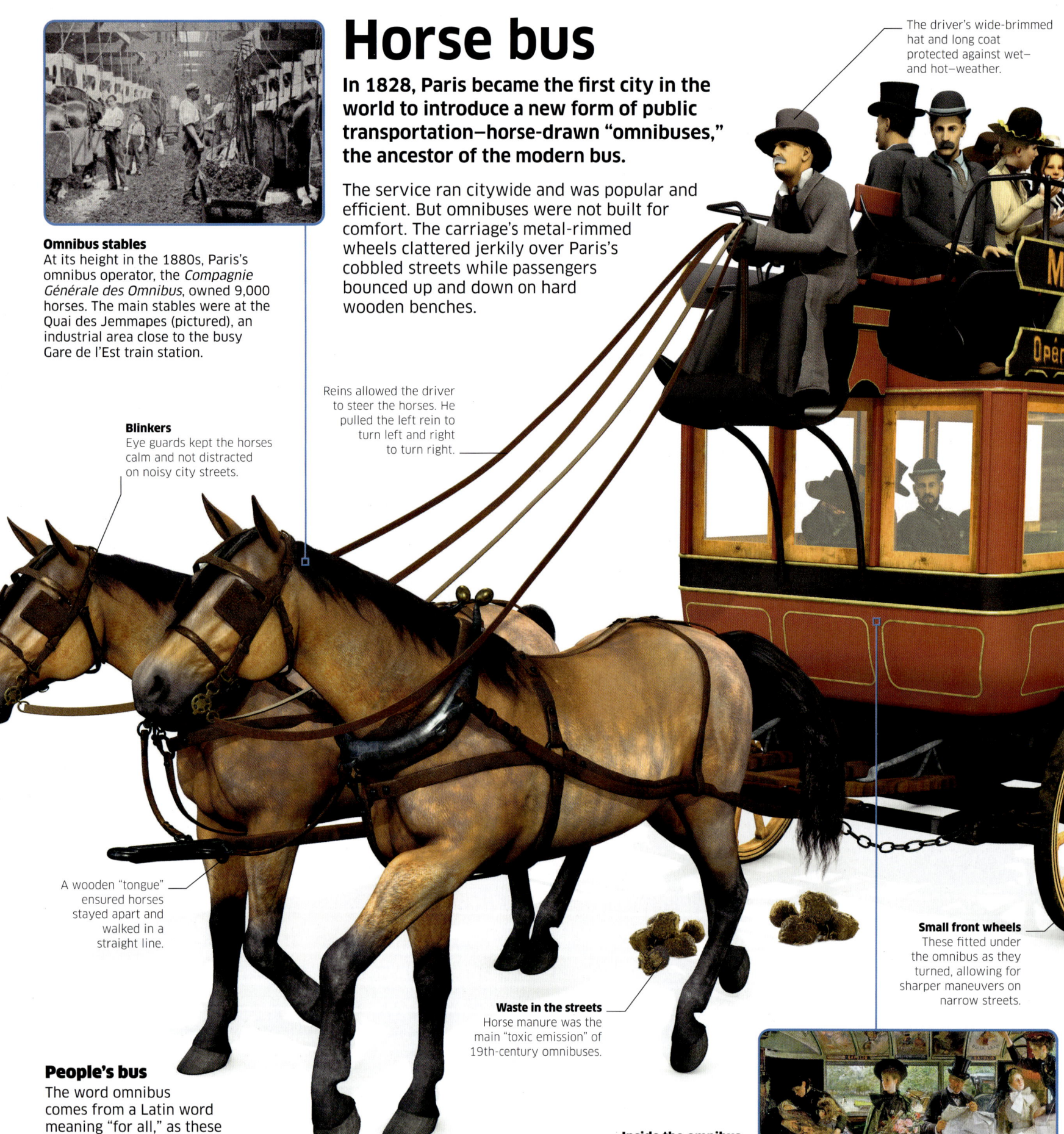

Omnibus stables
At its height in the 1880s, Paris's omnibus operator, the *Compagnie Générale des Omnibus*, owned 9,000 horses. The main stables were at the Quai des Jemmapes (pictured), an industrial area close to the busy Gare de l'Est train station.

The driver's wide-brimmed hat and long coat protected against wet— and hot—weather.

Reins allowed the driver to steer the horses. He pulled the left rein to turn left and right to turn right.

Blinkers
Eye guards kept the horses calm and not distracted on noisy city streets.

A wooden "tongue" ensured horses stayed apart and walked in a straight line.

Small front wheels
These fitted under the omnibus as they turned, allowing for sharper maneuvers on narrow streets.

Waste in the streets
Horse manure was the main "toxic emission" of 19th-century omnibuses.

People's bus
The word omnibus comes from a Latin word meaning "for all," as these buses were designed to be used by whoever could afford it. They were single-deck wooden carriages at first, with seats accessed by a ladder. Open-top omnibuses were introduced in 1853. Later, more sophisticated vehicles like the *Impériale* had fixed spiral staircases.

Inside the omnibus
Although unusual for the time, men and women sat together, eight passengers on each side. In theory, the omnibus was open to all, but the fare of 25 centimes was too expensive for Paris's poorer citizens.

IMPÉRIALE OMNIBUS

Type: Open-top double-decker

Location: Paris, France

Length (carriage): c.15 ft (4.6 m)

Year launched: c.1860

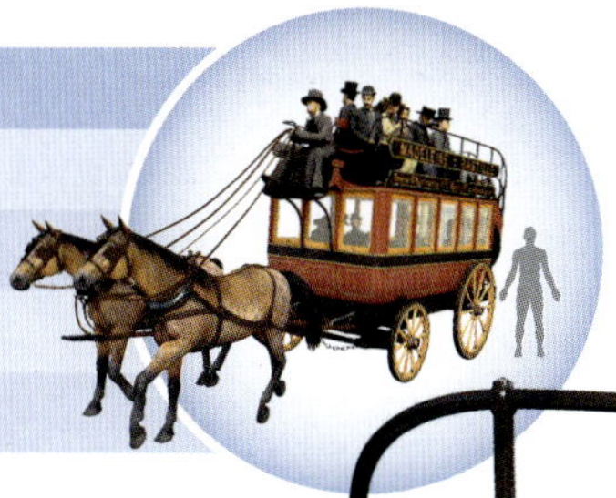

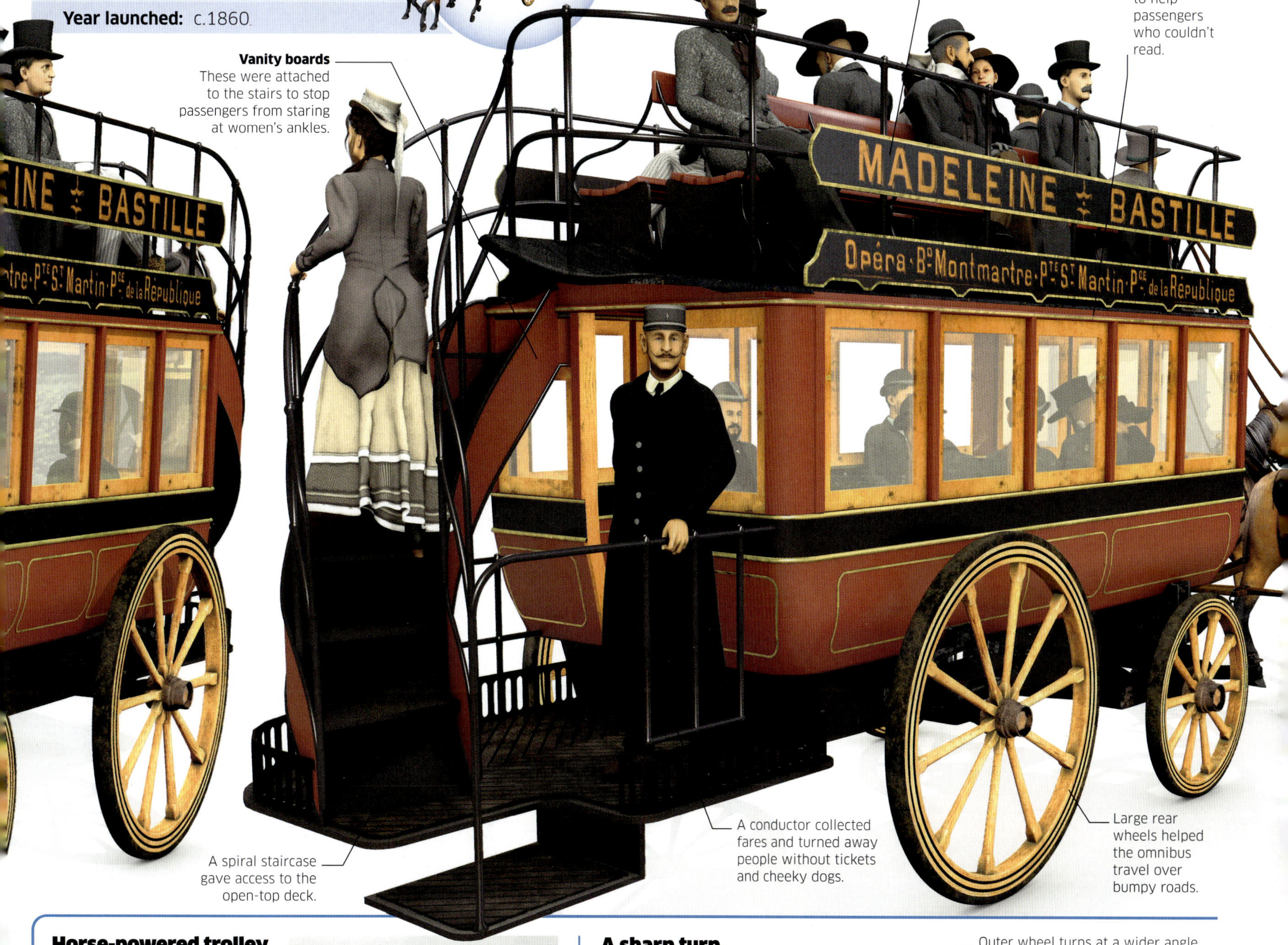

Vanity boards
These were attached to the stairs to stop passengers from staring at women's ankles.

Passengers on the top deck paid half the fare of those inside.

Route boards
Upper boards helped people identify the omnibus's route. Lower boards listed specific stops.

Omnibus carriages were color-coded to help passengers who couldn't read.

MADELEINE ⁑ BASTILLE

Opéra · Bᵈ Montmartre · Pᵗᵉ Sᵗ Martin · Pᶜᵉ de la République

A conductor collected fares and turned away people without tickets and cheeky dogs.

Large rear wheels helped the omnibus travel over bumpy roads.

A spiral staircase gave access to the open-top deck.

Horse-powered trolley

Horse trolleys were single- or double-decker carriages on metal rails pulled by two horses. They were popular in cities worldwide, such as London, UK (pictured). Horse trolleys were an improvement on the omnibus because their rails allowed horses to pull a greater load and created a smoother ride for passengers. In the 1890s, they slowly disappeared in favor of electric trolleys and railways.

A sharp turn

Until 1816, carriages struggled with tight street corners. That year, German inventor Rudolph Ackermann popularized an idea to give each wheel its own individual pivot, meaning they could turn around the same central point. This allowed carriages to turn more easily without their wheels slipping sideways. The steering on modern vehicles uses the Ackermann principle in a modified form that positions the wheels correctly for turns at higher speeds.

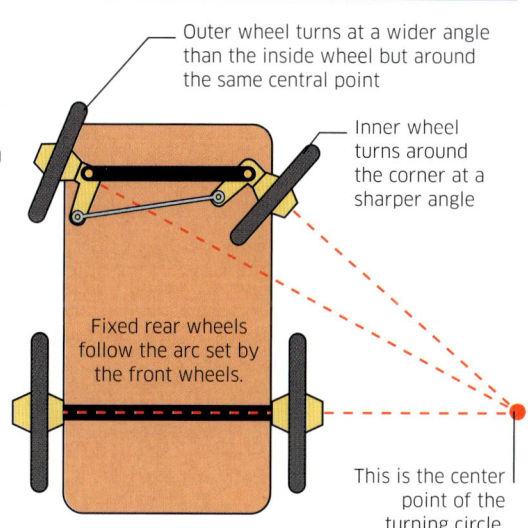

Outer wheel turns at a wider angle than the inside wheel but around the same central point

Inner wheel turns around the corner at a sharper angle

Fixed rear wheels follow the arc set by the front wheels.

This is the center point of the turning circle.

Road construction

From huge highways to small, single-track lanes, large networks of roads are a key part of any country's infrastructure and used by millions of vehicles every day.

Constructing a new road can take years and begins with an extensive surveying period to find land that is suitable. Depending on the landscape, tunnels, bridges, and other features may also need to be built to enable the path of the road. When completed, the road will need regular maintenance to ensure it can continue to support large volumes of traffic. In 2021, China alone spent more than $6 billion constructing and maintaining transport infrastructure.

Building a highway

Once a site has been selected, engineers begin to design the road, taking into account factors such as geography, soil conditions, and environmental impact. Construction then begins, and the road is built up in several layers. Drainage must be placed to the side and beneath the road so that water doesn't build up on its surface.

Construction workers
Experienced workers test the quality of the laid asphalt. When using concrete, they also use tools to smooth this down while it is wet.

3 Asphalt
Hot liquid asphalt is sprayed onto the subbase in layers, then compacted. Some roads use concrete instead, which typically lasts longer than asphalt but is more expensive.

Drainage
Networks of storm pipes carry water away. Pipes or cables for utilities may also be laid below the road.

4 Adding road markings
Once the asphalt has dried, markings, such as the center line, are painted on. Other road features can then be added, such as lights, barriers, and signage.

Highway
The biggest and often busiest roads, highways have multiple lanes and connect key cities and urban areas.

Line painter sprayer

Flyovers

When a road needs to pass over an obstacle or another road, a bridge called a flyover is constructed. Its foundation, supports, and the deck upon which the road sits are usually made out of concrete, with individual parts carefully lifted into place by a crane.

Asphalt paving machine
As the machine moves along, it spreads the asphalt smoothly and evenly. Its sprayers can be set to a range of different widths.

Edging stones
These border the road and have grills through which rainwater flows to the underground drainage systems.

2 Base and subbase
The next road layers are made of tough materials, usually a mix of crushed stone, gravel, additional soil, or crushed concrete. They provide strength and stability.

Base

9,024 miles (14,523 km)—the length of **Australia's Highway One**, the **longest continuous road** in the world, which **passes through every state** in the country.

17

Dump truck
Materials such as gravel are transported to the site by long dump trucks, which then tip out their load wherever it is needed.

Barriers
Railings and barriers made out of tough steel are designed to stop vehicles from accidentally going off the road.

Roadblock
This prevents vehicles from accessing the area until works are completed.

Embankment
Raised areas of earth are often built alongside a road to reduce noise and to make the road is less visible.

Road roller
Equipped with a large, heavy drum, a road roller moves over the hot asphalt, compressing it and removing any air bubbles that could cause cracks.

Excavator
Digging vehicles such as these can move large volumes of earth quickly, removing excess soil or creating trenches for below-road drainage.

1 Groundworks
Construction begins with earthworks—where the area is cleared and leveled and the remaining soil compacted so that it provides a supportive base for the rest of the road.

Bulldozer
As well as clearing any existing vegetation or debris, these large machines also flatten and level the ground.

Subbase

FOUR-STROKE ENGINE

At the heart of an engine are one or more fixed cylinders, each of which contains a movable part called a piston. Most internal combustion engines run on a cycle that involves two upward and two downward strokes, or movements, of each piston. These admit fuel and air into the engine, squeeze them together, ignite and burn the mixture, and finally push out the waste gases. When an engine is running, this cycle is repeated many times a second, as each piston goes through each stroke.

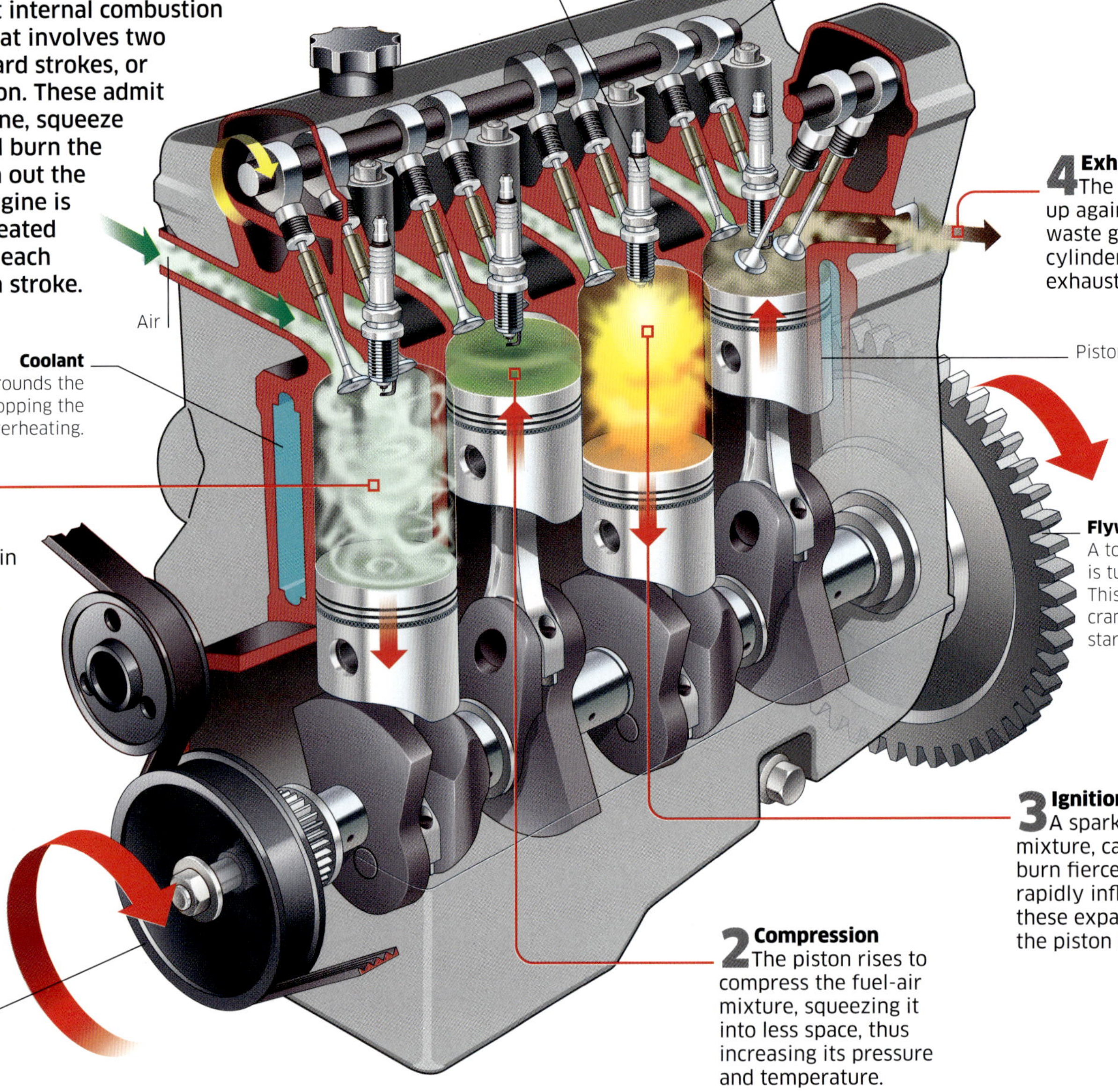

Spark plug
A spark plug produces a tiny spark of electricity to ignite the fuel and air.

Camshaft
Inlet and exhaust valves are opened and closed by the camshaft to let the fuel and air in and waste gases out.

Air

Coolant
This liquid surrounds the cylinders, stopping the engine from overheating.

4 Exhaust
The piston rises up again, pushing the waste gases out of the cylinder and into the exhaust system.

Piston

1 Intake
The piston contained in this cylinder descends, drawing fuel and air into the cylinder through an opening at the top.

Flywheel
A toothed flywheel is turned by the starter. This then turns the crankshaft and starts the engine.

3 Ignition
A spark ignites the mixture, causing it to burn fiercely and create rapidly inflating gases. As these expand, they drive the piston back down.

Crankshaft
This is turned by the pistons and keeps the four-stroke cycle going.

2 Compression
The piston rises to compress the fuel-air mixture, squeezing it into less space, thus increasing its pressure and temperature.

Inside the engine

Since their development in the 1870s, internal combustion engines have revolutionized the world. Packed into the centers of cars, motorcycles, and other vehicles, they generate large amounts of energy from either gas or diesel fuel.

Inside their closed cylinders, engines burn air and fuel together in a process known as combustion. This converts the chemical energy stored in the fuel into mechanical energy that can turn wheels or propellers and power a wide range of vehicles. Engines can contain varying numbers of cylinders, but the most common cars use between four and eight.

Diesel engines

Diesel engines do not have spark plugs. Instead, they compress air so much that it heats up to temperatures above 932°F (500°C), then the fuel is injected afterward, causing the explosion.

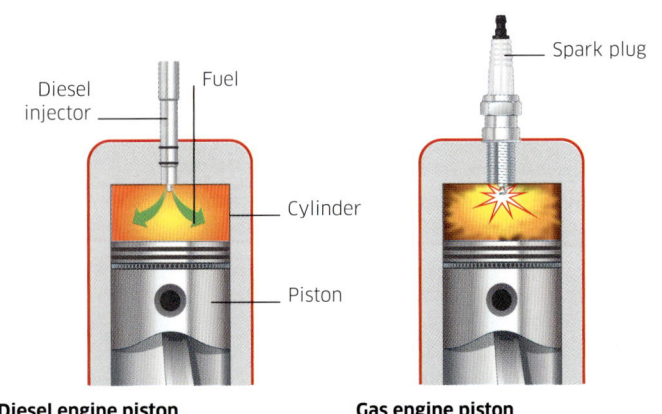

Diesel injector

Fuel

Spark plug

Cylinder

Piston

Diesel engine piston
Fuel injected into very hot air burns rapidly, creating downward force.

Gas engine piston
A spark causes the gas to burn rapidly, creating downward force.

1897 The year **Rudolf Diesel** successfully demonstrated **his own** version of the internal combustion engine—**the diesel engine.**

48 The record number of cylinders put into an engine, fitted **inside a motorcycle.**

19

STARTING A CAR

A series of electrical, chemical, and mechanical processes must all occur before a driver can successfully start their car and allow its engine to pull the vehicle away from a standstill.

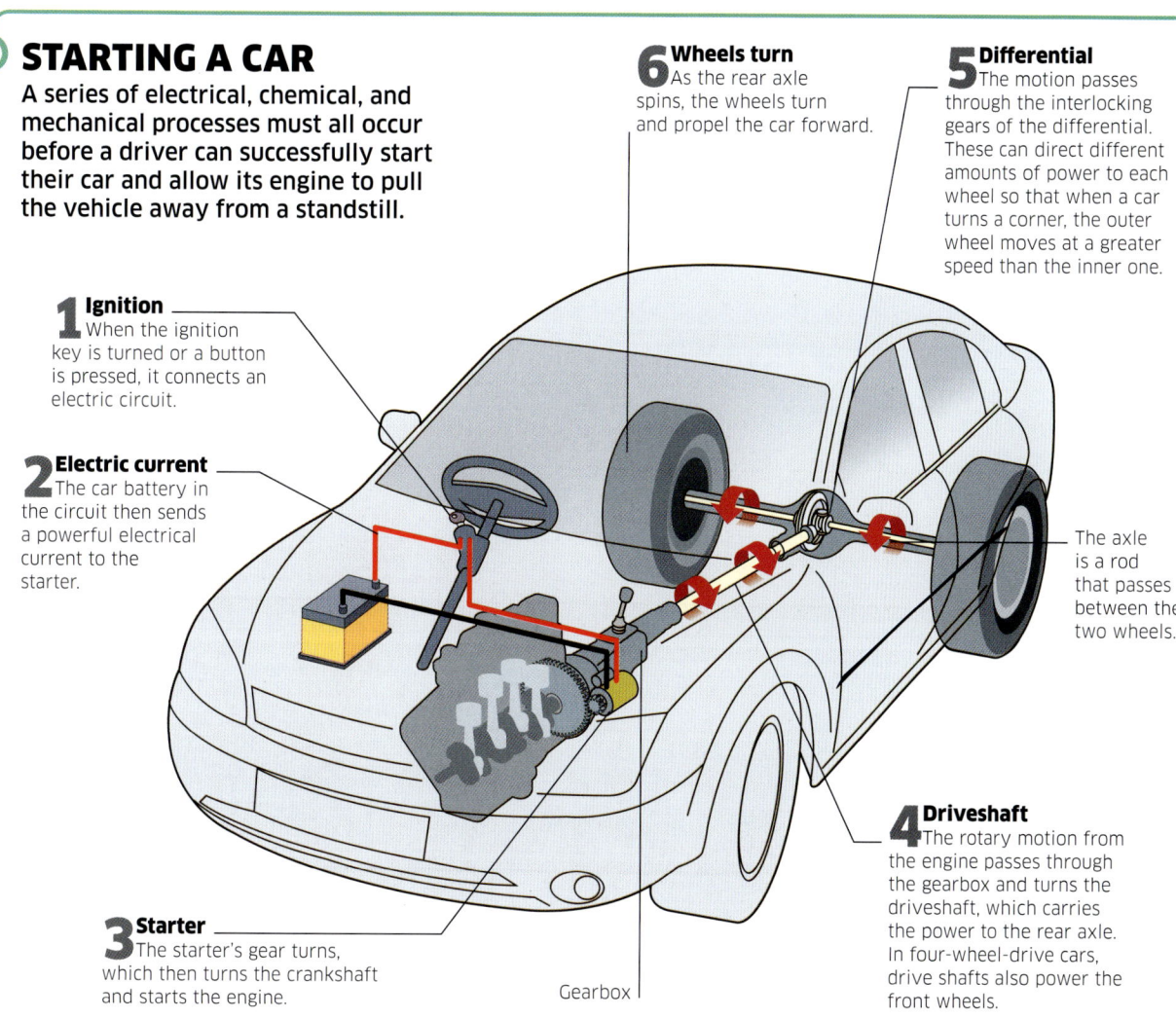

1 Ignition
When the ignition key is turned or a button is pressed, it connects an electric circuit.

2 Electric current
The car battery in the circuit then sends a powerful electrical current to the starter.

3 Starter
The starter's gear turns, which then turns the crankshaft and starts the engine.

Gearbox

6 Wheels turn
As the rear axle spins, the wheels turn and propel the car forward.

5 Differential
The motion passes through the interlocking gears of the differential. These can direct different amounts of power to each wheel so that when a car turns a corner, the outer wheel moves at a greater speed than the inner one.

The axle is a rod that passes between the two wheels.

4 Driveshaft
The rotary motion from the engine passes through the gearbox and turns the driveshaft, which carries the power to the rear axle. In four-wheel-drive cars, drive shafts also power the front wheels.

GEARBOX

Gears are toothed wheels that can mesh together so that as one gear turns, another does, too. When different-sized gears mesh together in a gearbox, they can alter the turning force, or torque, of the engine output. Lower gears provide more power, while higher gears are used at faster speeds.

Larger gear turns more slowly but with far greater turning force (torque)

Smaller wheel with less teeth turns four times faster but with less power

ENGINE LAYOUTS

Apart from smaller motors powering microcars and some motorcycles, most internal combustion engines contain two or more cylinders. These can be arranged in different ways.

Straight-four
Featuring four cylinders in a line, this is one of the most common arrangements in small and medium cars today.

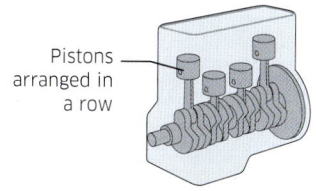

Pistons arranged in a row

V6
Found in racing IndyCars and some sports cars, these engines consist of two rows of three cylinders, angled against each other.

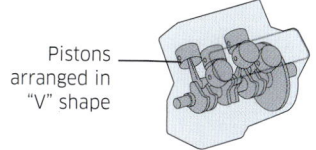

Pistons arranged in "V" shape

Flat-four
Two pairs of cylinders lie flat opposite each other, creating a wide but low-vibration engine.

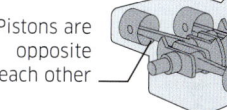

Pistons are opposite each other

Radial
This circular pattern of cylinders has powered many early aircraft and some modern acrobatic planes.

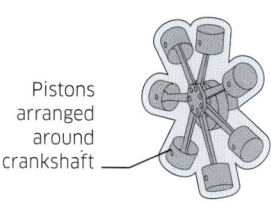

Pistons arranged around crankshaft

ENGINE POWER

Internal combustion engines power many of the world's vehicles. Many can generate large amounts of power, as the crankshaft turns many thousands of times in a minute—a measurement known as revolutions per minute (RPM). But power does not equal efficiency. Engineers are increasingly working to design engines that can travel the most distance on the least fuel.

Horsepower

The unit we use to measure a car's power was first developed by British engineer James Watt in the 1780s. He compared the force generated by engines to that of horses and defined 1 horsepower as the power of one horse lifting a 330-lb (149.7-kg) weight 100 ft (30.5 m) over one minute.

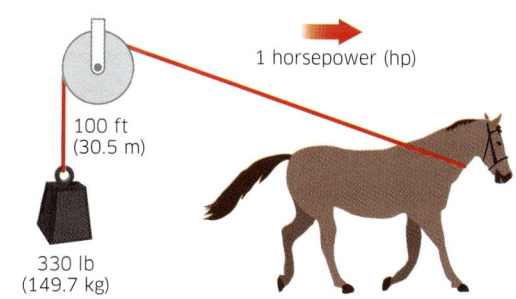

1 horsepower (hp)

100 ft (30.5 m)

330 lb (149.7 kg)

Engine efficiency

Vehicle designers are always looking for ways to get more power from engines while using less fuel. Adjusting engine components, aerodynamics, and onboard computers can all make a difference and lead to innovative new designs, such as those developed for the Shell Eco-marathon (left). Today, some smaller cars can now travel more than 51.7 mpg (22 km/l) of fuel.

1896 The date the world's **first speeding ticket** was **issued**, in **England**.

1.45 billion—the **total number** of **cars** in the **world** at the **end of 2023**.

On the road

Over the past 140 years, the typical car has evolved continuously, from "horseless carriage," through luxury item, to family essential.

Cars with engines were pioneered in Europe, but it was not until the US began to mass-produce vehicles on moving production lines that ordinary people could afford them. Increasing in power and sophistication while changing in style with the times, they became symbols of wealth and independence. Today, a wide range of cars are on sale all over the globe, but there are also efforts to reduce their impact on the environment, including developing electric cars.

OLDSMOBILE CURVED DASH

Origin: US
Year: 1901
Top speed: 20 mph (32 kph)

Named after its creator Ransom Eli Olds, this car was the first to be built in large numbers. The single-cylinder engine was positioned under the two seats, and a two-speed gearbox transmitted power to the rear wheels. The driver steered using a tiller—a long lever at the front.

Curved front
A rolled metal panel protected the car's occupants from mud and flying stones.

Soft tires
Rubber tires and built-in suspension cushioned the solid axles and wooden wheels.

FORD MODEL T

Origin: US
Year: 1908
Top speed: 42 mph (68 kph)

More than 15 million Model Ts had been made by 1927, using the world's first moving production line. It had high clearance (the gap between a car's body and the road) to tackle the era's rough roads.

CITROËN TRACTION AVANT

Origin: France
Year: 1934
Top speed: 85 mph (137 kph)

This family car was packed with new technology, such as front-wheel drive. Its excellent handling made it safer than other cars to drive fast, and it came with a choice of body styles and engines.

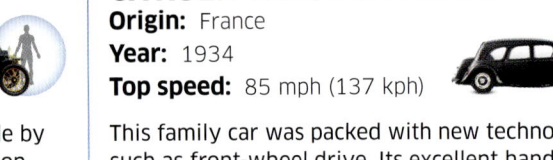

VOLKSWAGEN BEETLE

Origin: Germany
Year: 1938
Top speed: 70 mph (113 kph)

"Volks Wagen" means "People's Car" in German, and this rear-engined, air-cooled sedan was devised as cheap family transportation. More than 21 million Beetles were made—still a world record!

1959 The modern **three-point seatbelt is invented** by Swedish Volvo engineer **Nils Bohlin**—subsequently **saving over 1 million lives.**

3.1 seconds—the time a **Tesla Model 3** takes to reach 60 mph (97 kph) from a standing start.

21

CHEVROLET BEL AIR
Origin: US
Year: 1957
Top speed: 120 mph (193 kph)

With its extravagant rear fins, two-tone paintwork, and chrome trim, the Bel Air was a classic American convertible of the 1950s. A powerful engine and automatic gear changing came standard.

Tall tails
The tail fin styling was inspired by fighter jets and rockets.

MINI
Origin: UK
Year: 1959
Top speed: 100 mph (161 kph)

To fit four people inside, the Mini's engine was mounted transversely at the front—a first at the time. It was speedy and maneuverable, and sports models won the Monte Carlo Rally three times.

RENAULT 16
Origin: France
Year: 1965
Top speed: 105 mph (170 kph)

This car reshaped family motoring with its hatchback trunk and folding back seat, enabling it to carry cargo. Its soft suspension system absorbed road bumps and was ideal for highway driving.

HONDA CIVIC MKI
Origin: Japan
Year: 1972
Top speed: 90 mph (145 kph)

The Civic's innovative engine achieved "lean combustion" to reduce noxious emissions and meet new antipollution laws. Honda licensed this groundbreaking technology to other car makers.

JEEP WAGONEER
Origin: US
Year: 1963
Top speed: 95 mph (153 kph)

One of the first cars to offer the comfort of a family estate (a spacious car with a large trunk) with the offroad capability of a four-wheel drive, the Wagoneer was an early sport-utility vehicle (SUV).

SAAB 99
Origin: Sweden
Year: 1967
Top speed: 120 mph (193 kph)

The Saab 99 set new trends in crash protection, braking, visibility, and wet-weather handling. Other innovations included the heated driver's seat and headlight wipers.

TOYOTA PRIUS MKI
Origin: Japan
Year: 1998
Top speed: 99 mph (159 kph)

This small sedan made history as the first gas-electric hybrid on public sale. At low speeds, an electric drive ensured low emissions, while gas power took over on faster roads.

TESLA MODEL 3
Origin: US
Year: 2017
Top speed: 162 mph (261 kph)

Before the Model 3 appeared, Tesla had already shaken up the electric car market. Its 2017 offering was much more affordable and could cover more than 250 miles (402 km) between charges.

22   land ○ ELECTRIC CAR

201 mph (322 kph)—the **top speed of Formula E Gen 3 vehicles**—the fastest all-electric race cars.

Electric car

With no internal combustion engine (see pages 18–19), electric vehicles do not use gas or diesel to run. Instead, they are powered by electric motors—making them far quieter and less polluting than regular vehicles.

Approximately 14 million electric cars were sold worldwide in 2023. Unlike other cars, all-electric vehicles produce zero emissions when running, although the electricity they consume may have created emissions when being generated. They have far less moving parts than regular cars, with no fuel pump, ignition, or exhaust systems.

Driver's console
A touchscreen gives the driver access to many electronic driving aids, including different acceleration modes, parking assist, and cruise control.

Front traction motor
This powerful electric motor turns the front wheel axle with large amounts of torque in order to propel the vehicle forward.

Heat pump assembly
This powers a liquid cooling system, which removes heat from the motors and keeps the battery pack within 59–77°F (15–25°C).

Front trunk
The streamlined hood opens to access storage space that can hold up to 110 lb (50 kg).

Batteries
Thousands of rechargeable lithium-ion battery cells are located under the passenger cabin floor.

Negative terminal (cathode)

Positive terminal (anode)

Separator

High-end electric car

The first all-electric luxury sedan car, the Model S was originally rear-wheel drive only but now has two traction motors that power all four wheels. Hidden under the passenger compartment floor is its giant battery pack weighing 1,378 lb (625 kg)—almost a third of the car's total weight.

Inside a lithium-ion battery
An electric car's battery stores energy. When connected to a circuit, electricity flows from the battery's negative end (anode) to its positive end (cathode), passing through and powering the car's motor.

More than **one in three cars sold in 2030 will be electric**, according to the International Energy Agency.

The 2023 Model S Plaid has a "summon" mode where the vehicle can move itself out of parking spaces without a driver.

23

Side-impact protection
This includes strengthened bars to protect the car's occupants, as well as side airbags that inflate on impact.

Rear seats
The three rear seats are adjustable and can be folded down to create a large storage area, in addition to the trunk at the rear.

Rear traction motor
This powerful 252kW electric motor works with the front motor— allowing the car to reach speeds of up to 155 mph (249.5 kph).

TESLA MODEL S
Five-door luxury sedan car

Origin: US

Year: 2012

Length: 16⅜ ft (5.02 m)

Rapid supercharger
This fast curbside charging station can typically recharge batteries enough for 200 miles (320 km) of range in under 20 minutes.

Shock absorber
This part of the rear suspension system absorbs impact forces and can be adjusted for different driving conditions.

Charge port
The vehicle's port can be connected to supply-main electricity.

Chassis
The sturdy frame of the car is mostly made from lightweight aluminum.

Stacked packs
Large groups of batteries in the Model S give the car a maximum range of 348–402 miles (560–647 km) between charges.

The 2023 Tesla Model S Plaid can accelerate from a standstill to
60 mph (96 kph)
in just 1.99 seconds.

Regenerative braking
During braking, an electric car's motor no longer needs to supply power to the wheels. The motor instead becomes a generator— turning the movement energy of the spinning wheels into electricity that can be sent back to the car's battery pack. Gas or diesel cars are not able to store energy in this way.

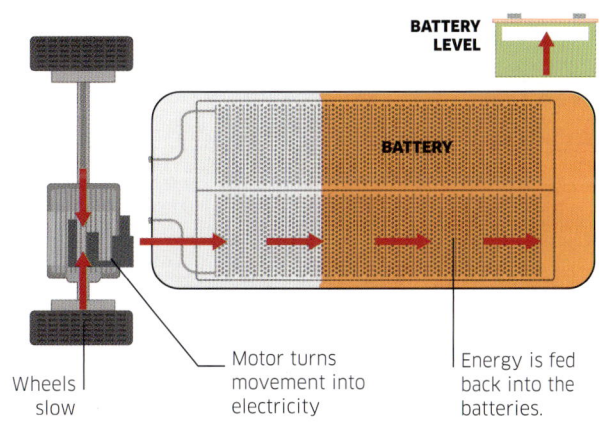

BATTERY LEVEL

BATTERY

Wheels slow

Motor turns movement into electricity

Energy is fed back into the batteries.

The first hybrid
Hybrid cars have both electric motors and an internal combustion engine. The first was the Lohner-Porsche Semper Virus in 1900. Its engine powered a generator that sent electricity to the wheels, giving the car a top speed of 21.8 mph (35 kph).

24   land ○ LUXURY CARS

100 ft (30.51 m)—the **length of the longest car**, a limousine that **can carry 75 people** as well as a pool and a helipad.

ROLLS-ROYCE SILVER GHOST
Origin: UK
Year: 1906
Top speed: 75 mph (121 kph)

The Silver Ghost gained its reputation as "the best car in the world" due to its mechanical refinement and near-silent running. Each car produced had an individually hand-built body.

BUGATTI TYPE 50T COUPÉ
Origin: France
Year: 1931
Top speed: 110 mph (177 kph)

Made by the manufacturer of many Grand Prix winners, the high-performance Type 50 had a responsive, eight-cylinder engine to match its distinctive Art Deco shape and bold colors.

BENTLEY R-TYPE CONTINENTAL
Origin: UK
Year: 1951
Top speed: 120 mph (193 kph)

Bentley produced this four-seater grand touring car for long, fast journeys. Its aluminum, aerodynamic body shape gave it excellent performance at high speeds. The seats were upholstered in rich leather.

Luxury cars

At the dawn of motoring, all cars were luxury purchases and playthings for the rich. But when motor racing demonstrated the reliability of high-end cars and mass production made them cheaper, they became an option for the everyday buyer.

Manufacturers such as Bugatti and Bentley proved the quality of their cars on the race track and used their experience to create high-spec production models. Meanwhile, increasingly advanced engineering created luxury cars that not only boasted powerful engines, but also offered comfortable, fashionable interiors.

CHEVROLET CORVETTE
Origin: US
Year: 1953
Top speed: 142 mph (229 kph)

There have been eight generations of Corvette two-seater sports cars—the one shown here is a 1959 version of the original type. They all have powerful eight-cylinder engines and lightweight fiberglass bodywork.

Whitewall tires

Two-tone color scheme

PORSCHE 911
Origin: Germany
Year: 1963
Top speed: 124 mph (200 kph)

Featuring a rear-mounted, six-cylinder engine and an attractive, tapering body, the Porsche 911 has gone through many designs. Its original air-cooled engine was inspired by Volkswagen cars.

PONTIAC GTO
Origin: US
Year: 1964
Top speed: 125 mph (201 kph)

The GTO is often considered the original American "muscle car"—a regular production car boosted with a more powerful engine and uprated brakes and suspension.

JAGUAR XJ12
Origin: UK
Year: 1972
Top speed: 150 mph (241 kph)

Jaguar put the V12 engine into mass production for this super-saloon. Special tires and suspension meant it handled like a sports car, giving smooth and effortless performance.

$142.7 million (£115.1 million)—the cost of the **world's most expensive car, the** Mercedes-Benz 300 SLR Uhlenhaut Coupé, sold at a secret auction in 2022.

25

AUDI QUATTRO

Origin: Germany
Year: 1980
Top speed: 135 mph (217 kph)

This rally-winning coupé brought together several technologies for the first time—a turbocharged engine, four-wheel drive, and antilock brakes. It had impressive grip and handling in many conditions.

FERRARI F40

Origin: Italy
Year: 1987
Top speed: 201 mph (323 kph)

With just two seats, this car had a windbreaking, lightweight body. It made history by becoming the first production car in the world capable of speeds more than 200 mph (322 kph).

ASTON MARTIN DB7

Origin: UK
Year: 1993
Top speed: 165 mph (266 kph)

More than 7,000 of this slick coupé were made, making it Aston Martin's most popular model. Supercharged versions with larger engines were also built alongside elegant convertible models.

LAMBORGHINI AVENTADOR

Origin: Italy
Year: 2010
Top speed: 217 mph (349 kph)

This Italian supercar delivered shattering performance from a V12 engine positioned behind the cockpit. It could accelerate from 0 to 62 mph (100 kph) in under 3 seconds.

MAYBACH 62

Origin: Germany
Year: 2002
Top speed: 155 mph (249 kph)

The name Maybach had been given to one of the most expensive cars in the 1930s, and it was revived for this limousine. A glass-roofed giant, it was often seen delivering celebrities to red-carpet events.

RANGE ROVER

Origin: UK
Year: 2022
Top speed: 155 mph (249 kph)

The fifth generation of this sport-utility vehicle is one of the most desirable luxury cars. As well as being very capable offroad, it is now available in both hybrid and all-electric models.

Scissor-style doors
Upward-opening doors have been a feature of Lamborghinis since the Countach in 1971.

Wedge-shaped bodywork
The low, dartlike profile of the Aventador makes it highly aerodynamic.

Air intakes
These funnel plenty of cold air into the V12 engine.

Air scoop
One scoop on each side of the car funnels large amounts of air into the engine. They can also protect the driver should the car roll over.

Speed demon
Named after French race car driver Pierre Veyron, this elite supercar was designed in Germany, assembled in France, and boasted a price tag of over $1.2 million (£1 million). The two-seater, two-door vehicle is sleek, highly streamlined, and built low to the ground, measuring just 47.4 in (1.2 m) tall.

Fuel guzzler
Made out of 250 hand-welded parts, the tank holds 22 gallons (100 liters) of fuel, enough for 155-249 miles (250-400 km) of driving.

Super spoiler
The angle of the Veyron's rear spoiler can be altered to produce 882 lb (400 kg) of downforce, helping keep the car low to the ground so it can safely reach even greater speeds. When braking sharply at high speed, the spoiler tilts at a 55° angle to act as an air brake, creating drag and slowing the car down.

Run-flat tires
These reinforced tires can stay rigid even if they get a puncture, and the car can still travel up to 31 miles (50 km) on a flat tire.

Wheels
The wide wheels have low-profile tires glued on. These need regluing once every 18 months and cost around $39,000 (£32,000) for a set of four.

Manufacturer marking
The Bugatti logo is embossed on the wheel hub.

How turbochargers work
The Veyron boasts four turbochargers—devices designed to increase engine power by admitting more air into the cylinders. This leads to a more rapid burning during combustion, producing greater engine output.

1. Exhaust gases released from the engine power a turbine.

3. The compressor draws in and compresses air before it enters the engine.

COMPRESSOR

TURBINE

2. The turbine is attached by a shaft to the compressor, turning it at the same time.

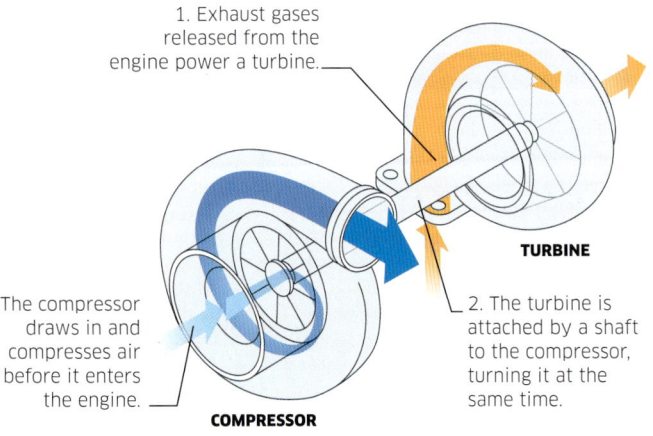

W16 Engine
The 16-cylinder engine is made of more than 3,500 parts and has a huge capacity of 0.28 cu ft (7,993 cu cm).

Gearbox
It took a team of engineers five years to perfect the Veyron's seven-speed gearbox, which has two clutches for fast action.

Monocoque
The car's inner shell is made of carbon fiber and forms a strong, stiff, and lightweight structure weighing 242.5 lb (110 kg).

BUGATTI VEYRON 16.4

Origin: Germany/France

Year: 2005

Length: 14⅝ ft (4.46 m)

Top speed: 253 mph (407 kph)

450 The **number of Veyrons built** between 2005 and 2015.

2.7 seconds—**the time** it takes the Bugatti Veyron **to accelerate from 0-62 mph (0-100 kph).**

27

Clever controls
A sleek dashboard houses the Veyron's controls, which allow the driver to monitor their speed and quickly switch gears. To access the supercar's maximum speed, drivers insert and turn a speed key in a slot beside their seat, which lowers the car and closes two underbody flaps.

Sports car

Supercars offer the ultimate in power, handling, and acceleration. Manufacturers invest vast amounts of money into designs to create vehicles that are not only fast, but stylish, too.

A top-of-the-line supercar, the Bugatti Veyron 16.4 has an engine that can deliver a phenomenal 1001 horsepower. A highly limited edition of the car, the Super Sport 16.4 debuted in 2010 and became the fastest road-legal production car in the world. Reaching an incredible speed of 267.86 mph (431.07 kph), it left its rivals in the dust.

Paddle shifters
Rapid gear changes, taking just 150 milliseconds, are made via quick shift paddles on the steering wheel.

Lightweight body
Carbon-fiber body panels provide toughness and rigidity.

Brakes
Eight titanium pistons power the brake discs, allowing the car to slow down swiftly. In testing, the discs were able to withstand temperatures as high as 2,012°F (1,100°C).

Radiator grille
This admits air in to help cool the Veyron's radiators, which draw heat away from the engine and other systems.

NASCAR

Fiery crashes, daring overtaking moves, and frequent changes of lead are common occurrences during a NASCAR race as speedy, specially built cars jostle for first place.

The National Association for Stock Car Auto Racing (NASCAR Inc.) was founded in 1948 and has since boomed to become America's favorite motorsport, with more than 1,500 sanctioned races a year. Cars rip around the track at speeds of more than 185 mph (300 kph). Many races end in desperately close finishes, with as little as 0.002 seconds separating the winner from the runner-up.

4.5 seconds—**the time it takes** an F1 car to accelerate from 0–124 mph (0–200 kph).

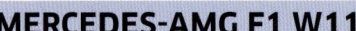

Formula One car

The pinnacle of motor racing, Formula One cars hurtle around twisting race circuits, pushing both vehicle and driver to the limit. The machines used are incredibly streamlined in order to reach the fastest possible speeds.

At the cutting edge of automotive technology, F1 teams produce a new design of vehicle annually, intent on outstripping their rivals. The shape of each car is designed to ensure air flows around it smoothly, reducing the force called drag, which pushes against the car to slow it down. Some of the new innovations used in these cars' materials, components, and technology later find their way into production road vehicles.

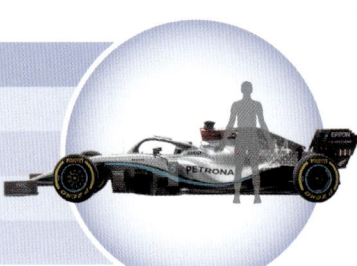

MERCEDES-AMG F1 W11	
Origin:	UK/Germany
Year:	2020
Length:	18⅞ ft (5.8 m)
Weight:	1,645 lb (746 kg)

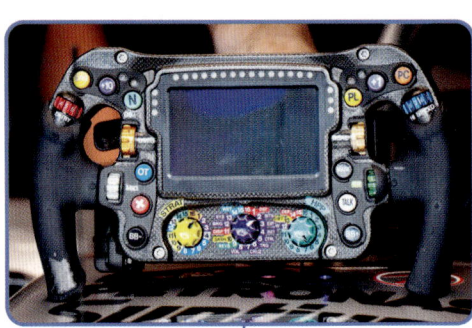

Steering wheel
More than 24 controls, as well as gear change paddles, are crammed into this detachable wheel. They enable the driver to communicate with their team and adjust braking, engine torque, and other factors that could affect the car's performance.

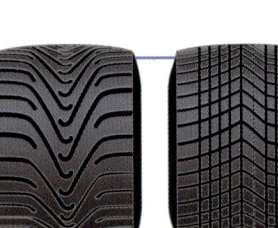

INTERMEDIATE FULL WET

Tires for all weathers
Eight different types of tires are used in F1 cars. In dry conditions, drivers can choose between softer, slick tires that offer more grip, or medium and hard tires that last longer. In wet weather, intermediate or full wet tires have tread patterns that disperse water.

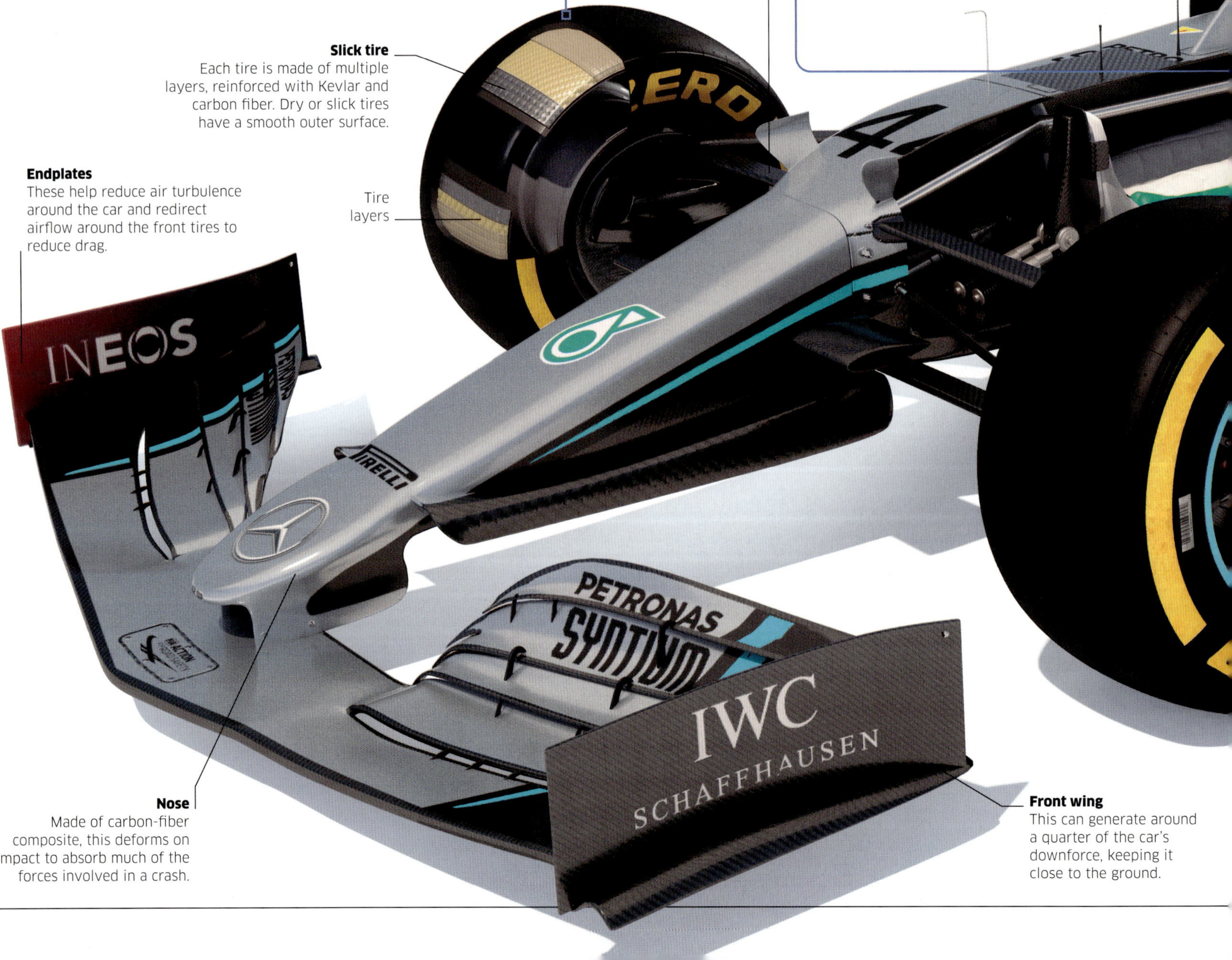

Front suspension
Adjusting the height and stiffness of the suspension affects how the car handles at high speed.

Antenna
Data from more than 300 sensors in the car is transmitted back to the race team for analysis via an antenna.

Slick tire
Each tire is made of multiple layers, reinforced with Kevlar and carbon fiber. Dry or slick tires have a smooth outer surface.

Tire layers

Endplates
These help reduce air turbulence around the car and redirect airflow around the front tires to reduce drag.

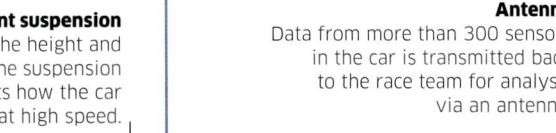

Nose
Made of carbon-fiber composite, this deforms on impact to absorb much of the forces involved in a crash.

Front wing
This can generate around a quarter of the car's downforce, keeping it close to the ground.

14,500 The **number of individual parts** that make up a **single F1 car**.

50 The number of **laps** an **F1 car tire** usually lasts.

F1 drivers **turning a corner** at high speed can experience **extreme G-forces** up to 6 G.

31

Downforce

An F1 car's unique shape directs air around it in specific ways, causing an area of low pressure beneath the vehicle. This produces downforce, which presses the car onto the track, giving it increased grip. After passing around and through the car, the air that flows behind it is turbulent, which can slow down other cars.

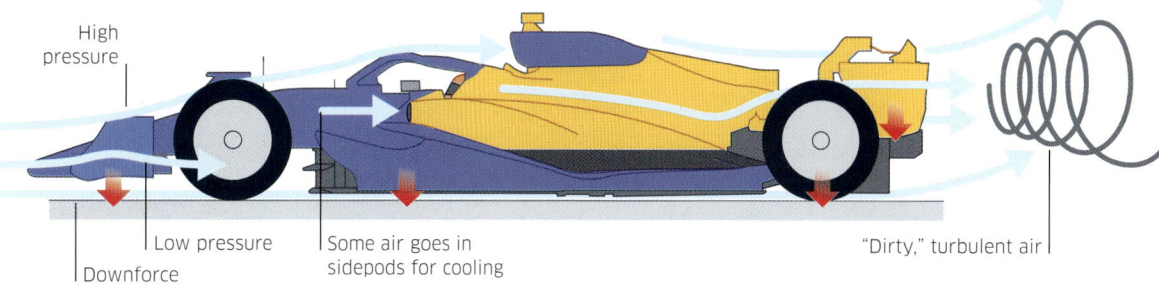

High pressure

Low pressure

Downforce

Some air goes in sidepods for cooling

"Dirty," turbulent air

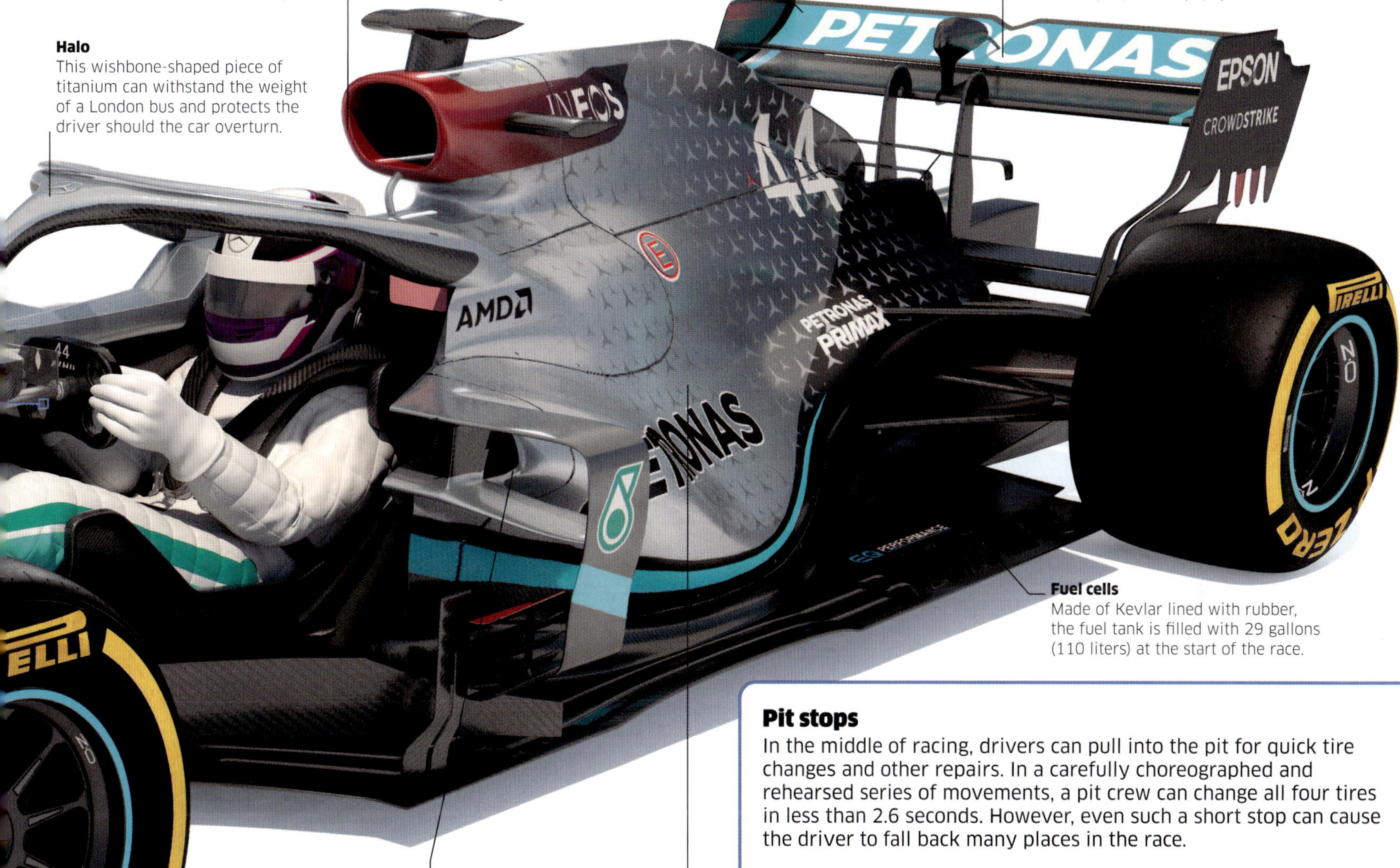

Engine air inlet
A humped inlet sucks air into the engine. It can also protect the driver's head if the car rolls or flips over.

Rear wing
The back wing exerts downforce as air passes over it, making sure the rear tires stay in contact with the track.

Drag reduction system
When activated at high speeds, this lowers part of the rear wing to reduce drag and give the car a 6.7–7.5 mph (10–12 kph) speed boost.

Halo
This wishbone-shaped piece of titanium can withstand the weight of a London bus and protects the driver should the car overturn.

Fuel cells
Made of Kevlar lined with rubber, the fuel tank is filled with 29 gallons (110 liters) at the start of the race.

Pit stops

In the middle of racing, drivers can pull into the pit for quick tire changes and other repairs. In a carefully choreographed and rehearsed series of movements, a pit crew can change all four tires in less than 2.6 seconds. However, even such a short stop can cause the driver to fall back many places in the race.

Sidepods
One on each side, the sidepods channel airflow around the car. They also house multiple radiators that draw heat away from the engine and other key parts that need to stay cool.

Engine
Behind the driver and beneath a tough shell lies the 98-in³ (1.6-liter) turbocharged engine. It can generate 840 horsepower and operate at up to 15,000 revolutions (turns of the engine around the crankshaft) per minute.

Medium tire
This tire is marked in yellow to indicate it uses a medium compound. Soft tires are marked in red and hard compound tires in white.

Winning W11

With aerodynamic sidepods and an innovative dual-axis steering system that helps heat tires up for greater track grip, the W11 broke circuit records. In this car, Lewis Hamilton won 11 of the 2020 season's 17 races—leading him to be crowned World Champion for the seventh time.

32 land • TESTING, TESTING

5,000 frames per second—the rate at which some high-speed cameras can film crash tests.

Testing, testing

Any new car goes through many years of development before it goes on sale. During this process, its features are rigorously tested—through computer modeling, physical challenges, and even by simulating serious crashes.

Testing is conducted on crucial individual car parts, digital and physical models, and full-sized vehicle prototypes. Its aim is to ensure that the vehicle and its systems work as expected in all environments. The data from each test is fed back to the vehicle's design and engineering teams, who seek out ways to make improvements. Even small alterations to single parts can result in major changes to the vehicle's overall performance.

HOW AN AIRBAG WORKS

Located in front of the driver and front seat passenger, airbags are designed to activate on impact. When sensors measure a sharp, sudden slowdown in speed, a signal is sent. This triggers a reaction that produces nitrogen gas in the bag, causing it to inflate, where it cushions the person's head and neck to protect them.

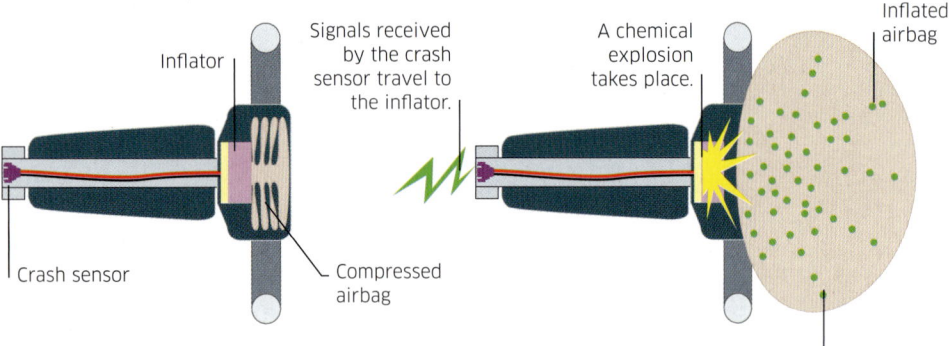

Inflator

Signals received by the crash sensor travel to the inflator.

Crash sensor

Compressed airbag

A chemical explosion takes place.

Inflated airbag

Nitrogen gas causes the bag to expand.

1 Sensing
A crash sensor detects sudden deceleration and sends a signal to the inflator housed in the center of the steering wheel.

2 Inflation
A small explosive charge in the inflator causes chemicals to mix, producing nitrogen gas, which inflates the bag in under 0.03 seconds.

TESTING INNOVATIONS

Many standard features on today's cars have come about as a result of testing. These include tire pressure monitors, airbags, and stability controls that reduce skidding when the brakes are applied.

Seat belts

Inspired by the harnesses of jet aircraft pilots, Swedish engineer Nils Bohlin developed three-point seat belts with shoulder and waist straps. Since their introduction into new cars in 1959, they have saved more than a million lives.

Crash test dummies

Used during crash tests to simulate human responses and injuries, these models are packed full of sensors and come in a range of sizes. Here, Pinocchio (right)—the first child crash test dummy—sits next to a modern one.

COMPUTATIONAL TESTING

Detailed digital models of vehicles or their parts can be constructed on powerful computers. These models are then put into lots of simulations, such as strength, impact resistance, and airflow tests. Changes to the digital model can be made easily, allowing engineers to try out many different solutions to a problem. The Cadillac CT6 luxury car was aided by 50 million hours of computer modeling and testing.

Computational Fluid Dynamics

These programs are used to simulate the airflow around vehicles, enabling engineers to predict temperatures, pressures, and speeds. They have led to great improvements in streamlining.

Wind tunnels

In this test, fans blow air under, over, and around a vehicle or model to simulate particular speeds or conditions, such as severe crosswinds. Analysis of how the air flows around cars can lead to changes that increase speed or allow greater fuel economy.

Fiat's Lingotto car factory in Turin, Italy, had a test track on its roof.

70 The number of different driving conditions cars can be tested in at China's Guangde Proving Grounds, which features 40 miles (65 km) of track.

33

CLIMATIC TESTING

To test how vehicles perform under stress, they are put into controlled environmental chambers. Here, they experience extreme weather, temperatures from -40°F to 176°F (-40°C to 80°C), as well as other conditions, such as high winds and high humidity.

Sudden snow
An electric car is tested inside the terraXcube extreme climate research facility in Bolzano, Italy, to see how its batteries perform in the cold.

TESTING TRACKS

Many manufacturers have huge testing tracks made up of a variety of turns, inclines, and driving surfaces. These thoroughly test out a vehicle's steering, suspension, and braking systems, as well as many other elements.

CRASH TESTING

New vehicles are subjected to a series of controlled crashes where hundreds of impact and force sensors inside the vehicle record data about the impact.

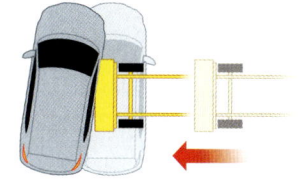

Side impact
A weighted trolley or sled strikes the side of the vehicle at 31 mph (50 kph) to see how the car's frame holds up.

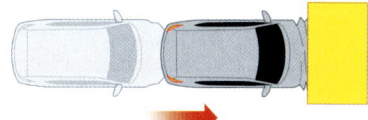

Full width frontal
A car is run into a rigid or collapsible aluminum barrier to simulate hitting another vehicle head on.

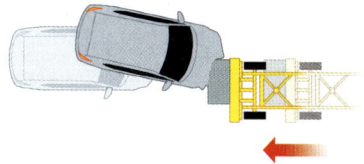

Frontal offset
Around 40 percent of the front of the car is struck by a trolley traveling at 40 mph (64 kph).

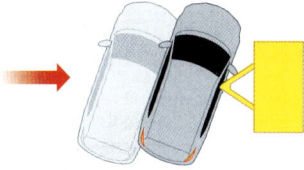

Oblique pole
This test mimics crashing into a tree or telephone pole, which punctures the side of the vehicle.

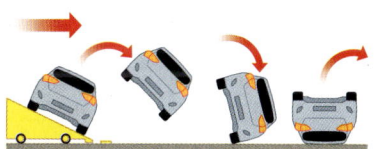

Rollover
A vehicle is driven up a ramp designed to tip the car onto its side and roof. This tests the safety of the passenger cabin.

EMISSIONS TESTING

Every car built must comply with global regulations on the levels of emissions they can produce. The amounts and types of polluting chemicals emitted, such as nitrous oxides and carbon monoxide, are measured by taking exhaust tailpipe gas samples while the car runs at different speeds. Here, the car is tested in red light to make it easier for the driver to concentrate on the screen that shows the various speeds they must test.

TIRE TESTS

New tire designs must be tested in both wet and dry conditions to measure their grip, handling, and rolling resistance. Less rolling resistance means better fuel economy for the vehicle. Braking distances in the wet and dry are tested at a range of different speeds, while the noise levels of the tires in action are measured by trackside sensors.

Safe swerving
A Jaguar XF undergoes tire testing on a wet track in Milan, Italy. A variable sprinkler system can cover the track with precise levels of water for different tests.

A new car rolls out of Hyundai's Ulsan plant in South Korea every 12 seconds on average.

2 Stamping machine
Each steel sheet is cut and punched into shape, then fed into a stamping machine. Excess steel may be used for other parts or recycled.

Plate   Flat sheet of steel   Steel bends

1 Rolled steel
Giant rolls of steel sheeting, some more than 1.2 miles (2 km) long, are unrolled ready for cutting. Increasingly, other materials such as aluminum and plastic are also used.

3 Parts collected
Stamped body parts are inspected and loaded onto a dolly or Automated Guided Vehicle (AGV) so they can be transported around the plant.

Steel press
The piece of steel is held between two plates. Hydraulics then press down with immense force, as much as 5,512 tons (5,000 tonnes), molding the steel to the shape of the plates.

Body assembly
Stamped parts are assembled by robots into a car unibody or chassis.

Conveyor
The body is carried around the line by an overhead gantry or rolling conveyor.

5 Dipping
Car bodies are dipped in a tank of anticorrosion chemicals to help prevent them from rusting.

AGV
These follow electric wires buried in the factory floor to travel to different production stations.

A picking robot loads parts onto a rack carried by an AGV.

Chemicals in the tank are attracted to the car's steel.

Spot weld
Each weld melts the metal of two different parts, so that they bond when solid.

Acid bath
An acid solution removes grease and dirt from the welding and stamping.

The car body is washed and rinsed before being dipped in tanks.

4 Welding
Teams of robot arms are programmed to make more than 2,000 different welds to build a car's frame or chassis in minutes.

Wiring harness
The car's electrical wiring, made up of hundreds of cables, is threaded through the vehicle.

The tailgate is fitted onto the vehicle body.

7 Adding parts
Major components are added, including the fuel tank, exhaust, and transmission systems.

Parking garage column
Volkswagen's Wolfsburg factory in Germany is one of the largest car-manufacturing plants in the world. Twin parking garage towers store up to 800 completed vehicles. Each one has robotic lifts and pallets that can collect, lift, and park cars in 110 seconds—all without human supervision.

23.4 million The number of **cars** that were produced in China alone in 2022.

1,653 tons (1,500 tonnes)–the amount of **sheet metal** used every day at **Volkswagen's** Wolfsburg car factory in Germany.

35

Modern car plant

A major automotive plant sprawls over a huge area, employs thousands of workers, and features numerous production and assembly lines. In the final stages, 2,000–4,000 parts are fitted into each vehicle.

Constructing cars

New cars are produced at a rapid rate. In 2022 alone, around 85.4 million rolled off assembly lines worldwide. They are produced in giant automotive plants that employ complex equipment.

Until production was revolutionized with the introduction of assembly lines, early cars were made individually by hand, which was a slow and laborious process. Now assisted by automated processes and hundreds of robots, a modern car-manufacturing plant is highly efficient, as the thousands of parts needed to make a car are brought together with perfect timing.

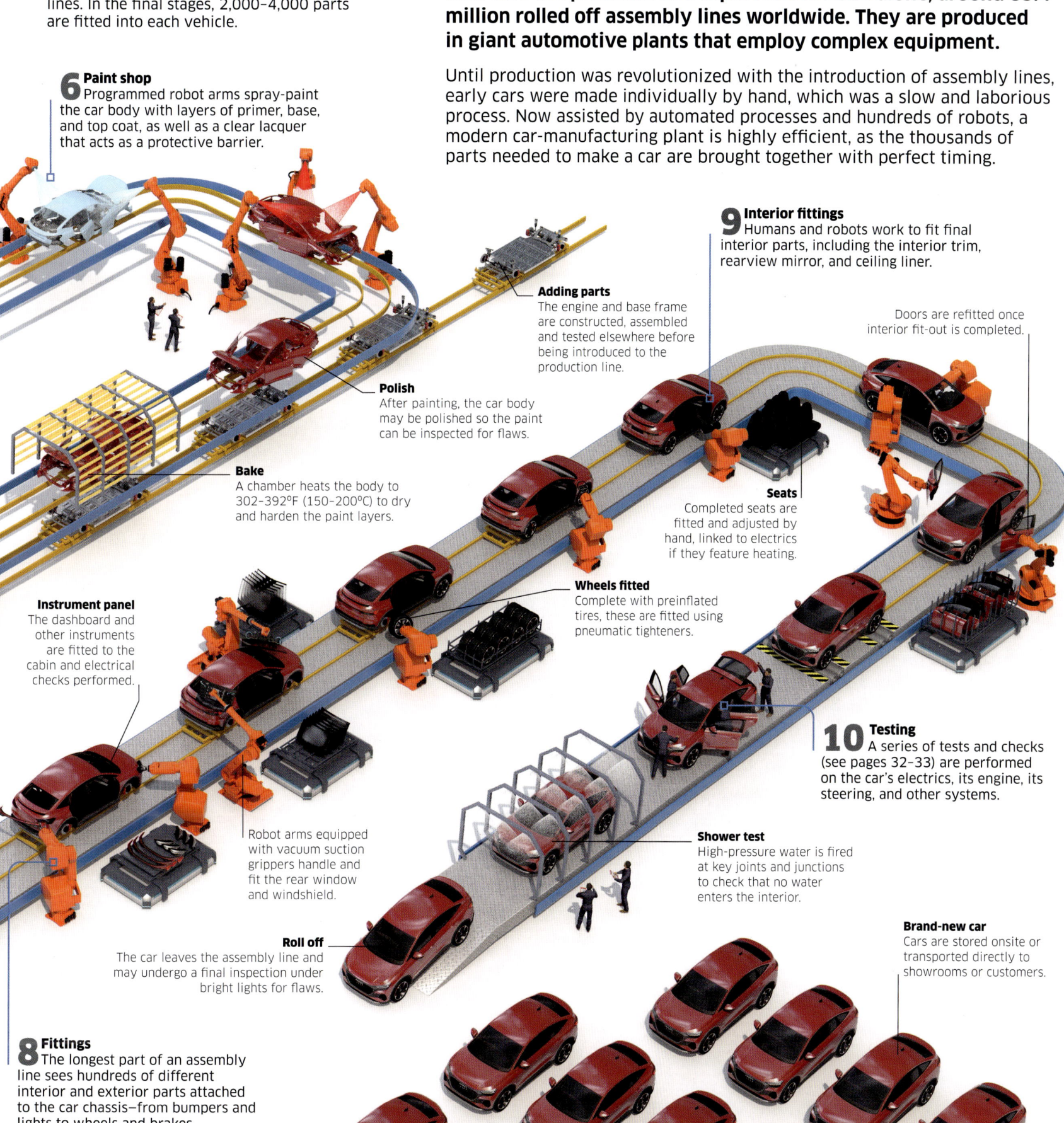

6 Paint shop
Programmed robot arms spray-paint the car body with layers of primer, base, and top coat, as well as a clear lacquer that acts as a protective barrier.

Adding parts
The engine and base frame are constructed, assembled and tested elsewhere before being introduced to the production line.

Polish
After painting, the car body may be polished so the paint can be inspected for flaws.

Bake
A chamber heats the body to 302–392°F (150–200°C) to dry and harden the paint layers.

9 Interior fittings
Humans and robots work to fit final interior parts, including the interior trim, rearview mirror, and ceiling liner.

Doors are refitted once interior fit-out is completed.

Seats
Completed seats are fitted and adjusted by hand, linked to electrics if they feature heating.

Instrument panel
The dashboard and other instruments are fitted to the cabin and electrical checks performed.

Wheels fitted
Complete with preinflated tires, these are fitted using pneumatic tighteners.

10 Testing
A series of tests and checks (see pages 32–33) are performed on the car's electrics, its engine, its steering, and other systems.

Robot arms equipped with vacuum suction grippers handle and fit the rear window and windshield.

Shower test
High-pressure water is fired at key joints and junctions to check that no water enters the interior.

Roll off
The car leaves the assembly line and may undergo a final inspection under bright lights for flaws.

Brand-new car
Cars are stored onsite or transported directly to showrooms or customers.

8 Fittings
The longest part of an assembly line sees hundreds of different interior and exterior parts attached to the car chassis–from bumpers and lights to wheels and brakes.

Impact!

Most high-speed collisions are not intentional, but this one has a purpose. As this life-size dummy smashes into the front of a car, it is helping manufacturers make their vehicles safer for both drivers and passengers.

Crash test dummies are accurate anatomical models made to represent people of different genders, ages, and sizes. Packed with up to 200 sensors, they can measure the acceleration, movement, and impact forces caused by a crash. The data they collect is analyzed by engineers to assess the effectiveness of a vehicle's safety systems, such as airbags.

Gas station

All vehicles require fuel to run, and a large majority use gasoline. This flammable liquid is made from oil extracted from under the ground (see pages 40-41), and then transported all around the globe to filling stations.

An average car's fuel tank can hold between 10 and 14 gallons (45 and 65 liters), and the driver must stop to refuel it regularly. Stations where fuel is available are found in towns and cities, as well as on more remote roads. They often also contain other facilities for vehicle owners—automatic car washes, vacuums for cleaning car interiors, and stores where wiper fluid and maintenance equipment can be purchased.

Signage and branding
Some gas stations are operated directly by fuel companies, whereas others are run by supermarkets or other companies. They usually have bright signs so that they are easily visible to motorists from afar.

Store
Often open 24/7 or for extended hours, these stores sell food, drinks, and a range of other supplies.

Air pump
A small hose extending out from this machine allows drivers to top up the air in a vehicle's tires.

Pump display
The screen at a pump has a running meter that shows the driver how much fuel they have put into their vehicle and its cost.

Dispenser hoses
After removing the fuel cap, drivers insert these nozzles into the side of their car, which leads to the fuel tank.

Underground pipes
Long systems of pipes carry fuel from the tanks to each of the pumping stations.

Transporting and storing fuel

A lot of the most important features of a gas station are located below ground. Oil tankers regularly deposit fuel into large underground tanks, which are equipped with safety features to prevent it from leaking into the environment.

Fuel safety

Gas can be a hazardous substance, and its vapors are harmful to inhale. One of the main risks in a gas station is of fire due to the fuel being very flammable. Spills must be reduced and ignition sources kept far away. Tanks must also be secure so that fuel doesn't leak into the environment. If not successfully contained, gas can leak through the ground and into waterways, harming fish and other wildlife in the area.

Oil spill
This 2020 oil leak off the coast of Mauritius had far-reaching effects on wildlife and ecosystems.

Gas vs. diesel

Two of the most common types of fuel are gas and diesel. They come from the same source—crude oil—but are refined differently, giving them different qualities. Diesel is heavier and more efficient—producing greater amounts of energy when burned. For that reason, it is usually used to fuel larger vehicles, such as trucks and container ships. Although both produce pollutants, gas cars are more environmentally friendly for short journeys.

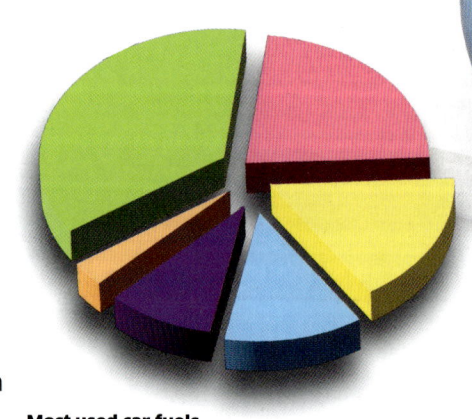

Most used car fuels

 Gas (36.4%)
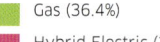 Hybrid Electric (22.6%)
Diesel (16.4%)
 Battery Electric (12.1%)
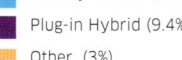 Plug-in Hybrid (9.4%)
Other (3%)

1,581⅞ miles (2,545.8 km)—the **longest distance driven** on a single **standard tank** of **fuel**, achieved in a **Volkswagen Passat 1.6 TDI BlueMotion** in 2011.

1 year—the estimated **storage life of gas** when in a sealed container.

39

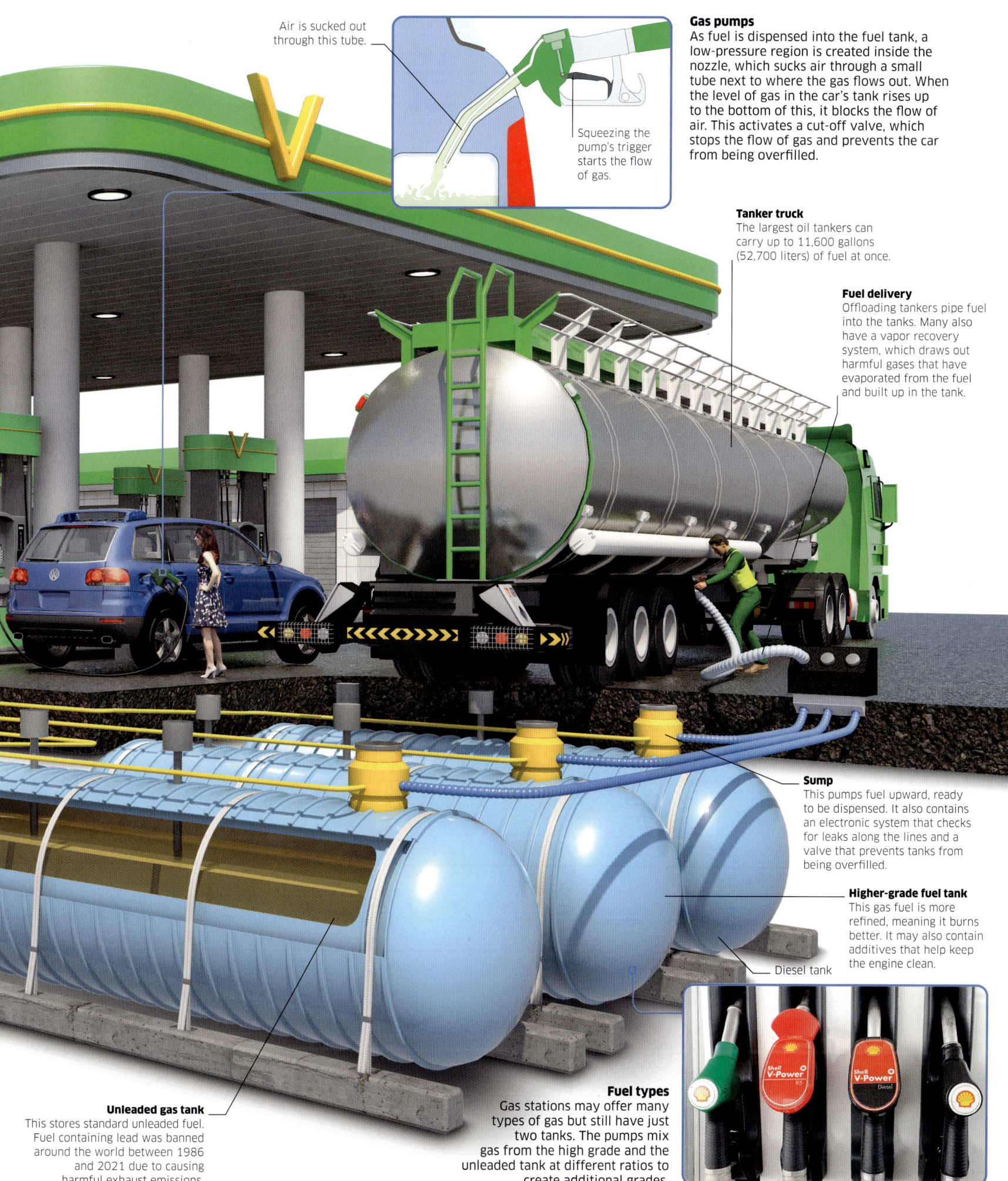

Air is sucked out through this tube.

Squeezing the pump's trigger starts the flow of gas.

Gas pumps
As fuel is dispensed into the fuel tank, a low-pressure region is created inside the nozzle, which sucks air through a small tube next to where the gas flows out. When the level of gas in the car's tank rises up to the bottom of this, it blocks the flow of air. This activates a cut-off valve, which stops the flow of gas and prevents the car from being overfilled.

Tanker truck
The largest oil tankers can carry up to 11,600 gallons (52,700 liters) of fuel at once.

Fuel delivery
Offloading tankers pipe fuel into the tanks. Many also have a vapor recovery system, which draws out harmful gases that have evaporated from the fuel and built up in the tank.

Sump
This pumps fuel upward, ready to be dispensed. It also contains an electronic system that checks for leaks along the lines and a valve that prevents tanks from being overfilled.

Higher-grade fuel tank
This gas fuel is more refined, meaning it burns better. It may also contain additives that help keep the engine clean.

Diesel tank

Unleaded gas tank
This stores standard unleaded fuel. Fuel containing lead was banned around the world between 1986 and 2021 due to causing harmful exhaust emissions.

Fuel types
Gas stations may offer many types of gas but still have just two tanks. The pumps mix gas from the high grade and the unleaded tank at different ratios to create additional grades.

Transportation and the environment

There are more cars, trains, ships, and planes traveling around the globe than ever before. This huge volume of traffic has a significant impact on the environment.

Most vehicles produce emissions—harmful gases that either cause air pollution or contribute to global warming. While modern technologies can reduce the level of emissions that spew directly into the air, the construction of vehicles also uses huge amounts of energy. However, along with producing and promoting alternative fuels, manufacturers are always developing new ideas to make the ways we travel more sustainable in the long term.

REDUCING EMISSIONS

A car's exhaust system is designed to process the waste gases produced in the engine and channel them away from the passengers. The exhaust manifold collects these from the engine cylinders and feeds them through a catalytic converter, which converts them into less harmful substances. The remaining gases pass through a muffler, which reduces the noise the vehicle produces, then they are expelled out the tailpipe at the rear of the car.

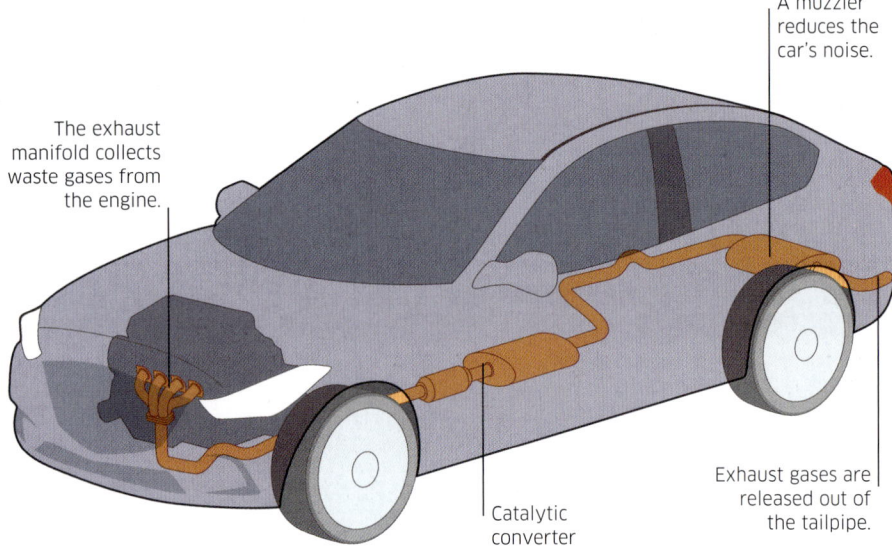

A muzzler reduces the car's noise.

The exhaust manifold collects waste gases from the engine.

Catalytic converter

Exhaust gases are released out of the tailpipe.

Catalytic converter

An invention developed in the middle of the 20th century, catalytic converters get rid of more than 90 percent of the most harmful substances a car produces. As pollutants such as carbon monoxide pass over the converter's sievelike surface, reactions happen where these substances are broken apart into cleaner and safer substances, such as carbon dioxide and nitrogen.

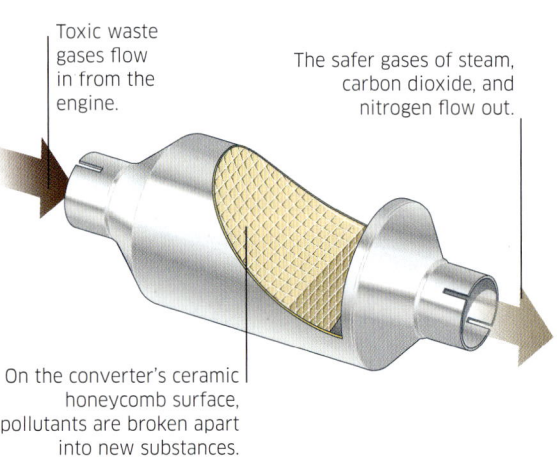

Toxic waste gases flow in from the engine.

The safer gases of steam, carbon dioxide, and nitrogen flow out.

On the converter's ceramic honeycomb surface, pollutants are broken apart into new substances.

HOW OIL IS MADE

Gas is derived from crude oil—one of several fossil fuels that emit greenhouse gases when burned. Oil is called a fossil fuel because it forms over millions of years from the remains of decomposed plants and animals. To access it, huge oil rigs at sea and on land drill down into the ground where the oil is buried.

1 ANCIENT CREATURES
Millions of years ago, tiny plants and animals, known as plankton, drifted in the ancient seas. When they died, their bodies sank to the bottom of the ocean and began to decompose.

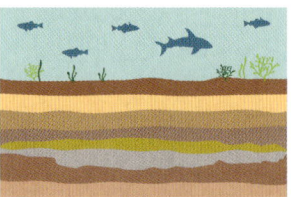

2 SEDIMENT BUILDS
Layers of sediment formed on top of the bodies, crushing them. As more and more layers built up, they put pressure and heat on the remains, which converted them into oil.

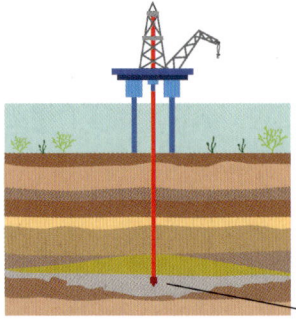

3 DRILLING FOR OIL
Millions of years later, oil and natural gas have formed and lie deep beneath the seabed. Oil rigs and drilling ships extend lines to depths of more than 39,000 ft (12,000 m) to extract the oil.

Oil deposit

Gas cars vs. electric cars

Electric cars are growing in popularity and are seen as a green alternative. Despite high emissions generated during production, their lifetime emissions are lower. The table below sets out the key differences between each type of vehicle.

	Electric	Gas
Production	Lots of energy used in production. Rare metals are needed for batteries.	Less energy used in production due to lack of battery.
On the road	Charging points are not widely available.	Gas stations are easy to access for refuelling.
Emissions & pollution	Use electricity off the grid that may come from fossil fuels	Emit carbon dioxide and pollutants and cause noise pollution
Cost to consumer	More expensive, but likely to get cheaper in the future	Relatively cheaper to purchase
Lifetime environmental impact	Overall lower lifetime emissions	Higher lifetime emissions

3.3 billion tons (3 billion tonnes)—the amount of **carbon dioxide emitted** by passenger cars in 2020.

1993 The year **catalytic converters became mandatory in all cars sold in the European Union** and the UK.

41

THE GREENHOUSE EFFECT

As the Sun warms the Earth, greenhouse gases trap some of its heat in the atmosphere—a process called the greenhouse effect. Emissions from transportation, industry, and many other sources release more of these gases, such as carbon dioxide, into the atmosphere, causing more heat to be trapped and gradually raising the temperature of the Earth.

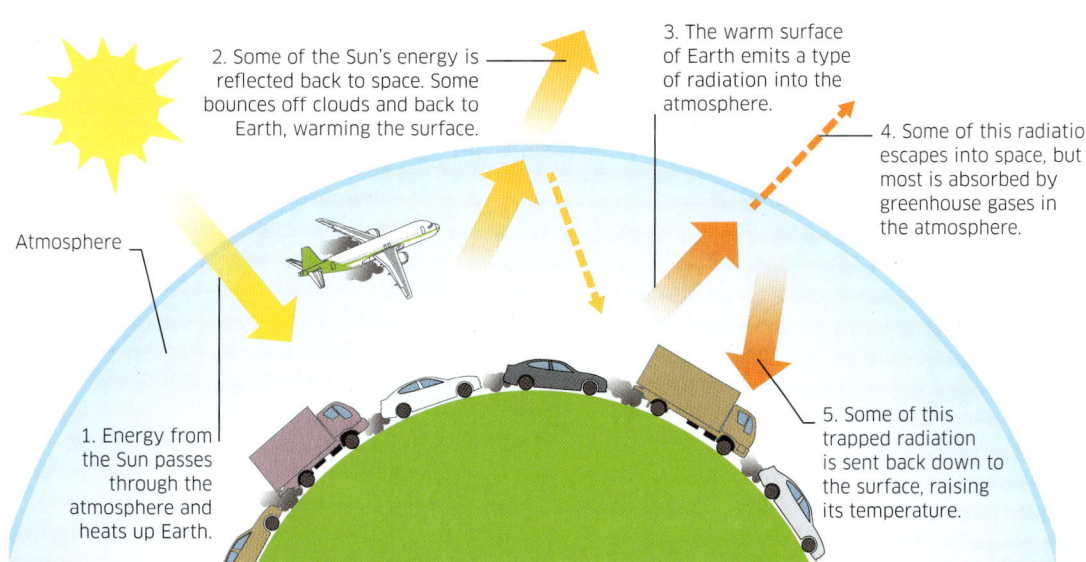

2. Some of the Sun's energy is reflected back to space. Some bounces off clouds and back to Earth, warming the surface.

3. The warm surface of Earth emits a type of radiation into the atmosphere.

4. Some of this radiation escapes into space, but most is absorbed by greenhouse gases in the atmosphere.

Atmosphere

1. Energy from the Sun passes through the atmosphere and heats up Earth.

5. Some of this trapped radiation is sent back down to the surface, raising its temperature.

TRANSPORTATION EMISSIONS

Around one-third of all carbon dioxide emissions come from transportation. As more people from around the globe come to own cars for the first time and to travel by train and plane, emissions have increased.

Emissions by transport type

Road transportation produces the vast majority of carbon dioxide emissions, with more of this coming from passenger transportation, such as cars and buses. In comparison, global rail produces smaller amounts of carbon dioxide.

Road (Passenger) 45.1%	Road (Freight) 29.4%	Aviation 11.6%		

GLOBAL TRANSPORTATION EMISSIONS IN 2018

Shipping 10.6%

Rail 1%

Other 2.2%

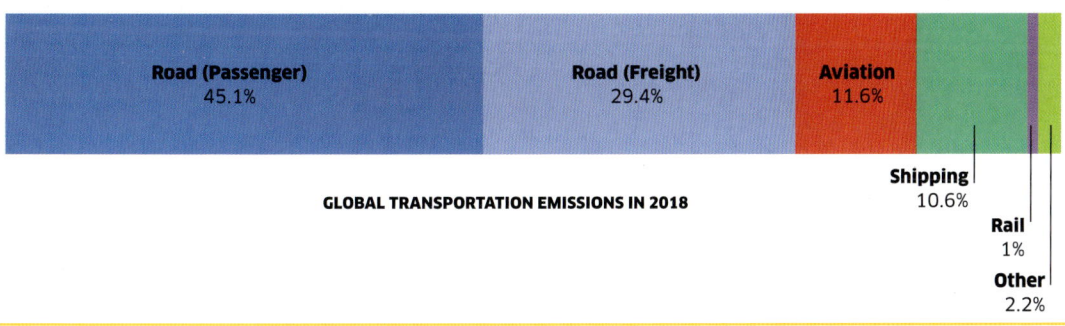

Short-haul flight

Long-haul flight

Gas car

Rail

Bus (Half-full)

Bike

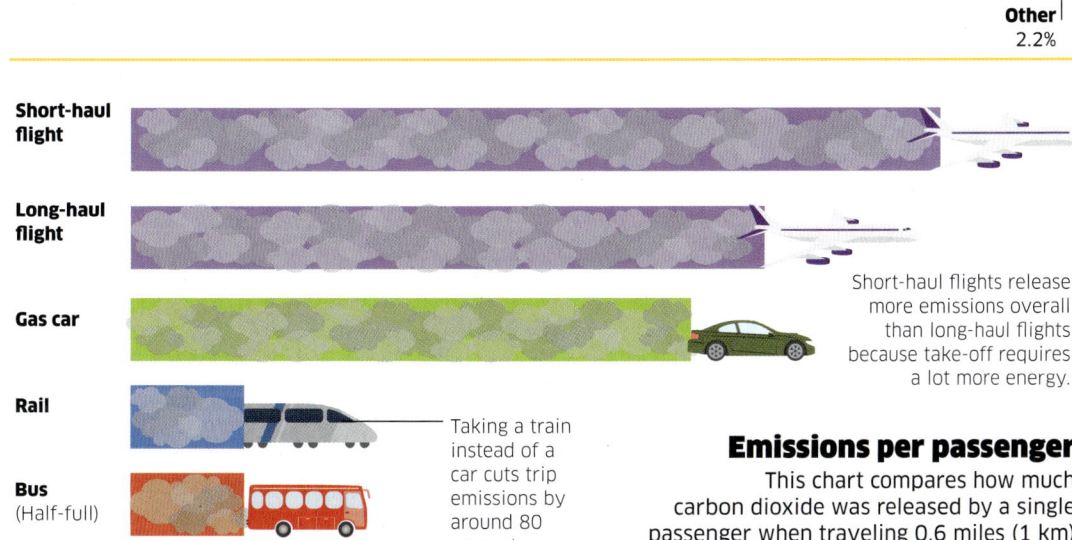

Short-haul flights release more emissions overall than long-haul flights because take-off requires a lot more energy.

Taking a train instead of a car cuts trip emissions by around 80 percent.

Emissions per passenger

This chart compares how much carbon dioxide was released by a single passenger when traveling 0.6 miles (1 km) in a variety of different vehicles. Using transportation methods that carry multiple people, such as buses or trains, reduces an individual's emissions.

SUSTAINABLE FUTURE

Around the world, governments, engineers, and urban planners are looking at ways we can reduce the impact of transportation on the environment. New technologies and alternative fuels, such as hydrogen or biofuels, are being developed to make cars cleaner. Promoting public transportation and other ways of reducing journeys can also have an impact.

Solar-powered rail

This railway tunnel in Belgium is equipped with solar panels, which help power the network. Scientists in India are also trialing solar panels on top of trains to provide power directly.

Biofuels

This bus traveling on a guided busway in the UK runs on biofuels—fuels made from plants. These fuels produce no emissions, but making them can involve using huge amounts of land.

Using AI

In the future, AI could help make transporting goods more efficient. By working out where to combine freight from different sources and what the best journeys are, it could cut emissions.

Driverless car

Self-driving cars are able to navigate without human input and are in development all over the world. They are predicted to become more widespread in the next 20 years and may use more environmentally friendly fuels, such as hydrogen.

Driverless, or autonomous, cars rely on sensors to identify road markings, signals, and lanes and to detect and track other vehicles, pedestrians, and obstacles. This data is acted upon by the car's controller, which adjusts wheel speed and angle to guide the vehicle to its destination. Future driverless cars are likely to be powered by fuel cells that convert hydrogen into electricity to power the car's motors.

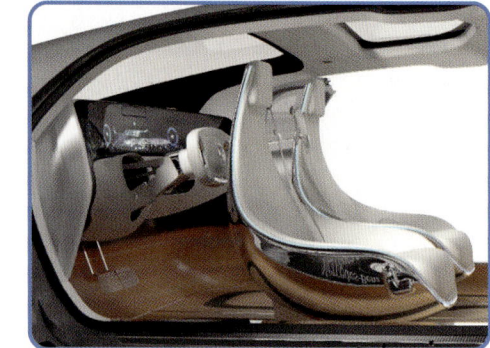

Clever controls
Virtual controls, gauges, and data panels are projected onto the driver dashboard. Even when the steering wheel is gripped by a human driver, the car's driverless features can operate to provide assistance and aid safe driving.

Sunroof
The roof is coated in nanoparticles that can adjust how much or how little light is let into the passenger cabin.

MERCEDES-BENZ F015

Origin: Germany

Year: Future concept

Top speed: 125 mph (201 kph)

Length: 17 ft (5.2 m)

Body panels
Light but strong carbon-fiber composite materials are fitted over an aluminum and steel frame.

Hood
With no bulky internal combustion engine, the car's front end can be sculpted low and aerodynamic so air flows smoothly past.

RADAR sensor
Mounted in the front grill, this device uses radio waves to detect moving vehicles and objects ahead of the car.

Air intakes
These suck air into the car's power system. The oxygen in the air is used by a fuel cell to produce electricity.

LED lamps
A cluster of small, bright LED lamps show whether the car is being driven by a person or is in autonomous mode.

Driverless saloon

This sleek saloon is designed as a high-end concept car. It seats four and is propelled by hydrogen fuel cells that generate electricity and produce water as the only emission. Computers and a sophisticated sensor package control and navigate the car in driverless mode. The vehicle can also be driven by a person in manual mode.

380 miles (612 km)–the **distance a Hyundai Nexo Blue** can **travel** on just 13.9 lb (6.3 kg) of **hydrogen fuel.**

1,000-19,000 gigabytes–the estimated **amount of data generated** by a future **driverless car's sensors every hour.**

43

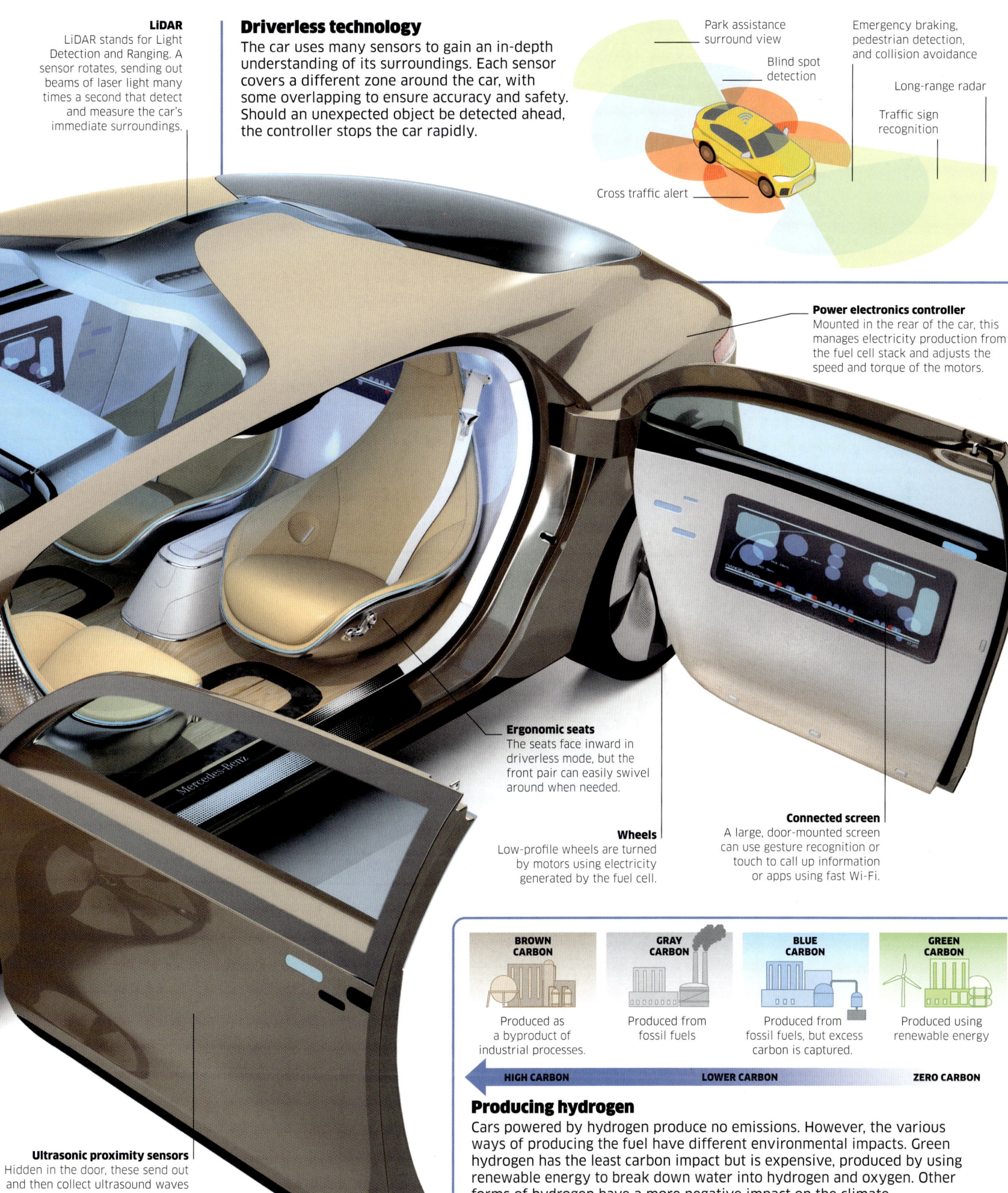

LiDAR
LiDAR stands for Light Detection and Ranging. A sensor rotates, sending out beams of laser light many times a second that detect and measure the car's immediate surroundings.

Driverless technology
The car uses many sensors to gain an in-depth understanding of its surroundings. Each sensor covers a different zone around the car, with some overlapping to ensure accuracy and safety. Should an unexpected object be detected ahead, the controller stops the car rapidly.

Park assistance surround view

Blind spot detection

Emergency braking, pedestrian detection, and collision avoidance

Long-range radar

Traffic sign recognition

Cross traffic alert

Power electronics controller
Mounted in the rear of the car, this manages electricity production from the fuel cell stack and adjusts the speed and torque of the motors.

Ergonomic seats
The seats face inward in driverless mode, but the front pair can easily swivel around when needed.

Wheels
Low-profile wheels are turned by motors using electricity generated by the fuel cell.

Connected screen
A large, door-mounted screen can use gesture recognition or touch to call up information or apps using fast Wi-Fi.

Ultrasonic proximity sensors
Hidden in the door, these send out and then collect ultrasound waves to measure the precise distance away from nearby objects.

BROWN CARBON
Produced as a byproduct of industrial processes.

GRAY CARBON
Produced from fossil fuels

BLUE CARBON
Produced from fossil fuels, but excess carbon is captured.

GREEN CARBON
Produced using renewable energy

HIGH CARBON · LOWER CARBON · ZERO CARBON

Producing hydrogen
Cars powered by hydrogen produce no emissions. However, the various ways of producing the fuel have different environmental impacts. Green hydrogen has the least carbon impact but is expensive, produced by using renewable energy to break down water into hydrogen and oxygen. Other forms of hydrogen have a more negative impact on the climate.

LAUFMASCHINE
Dandy horse
Origin: Germany
Year: 1817

The first bicycle to be sold in large numbers, the *Laufmaschine*, also called "dandy horse," had no pedals. Riders sat on the cushioned leather seat and kicked along with their feet to get up to speed.

BIANCHI PARIS-ROUBAIX
Classic lightweight
Origin: Italy
Year: 1951

This steel-framed racing bike was fitted with a revolutionary four-speed gearshift lever that was attached to the rear fork. It required the rider to reach backward to pull it!

Spare tire

Gearshift lever

BOWDEN 300
Cruiser
Origin: US
Year: 1961

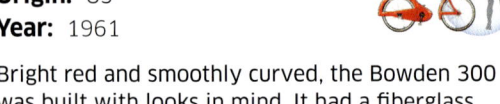

Bright red and smoothly curved, the Bowden 300 was built with looks in mind. It had a fiberglass frame and built-in lights that made it heavy on hill climbs but perfect for cruising by the beach.

THE LIGHT ROADSTER
Penny farthing
Origin: Germany
Year: 1892

Nicknamed the "penny farthing" after different-sized coins of the era, this odd-looking bike had a large front wheel (the "penny") that could easily roll over unpaved roads.

WOLFE AMERICAN ICEBIKE
Part bike, part sled
Origin: US
Year: 1920

Designed for use on frozen lakes and rivers, this icebike was adapted from a standard bicycle frame. The spikes on the rear wheel gripped the ice, while a sled replaced the front wheel.

Spikes

Sled steered by handlebars

Pedal power

One of the simplest and most popular means of transport, bicycles allow people to go four or five times faster than walking using the same amount of energy.

Most bikes use the same method of harnessing human power. Pushing the pedals drives a chain that transmits power to the rear wheel, which starts to rotate, moving the bike forward. Handlebars are used to turn the front wheel and change the bike's direction.

"Ape hangers"
These handlebars got their name because riders have to reach up and grab on like an ape hanging from a tree.

Gear shifter
Early Choppers had three gears, changed through a shifter styled like a car's.

RALEIGH CHOPPER
Wheelie bike
Origin: UK
Year: 1970s

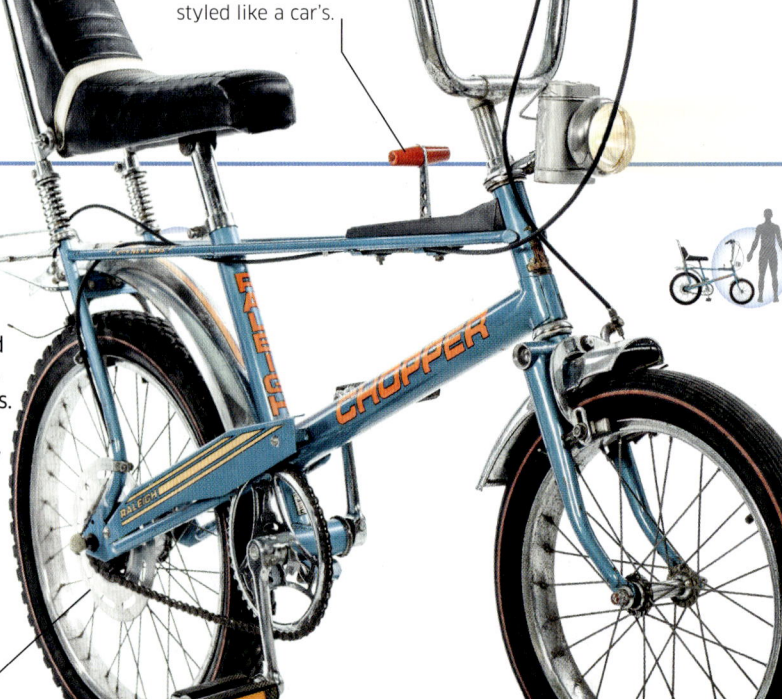

With its raised handlebars and padded seat, the Chopper was inspired by touring motorcycles. The saddle was mounted over the rear wheel, making it easy to lean back and lift the front wheel off the ground, earning this kind of bicycle the nickname "wheelie bike."

Spoke protector designed to look like a motorcycle disc brake

21 The number of day-long stages that make up the Tour de France, cycling's most **famous** race.

In **2016**, the city of **Copenhagen** announced that it had more **bikes** than **cars** on its streets.

More than **2 billion** bicycles are currently in use **around the world**.

45

SCOTT REFLEX
Mountain bike
Origin: Switzerland
Year: 2007

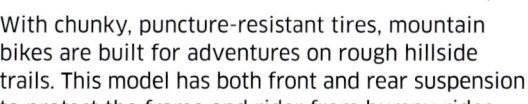

With chunky, puncture-resistant tires, mountain bikes are built for adventures on rough hillside trails. This model has both front and rear suspension to protect the frame and rider from bumpy rides.

GT PERFORMER
BMX
Origin: US
Year: 1985

BMX stands for "Bicycle Motocross," a sport where riders race on a dirt track. This model is a freestyle BMX designed for performing aerial tricks. Like all BMXs, it has a small, squat frame and a single gear.

Tucked-away cables
The front brake cable passed through the frame to avoid tangles when spinning the handlebars.

Three thick spokes
These wheels had an eye-catching design but lacked strength.

Freestyle BMX
was recognized as an Olympic sport in time for the 2020 Summer Olympic Games.

KINGCYCLE BEAN
Recumbent bicycle
Origin: UK
Year: 1990

Recumbent bicycles, where the rider lies down, are faster than ordinary bikes because they are more aerodynamic. This machine had a streamlined shell that allowed it to reach speeds of 57 mph (90 kph).

Windshield
This protective shield had to be taped on after the rider climbed in.

External shell made of fiberglass

BABBOE CURVE
Cargo bicycle
Origin: Netherlands
Year: 2015

With a large container on the front, cargo bicycles make it possible to transport heavy loads by bike. This model is often used in Dutch cities to take children to school.

CYFAC LE DUO CARBONE
Tandem
Origin: France
Year: 2015

First invented in 1890, tandem bikes allow a pair of riders to bring twice the power to a bicycle journey. Modern models are built from lightweight carbon fiber, so they require less effort to pedal.

Rear handlebars
The back rider's bars provide a place to hold on, but they do not steer the bike.

GOCYCLE G4I
Electric foldable bike
Origin: UK
Year: 2021

Bicycles fitted with an electric motor can take the strain off riders by supplying power to either the pedal cranks or the wheel hub. This model even folds for convenience.

Motorcycles and scooters

These nifty vehicles offer powered transport in a smaller, more convenient, and often more affordable form than cars as they are cheaper to buy and run.

The appeal and freedom of two- and three-wheel travel attracts millions of enthusiasts. Some run their motorcycles on and offroad for fun, leisure, adventure, or competition, whereas others use their vehicles for easy commutes to and from work. Bikes come in many different shapes and sizes—from small urban scooters to heavy long-distance cruisers and powerful performance sports motorcycles.

Handlebars
A single handlebar twist grip controlled both engine speed and the rear brake.

Saddle

DAIMLER REITWAGEN
Pioneering motorcycle
Origin: Germany
Year: 1885

Gottlieb Daimler and Wilhelm Maybach's machine featured a wooden frame and leather saddle. A leather belt transferred power from the engine to the rear wheel, giving it a top speed of 6.8 mph (11 kph).

Iron-clad wheels

Balance aid
Small outer wheels balanced the motorcycle.

HARLEY-DAVIDSON 8A
Vintage motorcycle
Origin: US
Year: 1912

A popular early motorcycle, this Harley-Davidson was nicknamed "Silent Gray Fellow" for the quiet running of its 494cc single-cylinder, single-gear engine and its muted paint scheme.

Fuel tank runs along top of frame

Underseat suspension

Rear mudguard

HONDA C100 SUPER CUB
Scooter
Origin: Japan
Year: 1958

Fitted with large plastic fairings and bigger-diameter wheels than traditional scooters, the world's most common motor vehicle remains popular today, especially in southeast Asia.

Fairings
A plastic shell covers the frame.

LAMBRETTA LD 125 INNOCENTI
Classic scooter
Origin: Italy
Year: 1951

More than 110,000 of these scooters were built, featuring sprung seats and an easily accessible foot platform. Switching between three gears, they could reach a top speed of 47 mph (76 kph).

Drum brake
This brake slows the bike by pushing on the inside of a drum within the wheel.

Smart use of space
Beneath the fairing that covers the rear-mounted engine is space for storage.

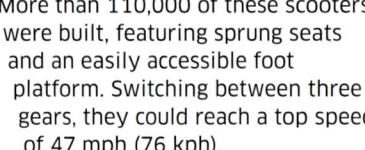

HONDA CB750
Road motorcycle
Origin: Japan
Year: 1969

With an upright riding style but racetrack levels of power, this popular motorcycle had a four-cylinder engine and five-speed gearbox. It possessed a top speed of more than 125 mph (200 kph).

The **GL1800 Goldwing** was the first motorcycle to **have an airbag.**

3 secs–the time it takes a **speedway motorcycle** to **accelerate from 0-60 mph** (0-96 kph).

When the **inventor's son Paul** first tested the **Daimler Reitwagen, its saddle caught fire!**

47

SUZUKI GSX-R1100 WR
Sports motorcycle
Origin: Japan
Year: 1994

This powerful motorcycle mimics track racing vehicles, with a low riding position and a high-performance engine delivering more horsepower than many small cars. Its top speed is 169 mph (272 kph).

KAWASAKI KX250
Offroad motorcycle
Origin: Japan
Year: 1990

These light but tough motorcycles can travel over all sorts of tough terrain on motocross tracks. Strong front and rear wheel suspension soaks up the impact forces of bumps and jumps.

Engine
The two-stroke engine delivers plenty of power.

ELECTRA GLIDE FLHS
Touring motorcycle
Origin: US
Year: 1987

This touring motorcycle with a 1337cc engine was designed for comfortable cruising over long distances. It had a windshield; panniers; a 5-gallon (19-liter) fuel tank; and a long, comfortable saddle.

AUTO RICKSHAW
Custom built
Origin: Varies
Year: 20th century

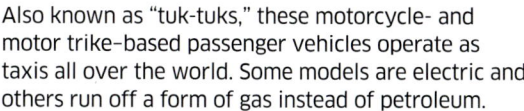

Also known as "tuk-tuks," these motorcycle- and motor trike–based passenger vehicles operate as taxis all over the world. Some models are electric and others run off a form of gas instead of petroleum.

PIAGGIO MP3 500
Motor trike
Origin: Italy
Year: 2006

Three wheelers offer greater stability than traditional motorcycles. This tilting trike has antilock braking, cruise control, and a suspension system that angles both front wheels when turning a corner.

HARLEY-DAVIDSON LIVEWIRE
Electric motorcycle
Origin: US
Year: 2019

With a lightweight aluminum frame and an engine delivering 105 horsepower, this electric motorcycle is speedy and maneuverable. It takes just one hour to fully recharge.

WESLAKE SPEEDWAY
Speedway motorcycle
Origin: UK
Year: 1981

Speedway motorcycles race without brakes around tight, oval-shaped dirt tracks. Weighing just 183 lb (83 kg), this world championship–winning motorcycle had fearsome acceleration and a top speed of 75 mph (120 kph).

Small engine
The 499cc single-cylinder engine is small but powerful.

Mudguards
These cover parts of both wheels and prevent debris from reaching the rider.

Rear wheel
This simple stripped-down wheel has no suspension at all.

Ridged tires
Knobby tread tires help give grip in the dirt.

Sports motorcycle

Able to accelerate and slow down very quickly, sports motorcycles are elite vehicles. They mimic the speed, handling, and overall performance of racing motorcycles but are safe and legal to use on the road.

Sports motorcycles tend to offer a powerful engine, sharp acceleration, and a streamlined body shell that allows air to flow smoothly around them. Riders often travel on these motorcycles in a tucked-down position–lowering their head and body toward the motorcycle to reduce the air resistance generated and achieve greater speeds.

YAMAHA YZF-R6

Origin: Japan

Year: 2015

Length: 80¼ in (2.04 m)

Top speed: 161½ mph (260 kph)

Motorcycle braking

Most motorcycles use a type of brake called a disc brake. When the brake lever on the handlebars is squeezed, fluid is pushed down a tube to force pads to press on a disc attached to the motorcycle's wheel. The pads generate lots of friction, slowing the wheel down.

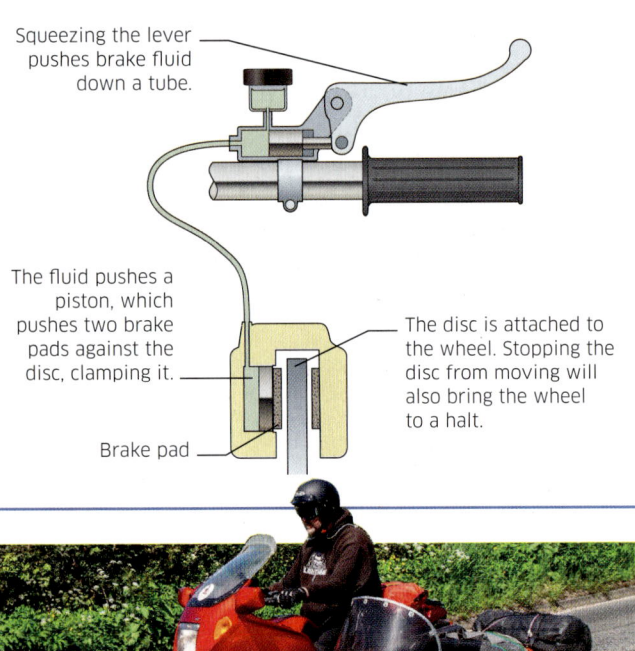

Squeezing the lever pushes brake fluid down a tube.

The fluid pushes a piston, which pushes two brake pads against the disc, clamping it.

The disc is attached to the wheel. Stopping the disc from moving will also bring the wheel to a halt.

Brake pad

Sidecars

Two wheels can become three with the addition of a sidecar, which can be attached to some motorcycles. Invented in 1893, single-wheeled sidecars can hold a passenger, goods, or–for military vehicles–a gunner and machine gun.

Rearview mirror

Front fairing
The protective cowling prevents riders from being buffeted by wind and streamlines the motorcycle.

Front fork
Holding onto the wheel, this fork provides suspension at the front of the motorcycle.

Front brake disc
Attached to the front wheel, this large disc slows the motorcycle when brake pads push on it.

53 million The estimated total **number of motorcycles sold** worldwide in 2021.

3 seconds—the **time it takes** for a YZF-R6 to **accelerate** from 0-60 mph (0-96 kph).

49

Yamaha YZF-R6

This supersport motorcycle pairs a sleek design with an ultra-high revving engine capable of delivering more than 87 kW of power. It has been upgraded every year since its introduction to include even more high-end features.

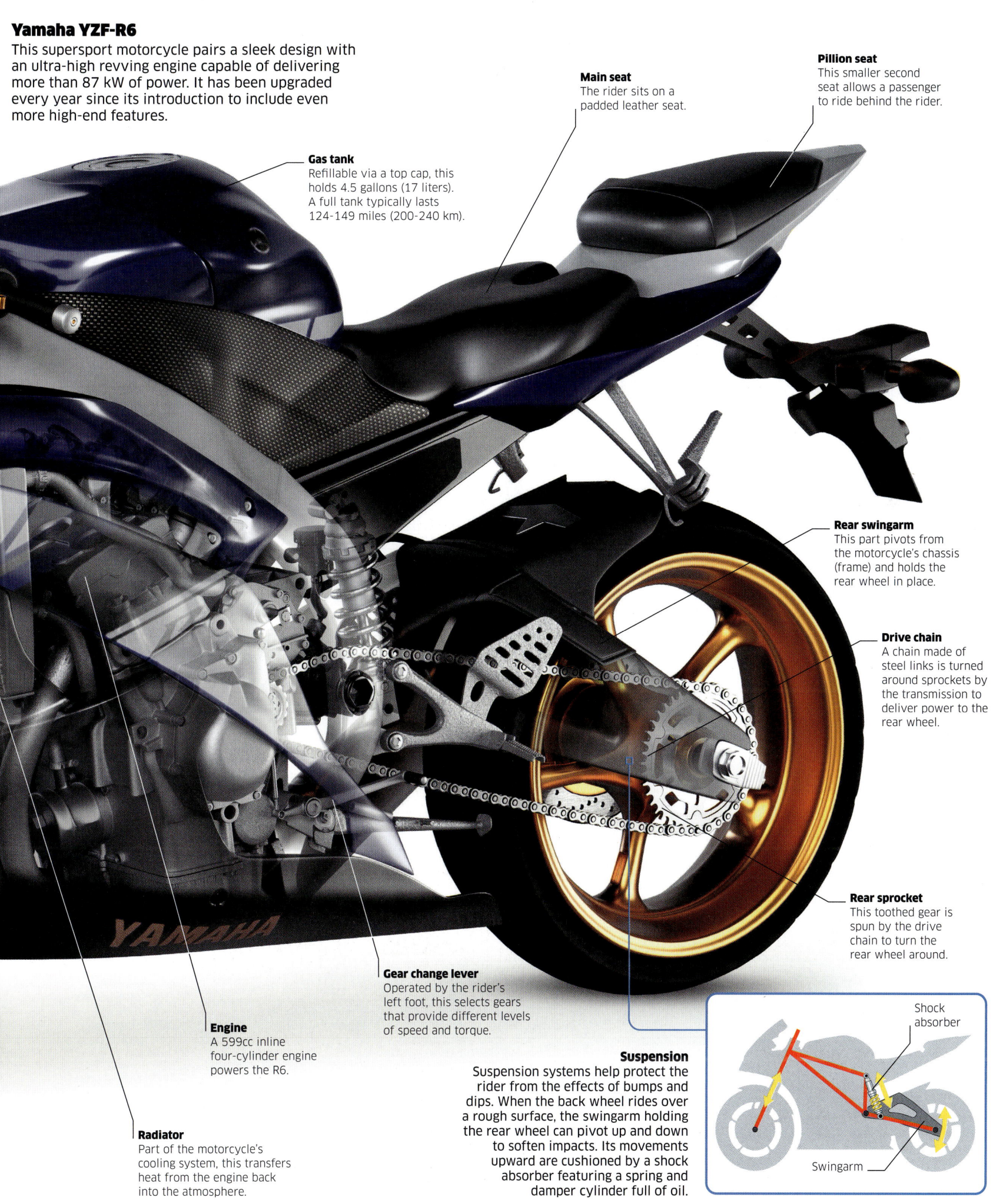

Main seat
The rider sits on a padded leather seat.

Pillion seat
This smaller second seat allows a passenger to ride behind the rider.

Gas tank
Refillable via a top cap, this holds 4.5 gallons (17 liters). A full tank typically lasts 124-149 miles (200-240 km).

Rear swingarm
This part pivots from the motorcycle's chassis (frame) and holds the rear wheel in place.

Drive chain
A chain made of steel links is turned around sprockets by the transmission to deliver power to the rear wheel.

Rear sprocket
This toothed gear is spun by the drive chain to turn the rear wheel around.

Gear change lever
Operated by the rider's left foot, this selects gears that provide different levels of speed and torque.

Engine
A 599cc inline four-cylinder engine powers the R6.

Radiator
Part of the motorcycle's cooling system, this transfers heat from the engine back into the atmosphere.

Suspension
Suspension systems help protect the rider from the effects of bumps and dips. When the back wheel rides over a rough surface, the swingarm holding the rear wheel can pivot up and down to soften impacts. Its movements upward are cushioned by a shock absorber featuring a spring and damper cylinder full of oil.

Shock absorber

Swingarm

Jacked-up body
This truck's body is based on a 2006 Chevrolet Silverado pickup and has been raised high above the wheels.

Trucks

These hardworking motor vehicles clock impressive distances as they travel along the world's roads, usually transporting a wide range of goods.

Trucks deliver millions of tons (tonnes) of cargo by road each day—from parcels to livestock, and chemicals to cement. Most use diesel engines, although a small but increasing number of trucks are powered by electricity, compressed natural gas, or biofuels. Some trucks perform specialized roles, such as cleaning streets or entertaining crowds.

MONSTER TRUCK
Viking Monster Trucks' Thor
Origin: US/Sweden
Length: Approx. 14⁶/₈ ft (4.5 m)

Heavily modified vans and pickup trucks can be fitted with huge suspension systems and tires. Powered by engines burning methanol, they race around arena circuits and perform stunts.

MOTORHOME
Newmar Dutch Star
Origin: US
Length: 38–43 ft (11.6–13.1 m)

These large recreational vehicles (RVs) contain living and sleeping accommodations for between two and eight people, often including kitchens and showers.

Terra tires
Tires of 66 in (1.68 m) in diameter make up around a third of the monster truck's weight, enough to crush these cars.

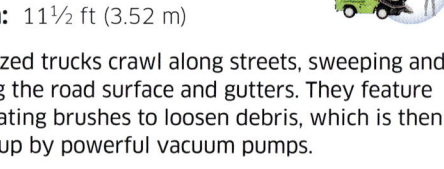

MINI TRUCK
Piaggio Ape truck
Origin: Italy
Length: 8¹/₈ ft (2.5 m)

Local deliveries and small building projects make use of these compact trucks with short wheelbases. Some, like this model, feature a single front wheel for tighter turns in crowded urban areas.

FORKLIFT TRUCK
Toyota Traigo 70
Origin: Japan
Length: Varies

These versatile lifters and shifters operate in many storerooms, delivery centers, and farms. Their forks rise up and down vertical masts to raise and lower loads that can weigh many tons (tonnes).

STREET SWEEPER
Green Machines 636
Origin: UK
Length: 11½ ft (3.52 m)

Specialized trucks crawl along streets, sweeping and cleaning the road surface and gutters. They feature fast-rotating brushes to loosen debris, which is then sucked up by powerful vacuum pumps.

237⅝ ft (72.42 m)—the **world record for the longest ramp jump** by a monster truck, set by **Bad Habit** in 2013.

More than **one-third** of all transport-related carbon emissions are produced by trucks.

51

POLAR TRUCKS
KamAZ-43114 offroad truck
Origin: Russia
Length: 26 ft (7.96 m)

Designed for work in Antarctica or the Arctic, these tough trucks are equipped with all-wheel drive and deep tread tires to help them maintain traction over slippery or icy ground.

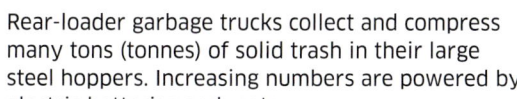

DROPSIDE TRUCK
Foton Miler
Origin: China
Length: 16⅝ ft (5.08 m)

Typically small- or medium-sized, dropside trucks are equipped with flatbed trailers surrounded by hinged sideboards and a tailboard that can fold down to enable easy loading and unloading.

RIGID BOX TRUCK
DAF LF Electric
Origin: Netherlands
Length: 35⅜ ft (10.8 m)

These unarticulated trucks feature a solid, roofed trailer body behind the driver's cab. Rigid box trucks carry a great variety of cargo, frequently stored on wooden pallets so it can be unloaded quickly.

GARBAGE TRUCK
Isuzu NQR 70
Origin: Japan
Length: 29⅛ ft (8.9 m)

Rear-loader garbage trucks collect and compress many tons (tonnes) of solid trash in their large steel hoppers. Increasing numbers are powered by electric batteries and motors.

TOW (RECOVERY) TRUCK
Mercedes-Benz 815 Atego 4x2
Origin: Germany
Length: 27⅛ ft (8.28 m)

Tow trucks can either tow a broken-down vehicle behind them or use a crane or winch to lift it up onto their flatbed, as here. They then transport the vehicle to a garage for repairs.

TANKER TRUCK
Renault Premium 460
Origin: France
Length: Varies

While some tankers are rigid trucks, many are articulated trucks pulling trailers. These can include tanker trucks with sealed tanks of liquid foodstuffs, chemicals, oil, or even sewage.

CAR TRANSPORTER TRUCK
Scania P230 DB
Origin: Sweden
Length: Varies

Cars, vans, and pickup trucks are driven onto a transporter trailer's rails and secured in place with wheel clamps or harnesses. The trailer can then be hauled by a truck.

Overhanging car
To fit the maximum amount of vehicles, one car sits above the driver's cab.

Two decks
A lower and higher platform both store multiple vehicles.

Refrigerated trailer
Also called a "reefer," this transports goods at a controlled temperature.

Front of trailer supported by semi truck's rear wheels

Cab communications
The latest trucks contain a range of electronic driving aids to help on long-haul trips. This includes lane assist and adaptive cruise control, which uses radar to keep the truck safe distances from vehicles ahead. Drivers are also able to communicate with other truckers via citizen band (CB) or VHF radio or with smartphone apps.

Opening up

Popular in Europe and Asia, cab-over trucks have the engine and front axle immediately below the driver. When the engine needs to be accessed, a hydraulic system tilts the cab up and forward. Trucks with cab-over designs have better visibility and a slightly shorter turning circle.

Semi truck

These tireless haulers form the backbone of the world's road freight transport. A typical semi truck travels 80,800 miles (130,000 km) a year—more than three times around the world.

Also known as tractor units or "big rigs," semi trucks can pull a wide range of different trailer types—from refrigerated box trailers for food and flatbeds that carry shipping containers, to specialized livestock or car transporters.

Engine cover

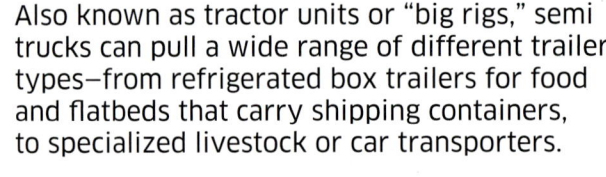

Australian road train

In Australia, trucks can be much longer and larger than most other places in the world. Vast road trains made of a single truck pulling three or four trailers travel across the country. These vehicles can be up to 175½ ft (53.5 m) long and weigh as much as 150 tons (136 tonnes).

Heavy-duty hauler

This large North American semi truck features a deep bay for its powerful diesel engine and 10 wheels split across three axles. Designed for long-distance trucking, the large sleeper cab contains living, sleeping, and storage space.

Fifth wheel

A large greased plate, the fifth wheel, sits above the truck's rear wheels. It is designed so that a kingpin—an attachment that sits underneath a trailer—can easily slot into it. Once coupled securely, the kingpin can rotate in the fifth wheel, providing a pivot point around which the truck and trailer can turn corners with ease.

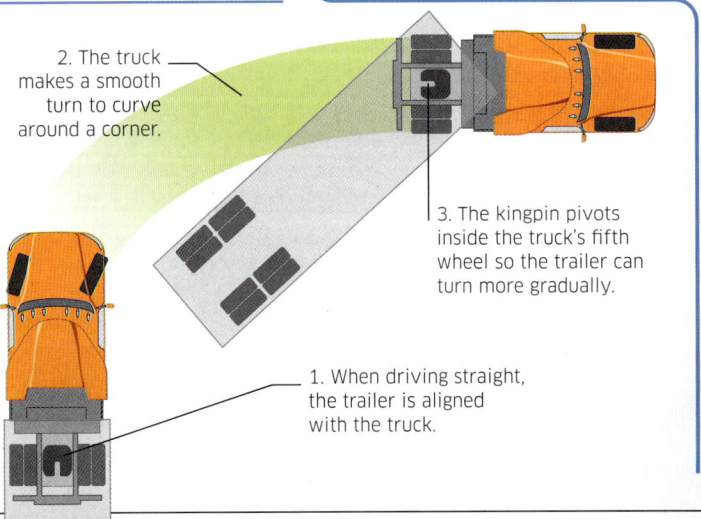

2. The truck makes a smooth turn to curve around a corner.

3. The kingpin pivots inside the truck's fifth wheel so the trailer can turn more gradually.

1. When driving straight, the trailer is aligned with the truck.

Air intake
A radiator behind the grille cools the engine.

Exhaust stack
This expels waste engine gases high and clear of the truck to avoid smoke clouding the road.

Gearshift
A typical semi truck has 10 forward gears and two reverse gears.

Wind deflector
The sloping cab roof helps reduce air resistance, which improves fuel efficiency.

Sleeper storage
As truckers can be away for weeks at a time, the cab contains storage for clothes and personal items.

Rear-mounted camera
A camera at the back of the cab streams live views of the area behind the truck for reversing or operating in tight spaces.

Cab porthole
These small windows admit light and ventilation into the living space to reduce reliance on electric lights and fans.

Bed
A short bed space allows the driver to get crucial rest when pulled into a rest stop or truck stop.

Fifth wheel

Storage area or "hold"

Truck fridge
Providing cool storage for the driver's food, the mini fridge is powered by the truck's batteries or auxiliary power unit.

Air suspension seat
Pressurized air cushions the driver's movements to provide a comfortable ride.

Fuel tank
This large 140-gallon (530-liter) fuel tank is one of two. Their vast size enables long-distance hauling with minimal stops.

FREIGHTLINER TRUCK

Origin: US

Year: 1942

Length: Varies

Hauling capacity: Up to 40 tons (36 tonnes)

FOWLER GYROTILLER
Tracked cultivator
Horsepower: 170 hp
Year: 1935

This machine had a rear cultivator that used two rotating rings to break up the soil for planting sugarcane. Long spikes called tines allowed it to penetrate deep into the earth.

Operator's seat

Directional arrow
The driver used this arrow, which turned with the wheel, to see where the vehicle was heading.

Two sets of four tines

Front wheel

PLEASE KEEP OFF

Down on the farm

Driven by powerful engines and mounted on gigantic wheels or tracks, farming vehicles handle a vast range of jobs that were once done by humans and animals.

Early tractors were towing devices that used chains to drag equipment around the farm. They were revolutionized in the 1940s with an invention that made it possible to attach machinery securely to a tractor. This "three-point linkage" meant the weight of a tool like a plow was transferred from the ground to the tractor's rear wheels. This was safer and more efficient. Today, there are a range of farming machines, from those that use tracks instead of wheels, to combine harvesters and compact garden vehicles.

FERGUSON TE-20
All-purpose fieldhand
Horsepower: 24 hp
Year: 1947

Ferguson tractors were the first to adopt the three-point linkage system that wheeled tractors still use today. This allowed farmers to attach and power different tools quickly and easily.

Large rear wheels

JOHN DEERE 5430I
Crop sprayer
Horsepower: 215 hp
Year: 2008

Farmers sometimes spray their crops with chemicals that protect them from weeds, insects, or disease. This tractor has long arms that extend to either side and are capable of spraying a wide area.

Extendable booms
The arms on either side of the tractor are called "booms." Chemicals travel down the boom from a central tank.

CATERPILLAR TH406
Forklift
Horsepower: 99 hp
Year: 2008

Forklift tractors are specialized for lifting and moving the hay bales that often serve as animal feed. This machine has an extending arm capable of stretching out more than 20 ft (6 m).

CASE IH QUADTRAC 9370
Articulated tracks
Horsepower: 360 hp
Year: 1997

Very heavy tractors run the risk of compacting the soil. This machine, with its set of four small tracks, distributes its weight over a wide area. It bends in the middle to allow easier steering.

JCB FASTRAC 185-65
Rapid transporter
Horsepower: 188 hp
Year: 1994

Built to transport loads at speed, the Fastrac can exceed 46 mph (75 km/h). Most tractors are uncomfortable when driven fast, but this machine has unique suspension to ensure a smooth ride.

153 mph (246 km/h)—**top speed** reached by the **JCB Fastrac 2, the fastest tractor in the world.**

-94°F (-70°C)—the **coldest temperature** in which a tractor has operated, **towing scientific equipment** on a **2013 expedition to Antarctica.**

55

JOHN DEERE 6210R
Fuel-efficient tractor
Horsepower: 210 hp
Year: 2011

To reduce running costs and limit emissions, modern tractors are designed to be fuel efficient. John Deere's 6210R has a power management system to limit fuel use at low speeds or when pulling lighter implements like this cultivator.

HAGIE 204SP
Corn detasseler
Horsepower: 173 hp
Year: 2013

This specialized machine is built with a body that sits high above its narrow wheels so it can pass over the top of cornfields, removing the "tassels" (flowers) from the corn as it goes.

Combo cultivator
This tool uses a variety of blades to break up soil, ready for planting.

Spacious cab
As well as large windows with wide-angle views, the cab has a passenger seat and fridge.

CHALLENGER MT865C
Rubber tracks
Horsepower: 583 hp
Year: 2009

Some farming machines run on tracks that help them tackle rough ground. This machine has rubber tracks, an improvement on the earlier metal tracks that ripped up hard road surfaces.

JOHN DEERE S690I
Combine harvester
Horsepower: 530 hp
Year: 2013

A combine harvester can do multiple jobs that once took many farm workers. This model can automatically change speed to slow down when traveling over rough terrain.

Unloading pipe
When the wheat grain is ready, it is sent through this pipe into a nearby trailer.

Reel
This metal frame rotates as the harvester moves, bending grain so it can be easily cut.

KUBOTA L3200
Compact 4x4
Horsepower: 32 hp
Year: 2014

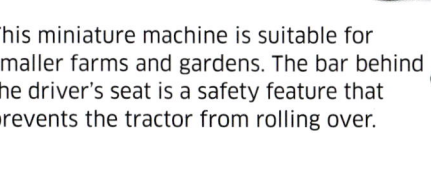

This miniature machine is suitable for smaller farms and gardens. The bar behind the driver's seat is a safety feature that prevents the tractor from rolling over.

NEW HOLLAND T9 700
Articulated tractor
Horsepower: 699 hp
Year: 2014

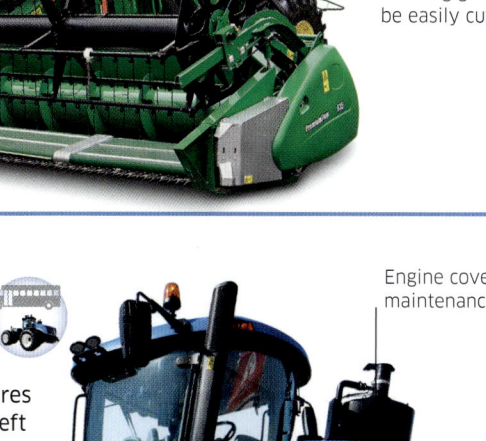

The larger the tractor, the more room it requires to turn, meaning crucial field space must be left empty of crops. To solve that problem, articulated tractors like this one bend around a central point, so they need far less space to maneuver.

Engine cover lifts up for maintenance access

1 Pick-up reel
A long, rotating header on the front of the combine gathers the crop, pulling it toward a cutter bar. Here, sharp blades cut the plant's stalks near their base, close to the ground.

2 Feeder system
A turning drum and moving conveyor belt draws the cut crop inside the combine harvester for processing. Fans help extract much of the dust that builds up, removing it from the cut crop.

Combine cab
The driver uses touchscreens to monitor the vehicle's systems, such as laser sensors and GPS that guide the vehicle on a preset path through the field.

3 Thresher drum
This 2½-ft- (75.5-cm-) diameter drum spins at 330–930 rotations per minute. The crop is beaten and rubbed between the drum and the curved surface below, shaking the grains free of their stalks. Around 70 percent of all the grain is collected this way.

CLAAS LEXION 8900

Origin: Germany
Year: 2019
Length: Varies
Top speed: 25 mph (40 kph)

4,755 gallons (18,000 liters)—the amount of **cereal crop** that can be stored in the Lexion 8900's **grain tank**.

26 The number of **work lights** the Lexion can turn on to **illuminate** its surroundings **when operating at night**.

57

6 Grain elevator
A turning screw called an auger transports the grain up and into the large grain tank.

Fan
This draws in fresh air through a sieve to filter out dust and plant particles. The air is then used to cool the engine compartment.

Crop collector
The Claas Lexion 8900 is one of the world's largest combine harvesters. Driven by a powerful diesel engine, it can deploy a 45-ft-(13.7-m-) wide header in large crop fields—harvesting as much as 105 tons (95 tonnes) of cereal crops, like wheat or rye, every hour.

7 Unloader
This long, pivoting arm is called an unloader. Inside, it is a long turning auger that transfers grain from the combine's elevator to a tank or hopper hauled by a separate truck. The Lexion 8900 can unload up to 40 gallons (180 liters) of grain every second.

Secondary separation
Long, spiral screw-like rotor paddles pull the separated stalks along, shaking out more grain. This falls into a chute below and travels on to be separated, cleaned, and stored.

Different headers
The attachment at the front of a combine—known as a header—can be swapped, depending on the crop. For crops such as corn, the header features spikes called snouts that travel between the rows. Gathering chains underneath snap the corn stalks, allowing the harvester to collect in the freed heads of corn.

Straw chopper and power spreader
The unwanted stalks, husks, and other plant matter is carried to a series of rotating blades, which cut the material up into short pieces. Chaff fans then blow this material out of a chute at the back of the combine.

4 Cleaning system
The grain falls through sieves and is carried to the cleaning system. There, a turbine fan produces air, which blows the chaff (the grain casing) and small straw stalks away, leaving the clean grain.

5 Collected grain
The grain in the collecting tank is transferred to the grain elevator.

Grain gain
Some combines are so efficient that they can fill their grain tanks within four minutes. To keep harvesting, these machines continually unload their grain into hopper trucks that run alongside the harvester. Cameras can monitor grain quality throughout the unloading process.

Combine harvester

Harvesting crops once involved lots of human and animal labor. Today, a single person in a combine vehicle can collect, clean, and separate an entire field's crop, hugely speeding up harvest time.

Modern combines are packed full of technology to optimize the harvesting process. Cameras and sensors adjust the front of the vehicle to ensure every pound (kilogram) of crop in the field is harvested, and sophisticated machinery inside separates the grains from the chaff.

Construction vehicles

A wide and varied range of rugged, heavy-duty vehicles enable the safe and speedy construction of all the structures we rely on in everyday life–from buildings and bridges to roads and houses.

The global construction industry uses hundreds of thousands of specialized vehicles, each of which is designed to be strong, reliable, and efficient. These vehicles help shape the construction site's landscape, dig trenches and foundations, remove obstructions, and transport building materials around the site wherever they are needed.

Rotating chain
The vehicle's chain can rotate at speeds of up to 730 ft (222.5 m) per minute.

CHAIN TRENCHER VEHICLE
Vermeer T555-III
Origin: US
Length: 21–25 ft (6.4–7.6 m)

Resembling a giant chainsaw on caterpillar tracks, this machine has a large rotating chain with broad steel teeth. These tear into the ground, removing earth to form deep trenches.

PIPELAYER
Caterpillar 583T
Origin: US
Length: 18 ft (5.4 m)

These specialized vehicles run on tracks and use a strong winch to lift, maneuver, and deploy sections of pipelines. Most have a counterweight at the rear to provide stability and balance.

Pipe sling
This can lift sections of pipe weighing up to 140,000 lb (63,504 kg).

Tough tracks
Caterpillar tracks allow the vehicle to easily travel over rough ground.

Extendable arm
Giant sections of the crane's "boom" can extend out up to 276 ft (84 m) high.

MOBILE CONSTRUCTION CRANE
Liebherr LTM 1500-8.1
Origin: Switzerland/Germany
Length: 70 ft (21.4 m)

Like many cranes, this vehicle has an arm that can extend out great distances, but that can also contract for easier transportation. The length of the crane arm and the weight of its load are counterbalanced by weights and beams called outriggers.

Driver's cab
The steel cab has safety glass windows to protect the occupants.

551 tons
(500 tonnes)–the maximum load that this LTM mobile crane can lift.

Long wheelbase
The distance between the front and back wheels stabilizes the crane.

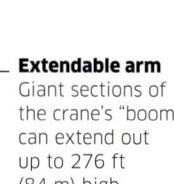

LOADING DUMP TRUCK
Thwaites 9 Tonne Front Tip
Origin: UK
Length: 15 ft (4.6 m)

Perfect for hauling away waste earth and rock, these vehicles range from small front-tip dumpers to larger vehicles capable of carrying loads of more than 298 tons (270 tonnes).

FRONT LOADER
Volvo BM loader
Origin: Sweden
Length: Varies

Large front loaders can carry more than 176.6 ft³ (5 m³) of earth or rock per trip. Powerful hydraulic pistons raise and lower the enormous broad bucket, or "loader," at the front.

Huge wheels
These turn with great torque but low speed.

BACKHOE LOADER
John Deere 310K EP
Origin: US
Length: 23 ft (7 m)

These versatile vehicles have both a broad excavator loader at their front and a bucket on a jointed boom known as a backhoe at their rear.

Strong scoop
Pistons powering the bucket allow it to dig to depths of 14¼ ft (4.34 m).

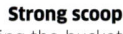

EXCAVATOR
John Deere 160D LC
Origin: US
Length: 29 ft (9 m)

Dedicated digging machines, these vehicles have a stout steel bucket with projecting teeth that dig into the ground. They help prepare foundations and pipeline trenches, as well as tear down structures.

ROAD ROLLER
Hamm HD140 roller
Origin: Germany
Length: 16 ft (5 m)

Usually used to compact the layers of a road, these machines have either one or a pair of heavy, large-diameter drums that exert great pressure on the material they roll over.

BULLDOZER
Caterpillar D9
Origin: US
Length: 18–23 ft (5.5–6.9 m)

Ridged caterpillar tracks and a powerful diesel engine give these earth-moving vehicles great traction. The large, hydraulically powered steel blade at the front pushes heavy loads around.

CEMENT MIXER
Western Star 4800TS
Origin: US
Length: Varies

These vehicles transport concrete in a large, rotating drum. A screw-shaped blade inside keeps the concrete churning, moving it around to keep it in a liquid state before it is poured.

Emergency vehicles

When disaster strikes, every second counts. Vehicles built to handle emergency situations are fast, powerful, and well-equipped.

From fire engines to ambulances, first responders make use of a range of specialized machines when dealing with a crisis. Sirens and flashing lights help warn other road users of the approach of fast-moving vehicles on their way from a fire station, hospital, or police station. Sometimes, air or water vehicles are also called in to support the teams working on the ground.

MOBILE WATER CANNON
Wasserwerfer 9000
Origin: Germany
Year: 1980s

In some countries, police use armored vehicles that fire high-pressure water jets to disperse crowds. In others, they are considered too dangerous.

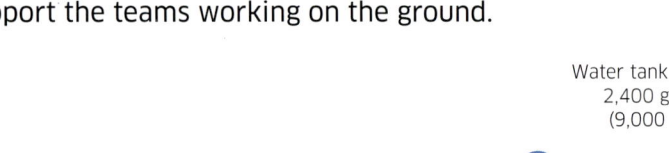

Water cannon
Each roof-mounted nozzle is individually operated.

Under pressure
Run-flat tires keep rolling even when punctured by broken glass.

Water tank holds 2,400 gallons (9,000 liters)

COMMAND SUPPORT
Enhanced Command Support Vehicle
Origin: UK
Year: 2008

Disaster scenes can be chaotic, so command vehicles are equipped with communication systems to help coordinate first responders.

Satellite housing antenna

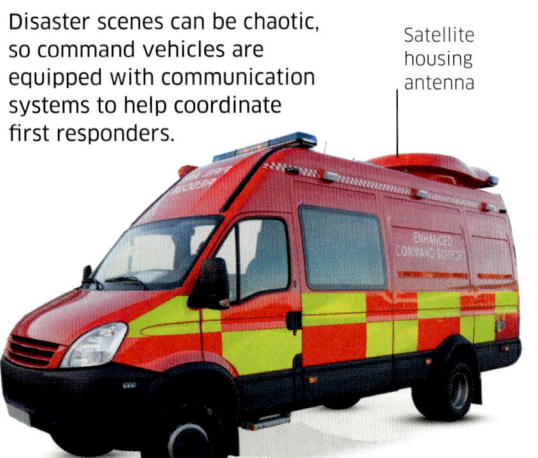

FIRE RESCUE
NATISK-3000 KS
Origin: Russia
Year: 1995 (civilian model)

With hoses, ladders, and sets of flameproof suits, these vehicles are equipped with tools to put out fires and rescue people trapped by the blaze.

Red paint alerts other road users

RESCUE HOVERCRAFT
Griffon 380TD
Origin: UK
Year: 2008

In coastal areas where land gives way to mud, marsh, ice, or water, the versatile hovercraft is a lifesaver. It skims the surface on a cushion of air held in by a neoprene rubber skirt.

FIRE-FIGHTING BOAT
Löschkreuzer "Weser"
Origin: Germany
Year: 1974

In areas next to water, boats with high-pressure water cannons are a valuable fire-fighting tool, capable of pumping thousands of liters directly from the sea, lake, or river onto the flames.

Light work
Two double searchlights extend upward for nighttime operations.

Super soaker
Each jet can hurl water mor than 330 ft (100 m).

AMBULANCE
Citroen Jumper
Origin: Greece
Year: 2002

Crewed by paramedics, ambulances are equipped to keep a patient in a stable condition on the way to the hospital, with oxygen, a defibrillator, stretchers, and life-saving medication on board.

The Tokyo Fire Department is the **largest urban fire department in the world**, with a total **staff of 18,600**.

33,000 gallons (125,000 liters)—the **volume of water a** Canadair CL-415 can **scoop and drop** per hour.

61

FIRE-FIGHTING AIRCRAFT
Canadair CL-415
Origin: Canada
Year: 1994

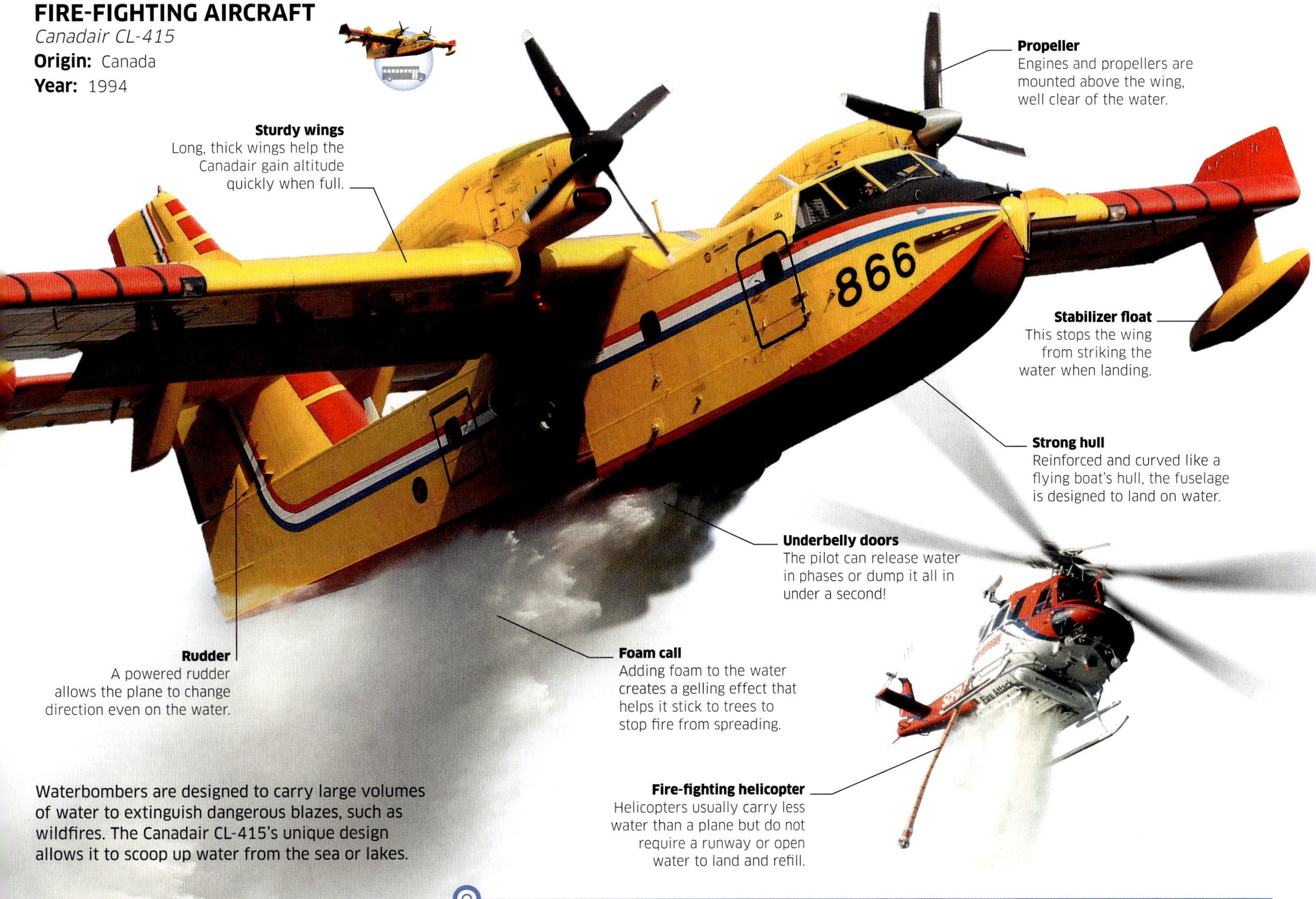

Propeller
Engines and propellers are mounted above the wing, well clear of the water.

Sturdy wings
Long, thick wings help the Canadair gain altitude quickly when full.

Stabilizer float
This stops the wing from striking the water when landing.

Strong hull
Reinforced and curved like a flying boat's hull, the fuselage is designed to land on water.

Underbelly doors
The pilot can release water in phases or dump it all in under a second!

Rudder
A powered rudder allows the plane to change direction even on the water.

Foam call
Adding foam to the water creates a gelling effect that helps it stick to trees to stop fire from spreading.

Fire-fighting helicopter
Helicopters usually carry less water than a plane but do not require a runway or open water to land and refill.

Waterbombers are designed to carry large volumes of water to extinguish dangerous blazes, such as wildfires. The Canadair CL-415's unique design allows it to scoop up water from the sea or lakes.

POLICE PATROL CAR
Opel Zafira Tourer
Origin: Germany
Year: 2012

Adapted from standard cars, patrol vehicles have a siren and blue lights, a radio link to the police station, and a secure area for suspects. Their very presence can also deter criminals.

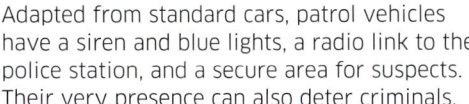

MOUNTAIN RESCUE
Land Rover Defender
Origin: UK
Year: 1990s

Rugged, all-terrain vehicles race up hillsides to assist lost or injured walkers and climbers. They carry ropes to reach tricky places and a medical kit, stretcher, and warm clothing to assist the injured.

Roofrack for large equipment

Detachable spotlight for nighttime rescues

Winch to pull stuck vehicles or lift trapped climbers

Hydrant to hose

Fire trucks can connect to the water in a water supply main through hydrants. When a hose is fitted, water from the hydrant reaches the fire engine's pump, which distributes it to spray hoses.

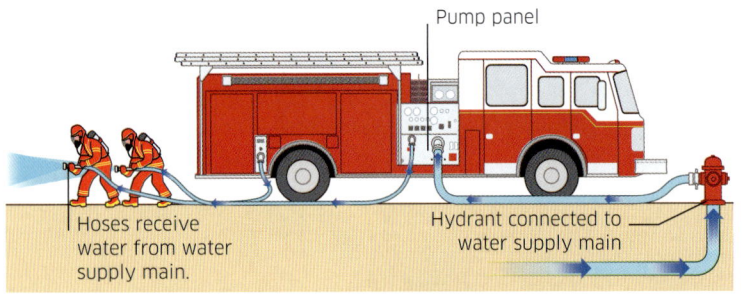

Pump panel

Hoses receive water from water supply main.

Hydrant connected to water supply main

Fire engine

Frequently the first vehicle to arrive at an emergency, fire engines are versatile, well-equipped trucks designed to tackle almost any problem firefighters face.

Fire engines transport firefighters, a wide range of tools (from concrete saws to thermal imagers), and key medical equipment such as defibrillators and oxygen tanks to the scene of a blaze or disaster. Usually powered by diesel engines, they contain powerful pumps to discharge water through turret guns and hoses.

Internal water tank
Metal baffles prevent the vast amounts of water in the internal tank from sloshing around.

Hard-suction hoses
Stored on the roof, these draw water in from an unpressurized water supply.

Foam proportioner
This controls the rate at which foam-creating chemicals are mixed with water.

Connected hose
This hose line is already fitted to a pump outlet so it can be used the moment the truck arrives at the scene.

Pump gauges
These display water flow rate and pump pressure from outlets such as hoses.

Control panel
Vital data about the truck is displayed here.

Water supply main inlet
A thick pipe carries water from the supply main into the fire engine.

Powerful fan
This large, portable fan can be used to disperse smoke and harmful gases at the scene.

Thick hoses
Strong reinforced hoses can be pulled inside a building or structure and operated by a single firefighter.

Connection point
Multiple outlets enable many separate hoses to be plugged in and supplied with pumped water.

223 ft (68 m)–the height of the **tallest fire engine** turntable ladder, found on a **Magirus M68L**.

3,434 gallons (13,000 liters)–the amount of **water that can be discharged every minute** by the most powerful fire engine **water turrets**.

63

Firefighters' cab
Up to eight firefighters wearing fire-retardant clothing, gloves, and hard helmets can fit into the cab. Strapped in and supported by cushioned seats, they can release their extensive harnesses with a single button in order to quickly spring into action.

ROSENBAUER AT

Origin: Austria

Wheelbase: 12–15 ft (3.6–4.55 m)

Crew: 2–8

Tank: 100–1,100 gallons (500–5,000 liters)

Pike
This long pole can snare objects or pull drywalling down to reach contained fires.

Ax
This can break through doors for access or to ventilate spaces from harmful smoke or fumes.

Bright lighting
LED lighting strips along the edge of the roof gallery illuminate the truck and its surroundings.

Water turret
This large electric water turret can be directed from the cab, where the operator can also alter its flow and spraying patterns.

Pressurized jet
The turret fires up to 926 gallons (3,500 liters) of water every minute.

Rosenbauer AT Pumper-Tanker
This Austrian fire engine is a generalist used by firefighters in more than 20 countries. It is a pumper-tanker type, meaning that it carries its own supply of water but can also access water supply mains when it needs more water to tackle bigger fires.

Warning lights
Bright blue flashing lights combined with loud sirens alert other vehicles to the fire truck's presence.

Aerial firefighting
Some blazes can be hard to subdue from the ground. When tackling tall fires or rescuing people, firefighters climb up giant telescoping ladders (carried on some trucks) or rise into the sky on a cherry picker's aerial platform (below right).

Generator
A portable electricity generator can be used to power electrical tools on site.

Hidden step
A fold-down metal ledge reveals lower equipment lockers and acts as step up to the tool locker and hose stores.

Driver's cab
Firefighters can control radio communications, sirens, powerful floodlighting, and a public address system from the cab.

RAIL NETWORKS

Railways were the first high-speed form of transportation, and today's modern electrified networks provide a fast, efficient, and relatively sustainable method of long-distance travel. Although cost-efficient to run, railways are expensive to build, putting them beyond the reach of many less wealthy countries.

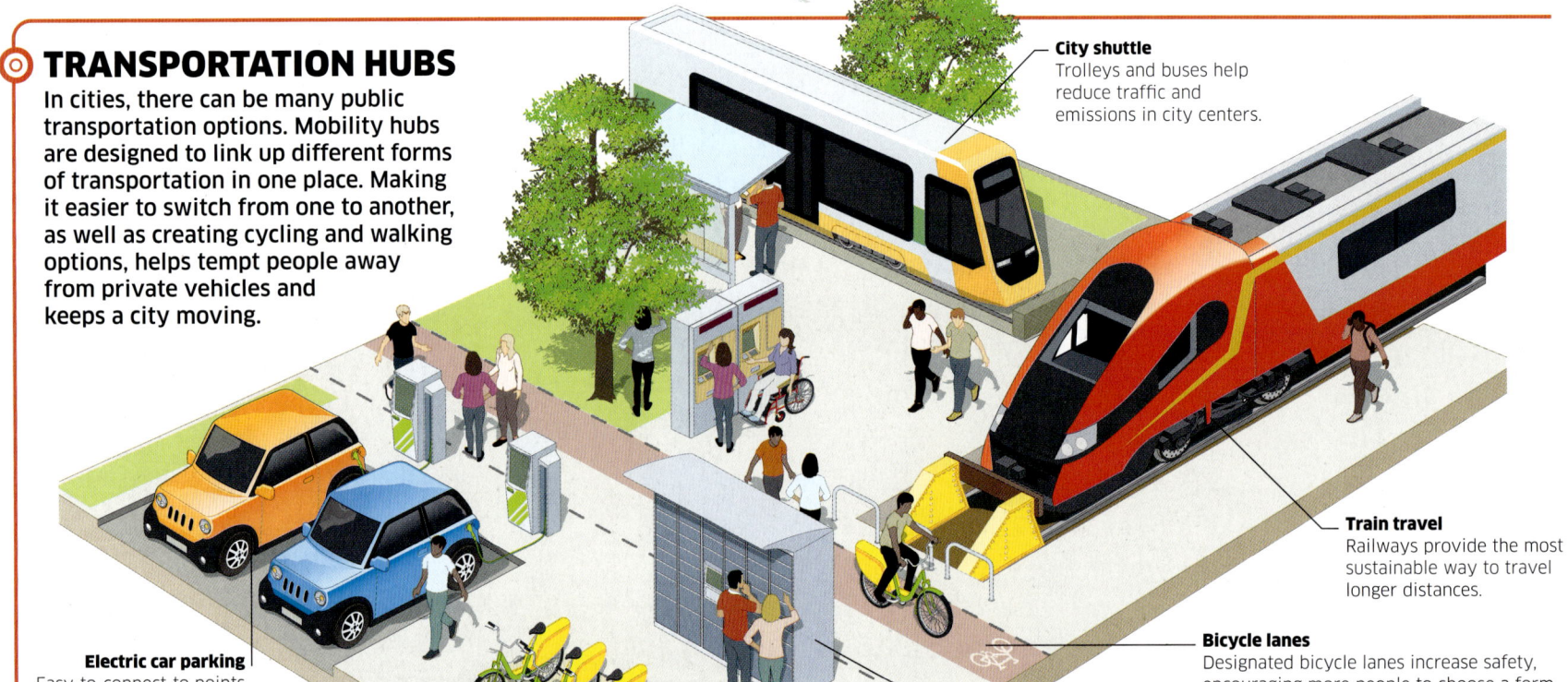

US
The world's largest rail network has over 136,000 miles (220,000 km) of track.

World rail network
This map shows the most extensive networks are in North America, Japan, and Europe.

China
China's railways are the world's busiest, and the country also has the longest high-speed rail network.

India
Rail travel is hugely popular in India, where fares are low and the network extensive.

TRANSPORTATION HUBS

In cities, there can be many public transportation options. Mobility hubs are designed to link up different forms of transportation in one place. Making it easier to switch from one to another, as well as creating cycling and walking options, helps tempt people away from private vehicles and keeps a city moving.

City shuttle
Trolleys and buses help reduce traffic and emissions in city centers.

Train travel
Railways provide the most sustainable way to travel longer distances.

Bicycle lanes
Designated bicycle lanes increase safety, encouraging more people to choose a form of transportation that keeps you healthy.

Package pickup
Package hubs are convenient for users and reduce the number of delivery vans on the road.

Electric car parking
Easy-to-connect-to points allow electric car users to charge their vehicles.

Bike and scooter rental
These provide a quick, convenient option for short journeys.

Transportation networks

Wherever we live, we all need to travel to work or for education, access stores and services, or visit friends and family. A range of public transportation methods make traveling accessible to most people.

Reliable, efficient public transportation is vital to a country's economy, is more sustainable than private vehicle use, and enhances the quality of life of people who live and work in cities. It is often funded by governments. However, in some hard-to-reach places, where conventional public transportation is not available, people have come up with other travel solutions.

SPECTACULAR STATION

A large rail network needs interchanges that can provide routes to multiple destinations and manage a huge volume of passengers. Grand Central Terminal in New York City has 44 platforms (the most in the world) and 67 tracks and is used by 750,000 people every day.

3.6 million The number of **people** that pass through Tokyo's **Shinjuku Station** in Japan every day, making it the **world's busiest railway station.**

1 bus produces the **same amount** of **emissions** as up to 30 cars.

65

MOVING PEOPLE

As well as being greener, public transportation is by far the most efficient way to move a lot of people through a route at the same time. This graphic compares transportation methods by their passenger capacity—the maximum number of people that can travel by that method on a given route for a set length of time.

Right on track
The graphic shows how many passengers could potentially travel along the same route in an hour. By far the lowest figure is for private cars and vehicles. By comparison, a suburban train can move 88,000 more passengers within the same time frame.

Key

👤 1,000 people

Mixed traffic | **Bus** | **Bicycle** | **Walking** | **Trolleys** | **Railways**

CREATIVE COMMUTES

When it comes to getting around town, some places pose their own unique challenges. Factors including location, layout, or climate rule out more traditional forms of public transportation—but undaunted citizens have devised imaginative methods of staying on the move.

Norry train, Cambodia
These homemade bamboo trolleys were made to run on disused rail lines. They carried people and goods between isolated communities, but they are now equally popular with tourists.

Monte toboggan, Madeira
Wicker baskets mounted on runners have been used on Madeira since the 1850s. Drivers push passengers on a 1-mile (2-km) winding mountain road, using their rubber-soled boots as brakes.

Rideau Canal Skateway, Canada
Every winter, a canal in the heart of Ottowa becomes a giant ice rink—and an express lane for the city's skating commuters. The skateway only opens when the ice is 1 ft (30 cm) thick.

Giant escalator, Colombia
The city of Medellín is built on a steep mountain. In 2011, an escalator opened that links the city center to one of its suburbs. Commuters make the 1,260-ft (384-m) ascent home in just six minutes.

PULLING POWER

Whether they are used to span the deep gaps between mountain peaks or to run up and down steep streets and hillsides, cables can provide efficient and inexpensive transportation solutions in some of the world's trickiest terrains.

Peak 2 Peak Gondola, Canada
Linking the Whistler and Blackcomb mountains, the Peak 2 Peak is the highest cable lift of its kind and can carry 4,100 people every hour.

Cable cars, US
The city of San Francisco runs the world's last hand-operated cable trolleys. Cars are hauled along the steep streets by constantly moving cables.

AIRBORNE TAXI

South Korea's Urban Air Mobility (UAM) project is set to make flying taxis a reality. Electric, remotely piloted aircraft will beat traffic jams in the capital, Seoul, by ferrying passengers between the airport and city center. The vertical take-off and landing drones can cover the 7.5-mile (12-km) route in just three minutes.

On the bus

A staple of both short- and long-distance travel, buses, trolleybuses, and coaches are found in countries around the world. Large numbers of passengers hop on and hop off every single day.

A bus is defined as a self-propelled, wheeled vehicle carrying 10 or more passengers. While trains, trolleys, and subways bear much of the load of public transportation, buses provide a convenient, cost-effective way of making journeys using existing roads or dedicated bus lanes. Hundreds of billions of passenger journeys are made by bus each year. In the European Union alone, there are more than 684,000 buses in operation.

Standing space
Room for people to stand in addition to seating means the bus can carry up to 44 passengers.

DESIGNLINE INTERNATIONAL TINDO

Electric bus
Origin: New Zealand/Australia
Length: 34⅛ ft (10.42 m)

In 2007, New Zealand's Tindo became the first electric bus to recharge its batteries using solar power. Like other electric buses, it produces no harmful exhaust gases and runs very quietly.

Roof window ensures natural light can reach the rear of the bus

BREDAMENARINIBUS AVANCITY+

Trolleybus
Origin: Italy
Length: 35 ft (10.7 m)

These electric vehicles run on regular roads but gain power from overhead electricity cables instead of an engine. Their motors offer good power from a standstill, so they operate well in hilly locations.

Connecting pole
Spring-loaded trolley poles connect with overhead electricity power lines.

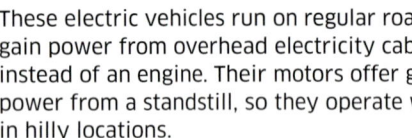
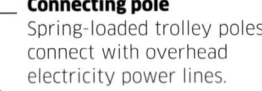

TOYOTA COASTER B40

Minibus
Origin: Japan/Tanzania
Length: 22⅜ ft (6.8 m)

Minibuses act as small-capacity shuttle buses over short distances. Some are built-up cabs on small truck chassis, while others—like the Toyota Coaster—are custom designed and built.

Luggage space on top

MARRAKECH CITY TOUR BUS

Open-topped bus
Location: Morocco
Length: 36 ft (11 m)

Converted double-decker buses with their roofs removed become the perfect sightseeing vehicles for tourists. Screens and sound systems can be installed for audio commentary, and seats on the top deck give excellent views.

Louver windows
Windows with angled slats open to provide ventilation for passengers sitting in the lower deck.

Built-in audio
Flags indicate the eight languages in which an audio commentary of the tour is available.

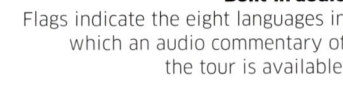

JEEPNEY
Jeepney
Origin: Philippines
Length: Variable

In the Philippines, leftover World War II military jeeps have been converted into small buses. Modern jeepneys are based on a great variety of base vehicles, including vans and small trucks. They tend to carry 8-18 people, with more sometimes hanging onto the outside.

Eye-catching exterior
A typical jeepney has a brightly colored paint job and is ornately decorated.

SETRA SIX WHEELER DOUBLE DECK COACH
Double-decker coach
Origin: Germany
Length: 47¼ ft (14.4 m)

Designed to travel long distances, diesel-powered coaches often have comfier seating than ordinary buses. Double-decker coaches carry more passengers and can even include cooking facilities and showers.

TORSUS PRAETORIAN
All-terrain bus
Origin: Czechia
Length: 28½ ft (8.7 m)

These rugged buses feature all-wheel drive, heavy grip tires, and a raised body all mounted on a strong suspension system. They can go offroad, clambering over steep slopes and rough ground.

US SCHOOL BUS
School bus
Location: US
Length: Up to 45 ft (13.7 m)

These bright buses are painted yellow for visibility. They come in four classes built by a range of different manufacturers and operate all over the United States.

VAN HOOL EXQUI.CITY 24 HYBRID
Bi-articulated bus
Origin: Belgium/Sweden
Length: 78 ft (23.8 m)

Most "bendy" buses have one articulation point (pivot) in the middle to travel around corners. Bi-articulated buses are longer and require two bends. This long bus can carry 140 passengers.

Connected sections
Walkways in the bendy parts connect each section of the bus.

Articulation point
Rubberized bellows expand and contract as the bus turns around a bend.

In 1932, the world's first ever
triple-decker bus
ran between Rome and Tivoli in Italy, carrying 88 passengers.

45 mph (72 kph)—the **top speed** of a **Routemaster.**

Top deck
Up to 40 passengers could get a seat and a view upstairs.

Destination
The replaceable blind showed the bus route number, main stops, and final destination.

Air inlet
A grill allowed cool air into passenger decks or used heat from the engine to warm the air via a radiator behind the grill.

Driver's cab
A driver sat in a half-cab alongside the engine bay. They drove using power steering, hydraulic brakes, and an automatic transmission.

Diesel engine
The 9.6-liter engine had a fuel economy of around 10 mpg (4.25 km/l). It worked well for stop-start driving in a city.

Hydraulic cylinder
Filled with hydraulic fluid, this device provided the force for the bus's power steering system.

AEC ROUTEMASTER RML

Origin: UK

Produced: 1954–1968

Length: 30 ft (9.12 m)

Capacity: 72 passengers

Sturdy frames
The main body of a Routemaster consisted of a light aluminum frame. This could be easily slotted over two steel subframes. Subframe A held the engine, while B carried the rear axle and suspension.

29 gallons (132 liters)—the amount of **diesel** that could be **held** in a **Routemaster's fuel tank**.

28 million miles (45 million km)—the **distance traveled by the Routemaster fleet** during its **first five years** of service.

69

Good overview
An angled mirror allowed the conductor to view the stairs and top deck from their position on the lower deck.

Seat patterns
The seat fabric was designed to match the interior colors.

36 Paddington Marble Arch
Victoria Camberwell
Peckham New Cross

Hop on
Passengers could hop on and hop off the open platform even as the bus took off.

London legend
Routemaster RMLs were first built in 1961, five years after the Routemaster RM (which seated 64) entered service. Both had no passenger doors but an open platform at the rear. Conductors had to remind passengers to "hold on tight!"

Rear suspension
A coiled spring rear suspension gave the bus a smooth ride over bumps or potholes.

Double-decker bus

This innovative two-level bus entered service in 1956 to help satisfy London's growing public transportation demand while taking up less space on the road.

With its aluminum body and rugged running gear, this much-loved bus far outlasted its predicted lifetime. Many traveled more than 1 million miles (1.6 million km). The last Routemasters on regular services in London were only withdrawn in 2005. Some vehicles continued on heritage routes until 2019 and a few still run as tourist buses.

Tickets, please!
Passengers bought tickets from a roving conductor who wore a ticket-printing machine. Its ink-covered roller could print up to 75,000 paper tickets. The conductor also controlled heating and ventilation inside the bus.

Power steering
Driving a heavy bus was made easier due to the Routemaster's power steering system. As the steering wheel was turned, a pump driven by the bus's engine pushed hydraulic fluid along fluid lines to move a piston. The piston assisted the steering rack and pinion gearing, angling the wheels so the bus could turn.

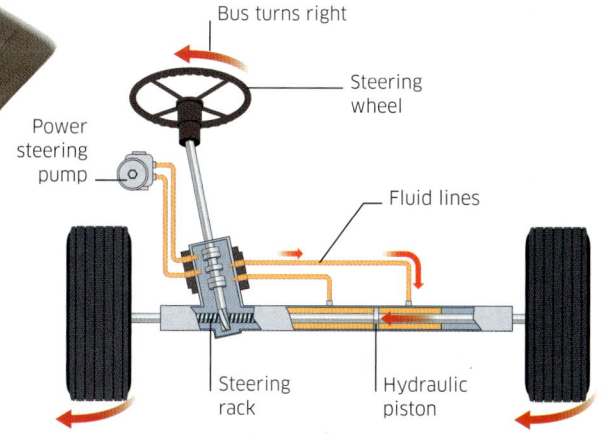

Bus turns right

Steering wheel

Power steering pump

Fluid lines

Steering rack

Hydraulic piston

Tilt test
All Routemasters underwent stringent safety testing. In tilt tests, the bus was angled sideways on a ramp while filled with sandbags mimicking the weight of a full load of passengers. To pass, all four tires had to stay in contact with the ground even when the bus tilted to 28 degrees.

A hydraulic ram tilts the ramp holding the bus.

70   land ○ ON TRACK

303 miles (487 km)—the length of the **longest perfectly straight rail track**, across Australia's **Nullarbor Plain.**

BUILDING RAILWAYS

Constructing a new line is a vast undertaking, beginning with land surveys and the selection of the ideal route. A line requires as little gradient as possible for fast, efficient running, so parts of the route are leveled or built up, while tunnels and bridges may be needed to overcome obstacles. Each section of track begins with a base layer, covered in sand or gravel.

Parts of a track

The main features of a track are two long rails on top of lots of rocky ballast, which bears the heavy load of the train as well as allowing drainage. Wooden sleepers lie at regular intervals under the rails, each attached to rail pads by a small fastening.

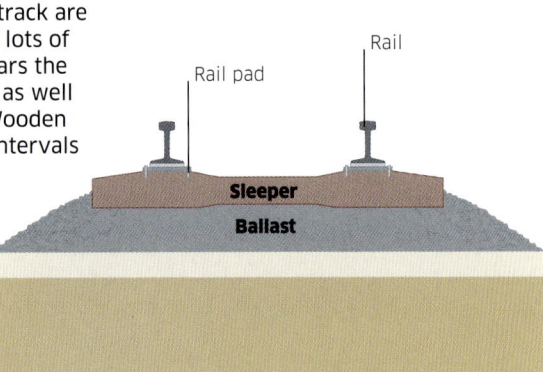

Rail pad | Rail | Sleeper | Ballast

Track laying

Despite modern cranes and automatic track-laying machines, constructing track remains labor intensive. It must be continually checked for accuracy, alignment, and smooth fitting. Most fast lines use continuous welded rail, where long lengths of steel rail are welded together.

First Transcontinental Railway
More than 20,000 workers laid 1,774 miles (2,885 km) of track across the US in the 19th century.

Laying light rail
Today, steel rails on a light rail line in Colorado need to be finished and fitted smoothly.

On track

Rail has long been a fast, affordable, and energy-efficient mode of transportation. Today, there is enough railway track crisscrossing the continents to circle Earth around 25 times!

The first railways ran during the 16th and 17th centuries. Horses, mules, or people pulled carts or wagons out of mines on wooden rails. These early rails were later replaced by wood covered in iron sheets, then cast-iron rails and, finally, steel.

TRACK GAUGES

The distance between the inner sides of a track's two rails is its gauge. Railways of different gauges have been constructed throughout the world, creating problems when lines of different gauges meet.

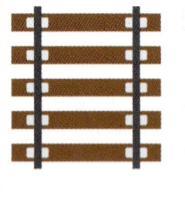

Standard
This 4-ft 8½-in (1,435-mm) gauge is used on 55 percent of the world's railways, including most in the US and UK.

Russian
The world's second most common gauge of 4 ft 11⅞ in (1,520 mm) is found in Russia, Finland, Ukraine, and Belarus.

Cape
With a narrow gauge of 3 ft 6 in (1,067 mm), this track is found in South Africa, New Zealand, Indonesia, and Japan.

Two foot
An ultra-narrow gauge of 2 ft (610 mm) is found mostly at industrial plants and amusement parks.

Bogies

Also known as trucks, these are the sets of axles, wheels, and suspension systems on which the body of a train sits. Trains on the Trans-Mongolian Railway have their bogies changed at the China-Mongolia border to swap to a different track gauge.

RAIL NETWORKS

The rise of steam-powered locomotives in the 19th century saw early railways boom. The UK rail network alone grew from 98 miles (157 km) to 10,373 miles (16,693 km) between 1830 and 1860. Huge lengths of track now stretch across many countries. The three longest national rail networks today are shown below.

US   92,307 miles (148,553 km)

China   68,206 miles (109,767 km)

Russia   53,155 miles (85,544 km)

Longest routes
- Moscow to Vladivostok
- Toronto to Vancouver
- Shanghai to Lhasa
- Sydney to Perth
- Dibrugarh to Kanyakumari
- Emeryville to Chicago

Epic journeys

It is now possible to travel extremely long distances on just a single train. The longest single rail line is the Trans-Siberian Railway, which crosses seven time zones as it travels 5,711 miles (9,288 km) across Russia from Moscow to Vladivostok.

8⅝ ft (2,642 mm)–the **track gauge** used by the **Guangzhou Metro APM Line.**

315 miles (507 km)–the **length of the track** at the world's **largest railway** yard–Bailey Yard in Nebraska. It handles up to 14,000 wagons a day.

71

WHEELS

With the exception of magnetic levitation systems (maglev trains), all trains run on wheels. As the only part of the train in contact with the rail, wheels play a vital role in keeping the train on track. They were originally wooden, then gained a rim made of iron or steel before eventually becoming entirely steel.

Wheel design

Train wheels are smaller on one side, forming a conical shape. This helps center the wheels on the rail when traveling straight ahead. When turning, one of the wheels will be forced slightly away from the rail. To keep them on track, flanged wheels were developed.

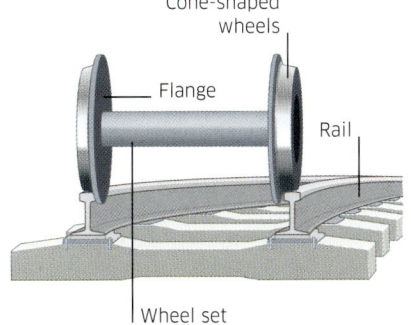

Cone-shaped wheels

Flange

Rail

Wheel set

Flanged wheels

A flange is a large lip, first developed by English engineer William Jessop in 1789. It helps keep the train on the track at all times, even when traveling around a bend.

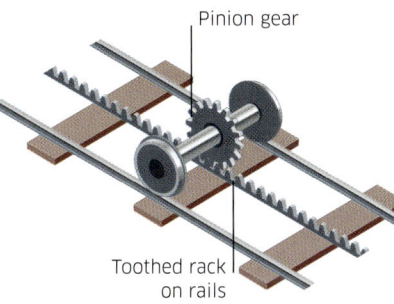

Pinion gear

Toothed rack on rails

Traveling uphill

Normal trains are not designed for traveling uphill. Mountain railways have an extra pinion gear cog between the wheels. This meshes with a toothed rack to drive the train up the steeper gradient.

Wheel arrangement

In the early 20th century, US engineer Frederick M. Whyte developed a way of classifying steam locomotives by the number of leading, driving, and trailing wheels they featured. Driving wheels were powered by the engine, while leading and trailing wheels were usually unpowered, supporting the locomotive and helping it negotiate bends in the track.

4-4-0 American

Four large driving wheels propelled this 1866 locomotive, which was employed on the early lines of the London Underground, UK.

Two driving wheels are on each side.

Four leading wheels sit at the front of the locomotive.

4-6-2

Three powered driving wheels on each side propelled the world's fastest steam locomotive, the *Mallard*.

Driving wheels

POINTS AND SWITCHES

These junctions connect one set of track rails with another diverging set, enabling trains to switch tracks. The new track may branch away to form another complete rail route or may be a small passing loop that a slower train travels along in order to let a faster train through.

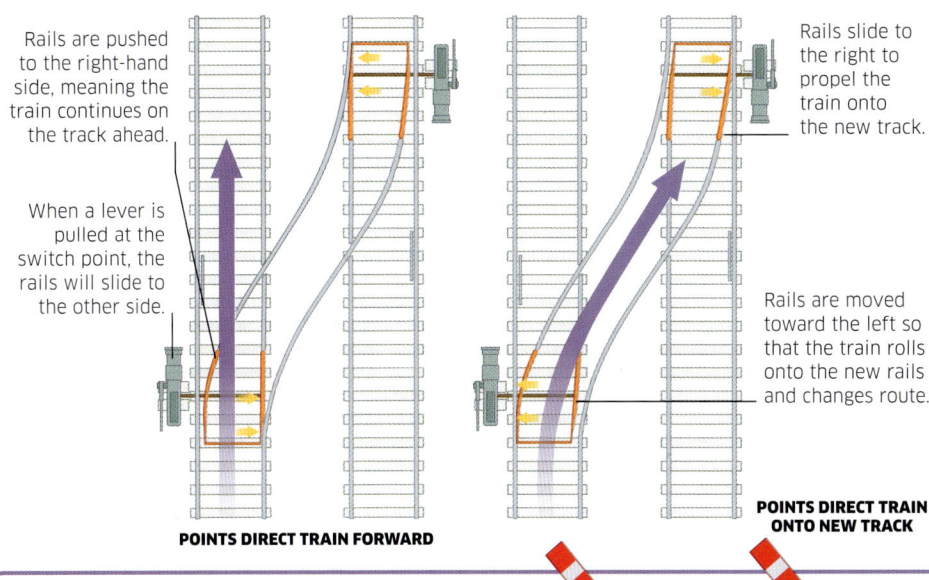

Rails are pushed to the right-hand side, meaning the train continues on the track ahead.

When a lever is pulled at the switch point, the rails will slide to the other side.

Rails slide to the right to propel the train onto the new track.

Rails are moved toward the left so that the train rolls onto the new rails and changes route.

POINTS DIRECT TRAIN FORWARD

POINTS DIRECT TRAIN ONTO NEW TRACK

SIGNALING

With long stopping distances, trains could easily collide or derail without an extensive signaling system. Modern trains receive track and traffic data electronically, but trackside signals featuring lights and signs are still used to inform drivers of the route ahead. A process called the semaphore system uses lights and pairs of pivoting arms to convey key information.

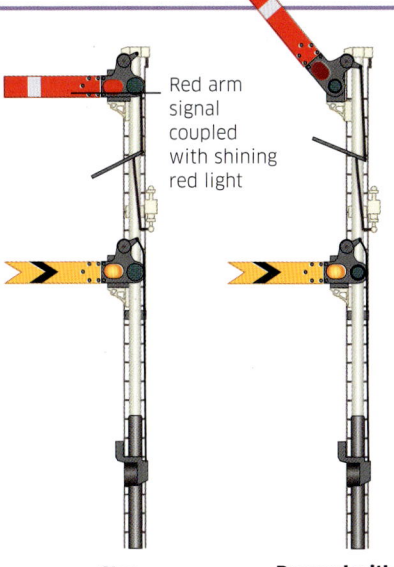
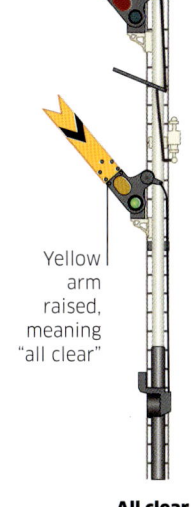

Red arm signal coupled with shining red light

Yellow arm raised, meaning "all clear"

Stop
Both arms horizontal indicates that the brakes should be applied.

Proceed with caution
This signal tells the driver to keep the train moving, but be prepared to stop.

All clear
Both arms raised means the train can continue at its normal speed.

EXTREME RAILWAYS

Some railways operate in extremely harsh conditions. The Mauritania Railway runs through the hot, sparse climate of the Western Sahara Desert. Trains hauling iron ore to the coast use sand plows and filter systems and can measure more than 1.9 miles (3 km) in length.

Mauritania Railway
Trains often haul 200 wagons on this 437-mile- (704-km-) long track. Sometimes, passengers sit in the wagons.

Freight trains

Railroads carry around 7 percent of all the world's transported cargo. Locomotives haul wagons and railcars carrying containers full of grain, chemicals, and other goods.

A typical freight locomotive can haul up to 200 times more cargo than a truck on the road and is three to four times more fuel-efficient than a road vehicle. Freight locomotives also work at mines and factories, as well as railway yards where they shunt carriages around to form long trains ready to be pulled by a locomotive (see pages 70–71).

STADLER GEAF 2/2
Electric shunter
Origin: Switzerland
Length: 27½ ft (8.4 m)

This locomotive can shift up to 551 tons (500 tonnes) when pushing wagons and carriages around a rail yard. A rechargeable battery pack allows it to operate on nonelectrified track for short time periods.

EMD SD60
Diesel-electric hauler
Origin: US
Length: 71¼ ft (21.7 m)

A 3,600-hp diesel engine generated the electricity required to power this locomotive's six electric traction motors. More than 1,100 SD60s and variants were built, 85 of which were used to haul heavy freight trains on the Union Pacific railroad.

Drive wheels
Twelve 40-in- (101.6-cm-) diameter wheels propelled the train along the track.

WABTEC FLXDRIVE
First battery-powered locomotive
Origin: US
Length: Variable

Built for heavy shunting since 2019, this model produces no emissions. Its lithium-ion battery cells are recharged by regenerative braking (see page 23) and by chargers fitted beside the tracks.

VOSSLOH G6
Industrial shunter
Origin: Germany
Length: 34 ft (10.35 m)

Short, squat shunter locomotives like this diesel-hydraulic-powered vehicle operate over short distances. They push and pull goods wagons around factories, docks, and industrial complexes.

CGR CLASS 7
Steam locomotive
Origin: Scotland/South Africa
Length: 51⅝ ft (15.7 cm)

Powerful pistons moved by expanding steam propelled these trains up to a steady top speed of 34.8 mph (56 kph). They were very reliable, operating from 1892 to 1972.

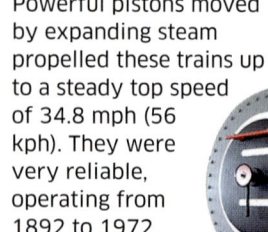

Almost **70 percent of all rail freight** is carried in just three countries: the **US, China, and Russia.**

83 million The **number of road trucks** that would be needed to replace all the freight hauled by rail in the US.

73

Flatbed wagon
This holds one or two trucks or trailers, each attached securely in place for the rail journey.

In 2001, an Australian train of
8 locomotives
and 682 wagons extended for more than 4 miles (7 km).

ROLLING HIGHWAY TRAIN
Cargo truck transporter
Origin: Europe and elsewhere
Length: Variable

These vehicles are formed when cargo-carrying road trucks drive onto low wagons via a ramp. They are then carried long distances by train–using less fuel and causing less congestion than road transportation.

Passenger car
Drivers rest here while their trucks are hauled. Some trains feature sleeper cars for overnight journeys.

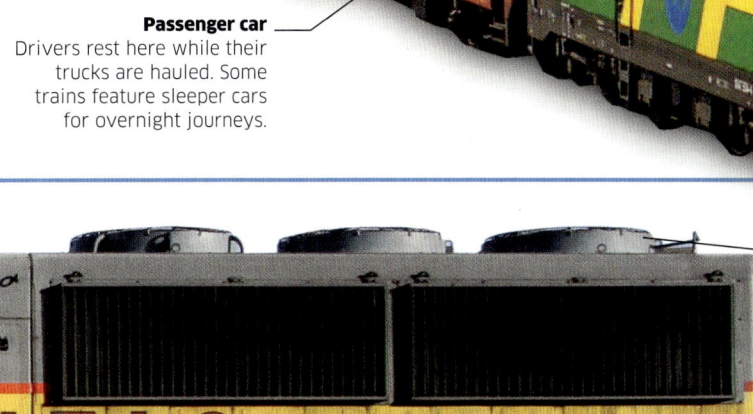

Cooling fans
Spun by electric motors, these drew heat away from the engine and other machinery.

UNION PACIFIC BIG BOY
Mountain range hauler
Origin: US
Length: 85¼ ft (26 m)

One of the biggest steam locomotives ever built, these operated over steep terrain in Utah and Wyoming. Vast quantities of coal had to be burned to power the 16 driving wheels.

SIEMENS VECTRON
Versatile dual-fuel locomotive
Origin: Germany
Length: 62¼ ft (19 m)

Not all freight railways are electrified, so this electric locomotive can also switch to a diesel engine fed by a 660.4-gallon (2,500-liter) fuel tank. It has a top speed of 100 mph (160 kph).

BOMBARDIER/ADTRANZ IORE
Powerful electric mine train
Origin: Canada/Germany
Length: 75⅛ ft (22.9 m)

Weighing 198 tons (180 tonnes), this powerful locomotive generates 10,500 hp. One is fitted at either end of a mine train to haul 68 heavy wagons filled with rocky ore.

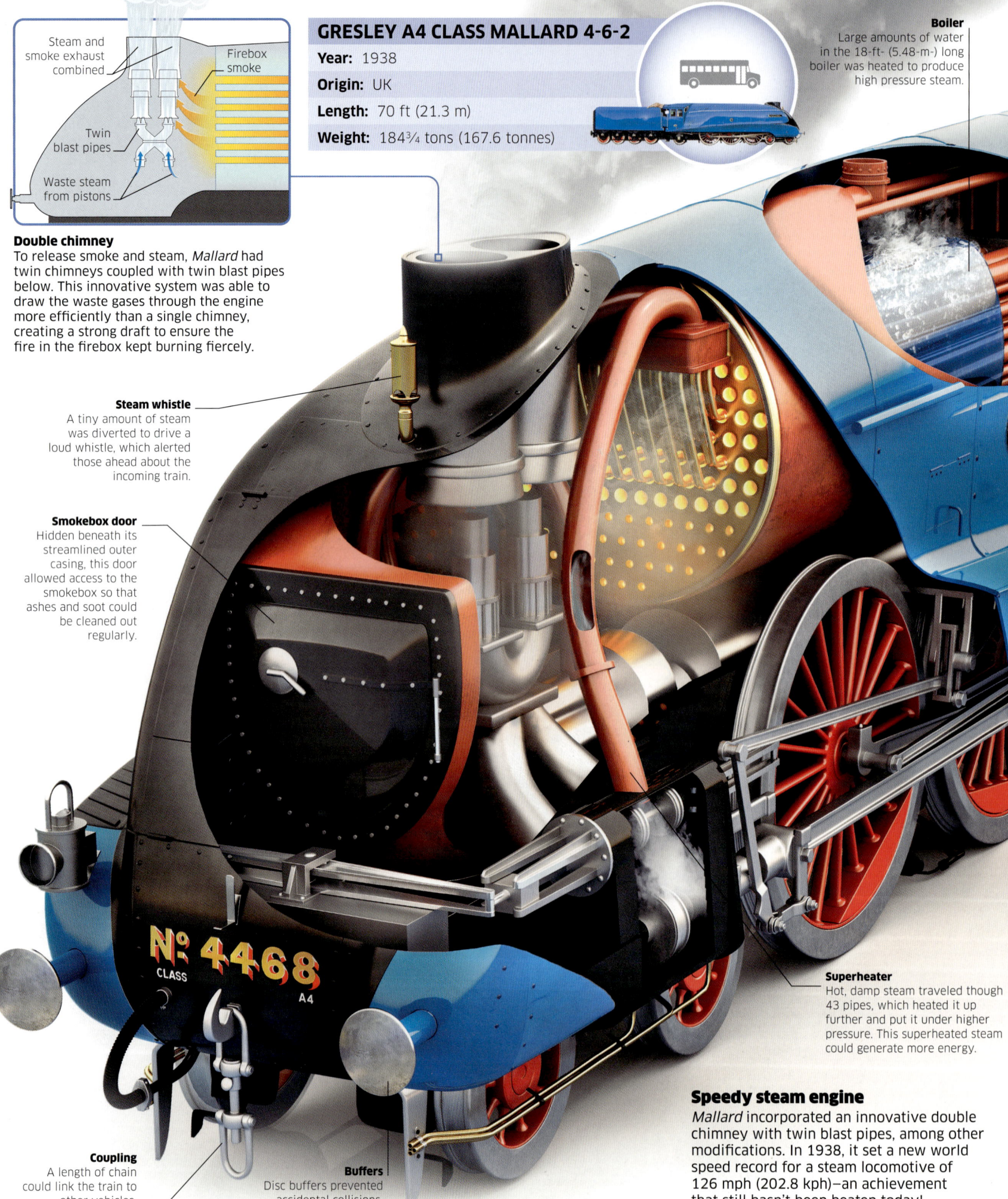

Double chimney
To release smoke and steam, *Mallard* had twin chimneys coupled with twin blast pipes below. This innovative system was able to draw the waste gases through the engine more efficiently than a single chimney, creating a strong draft to ensure the fire in the firebox kept burning fiercely.

GRESLEY A4 CLASS MALLARD 4-6-2

Year: 1938

Origin: UK

Length: 70 ft (21.3 m)

Weight: 184¾ tons (167.6 tonnes)

Boiler
Large amounts of water in the 18-ft- (5.48-m-) long boiler was heated to produce high pressure steam.

Steam and smoke exhaust combined

Firebox smoke

Twin blast pipes

Waste steam from pistons

Steam whistle
A tiny amount of steam was diverted to drive a loud whistle, which alerted those ahead about the incoming train.

Smokebox door
Hidden beneath its streamlined outer casing, this door allowed access to the smokebox so that ashes and soot could be cleaned out regularly.

Superheater
Hot, damp steam traveled though 43 pipes, which heated it up further and put it under higher pressure. This superheated steam could generate more energy.

Coupling
A length of chain could link the train to other vehicles.

Buffers
Disc buffers prevented accidental collisions.

Speedy steam engine
Mallard incorporated an innovative double chimney with twin blast pipes, among other modifications. In 1938, it set a new world speed record for a steam locomotive of 126 mph (202.8 kph)—an achievement that still hasn't been beaten today!

1830 The year the **world's first steam-powered intercity railway** line opened, running **between Manchester and Liverpool in the UK.**

6,005 gallons (22,730 liters)—the **maximum amount of water** *Mallard* could carry.

75

Coal wagon

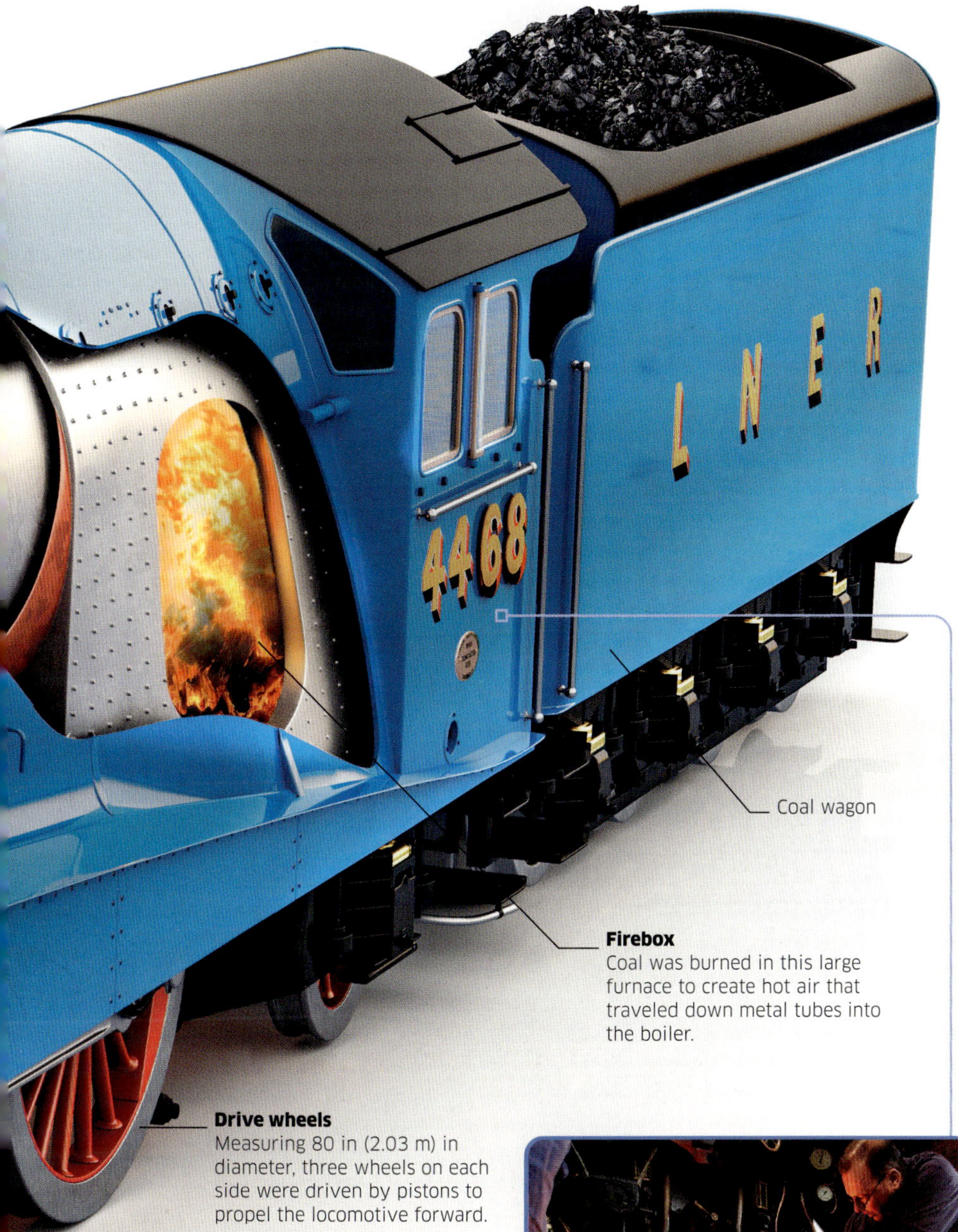

Firebox
Coal was burned in this large furnace to create hot air that traveled down metal tubes into the boiler.

Drive wheels
Measuring 80 in (2.03 m) in diameter, three wheels on each side were driven by pistons to propel the locomotive forward.

Feeding the firebox
In the cramped space between the firebox and the coal wagon, workers known as firemen had to shovel coal into the firebox constantly to keep the train running. They were also responsible for checking the boiler had enough water.

Steam locomotive

The first trains burned coal, oil, or wood to heat water in boilers—turning it into rapidly expanding steam. The steam then powered pistons that drove the train's wheels around.

By the 1920s, advances such as streamlining and using bigger, higher pressure boilers saw express steam trains reach speeds of 100 mph (160 kph). The quest for faster travel on long-distance UK rail routes resulted in the building of 35 powerful A4 class locomotives. They were introduced in 1935, and the fastest one of all was called *Mallard*.

Long-haul legend
Mallard remained in operation for many years after its record-breaking run, hauling passenger on routes such as London to Edinburgh. After 25 years of solid service and after traveling almost 1.5 million miles (2.4 million km), it was retired and is now on display at the National Railway Museum, York, UK.

Piston power
When a locomotive's boiler produces steam, it forces a piston along a cylinder. The piston is joined to a connecting rod and crankshaft, which converts the piston's straight-line movement into turning movement. A constant supply of steam is needed to keep the piston moving so that the train does not stall or stop.

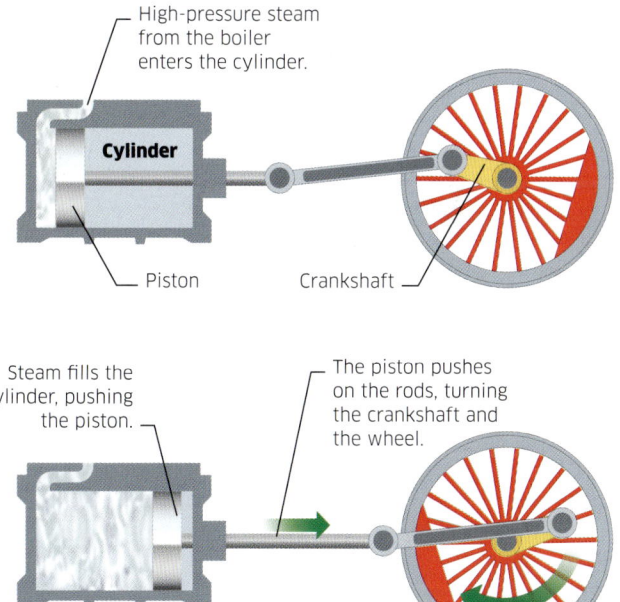

High-pressure steam from the boiler enters the cylinder.

Cylinder

Piston

Crankshaft

Steam fills the cylinder, pushing the piston.

The piston pushes on the rods, turning the crankshaft and the wheel.

Sandboxing
Many locomotives carried a sandbox full of dry sand, which was released on the track ahead of the wheels to produce extra grip. To ensure a plentiful supply, sand was dried in large ovens in engine sheds.

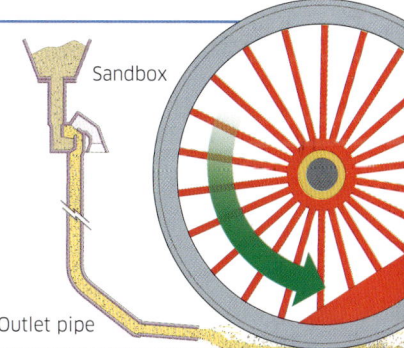

Sandbox

Outlet pipe

3.6 million The number of passengers that pass through Shinjuku Station in Tokyo, Japan, every day.

Passenger trains

In 1808, the world's first fare-paying train passengers paid 1 shilling (12 pence) each to travel at 12 mph (19 kph) around a small, circular track in London—the beginning of commercial rail travel.

Since then, railway networks have boomed. More than 621,000 miles (1 million km) of track spread across continents, and steam power gave way to diesel or electricity. Locomotives—self-powered engines that pull trains—or motorized rail cars now move hundreds of millions of passengers each day.

LNER A3 FLYING SCOTSMAN

Origin: UK
Year: 1923
Length: 70 ft (21.3 m)

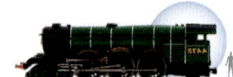

This much-loved long-distance locomotive hauled British trains between London and Edinburgh. In 39 years of high-speed service, it traveled more than 2.08 million miles (3.3 million km).

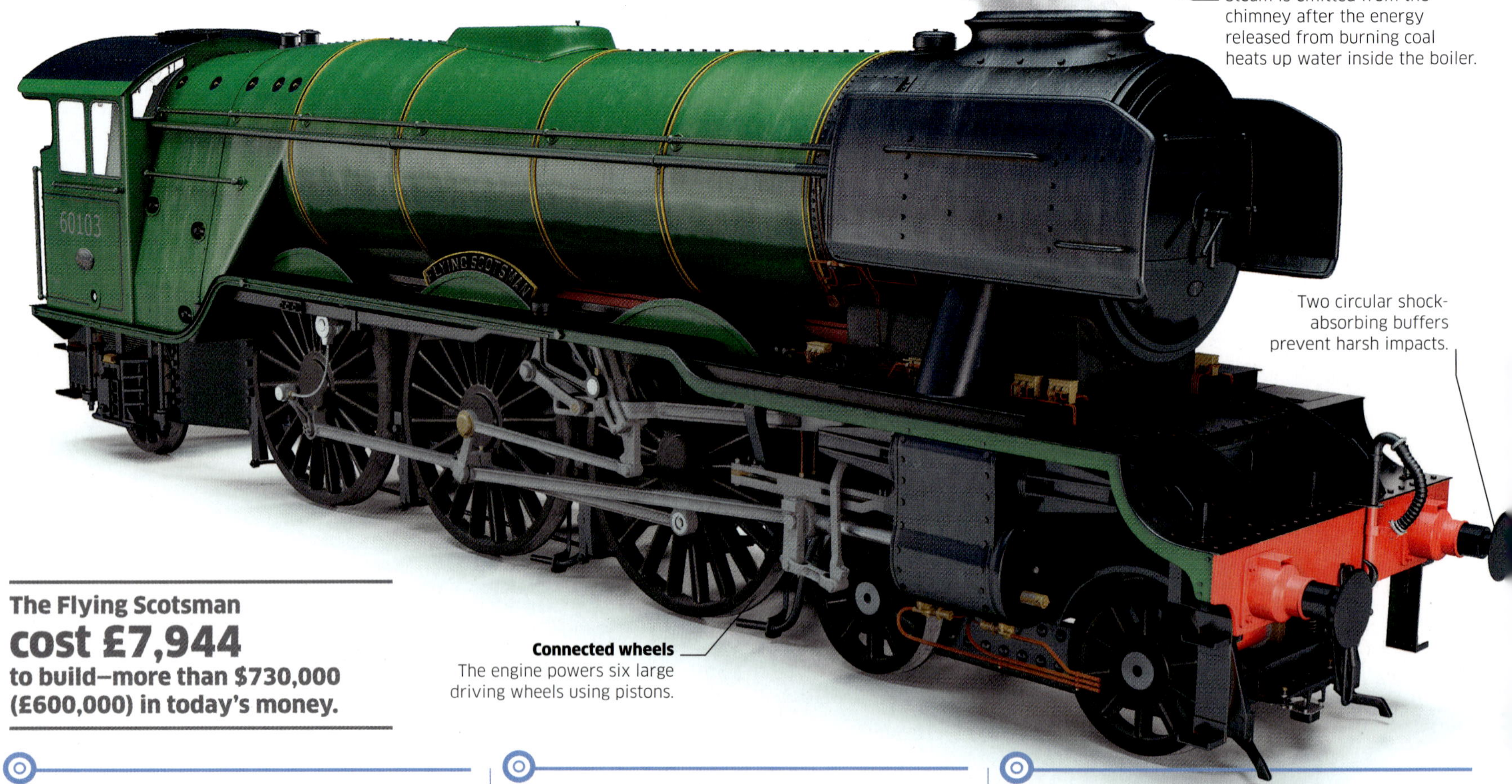
Steam is emitted from the chimney after the energy released from burning coal heats up water inside the boiler.
Two circular shock-absorbing buffers prevent harsh impacts.
60103
FLYING SCOTSMAN
Connected wheels The engine powers six large driving wheels using pistons.

The Flying Scotsman cost £7,944 to build—more than $730,000 (£600,000) in today's money.

DHR CLASS B

Origin: UK/India
Year: 1892–1925
Length: Varies

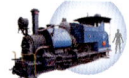

Small but powerful, these steam locomotives hauled passengers along the narrow-gauge Darjeeling Himalayan Railway, which rises more than 6,562 ft (2,000 m) through the mountains.

V&TRR NO.20 "TAHOE"

Origin: US
Year: 1875
Length: 51 ft (15.5 m)

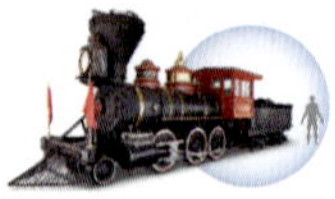

With a top speed of around 30 mph (48 kph), this steam locomotive ran for 51 years. Its large smokestack caught fire-causing sparks before they escaped, and it burned wood, coal, or oil.

ZEPHYR PIONEER

Origin: US
Year: 1934
Length: 197 ft (60.1 m)

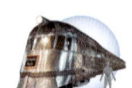

The first streamlined diesel railcar with a stainless-steel body was named after an Ancient Greek god of wind. It carried 72 passengers plus 25 tons (22.7 tonnes) of freight in a three-carriage train.

Front headlight, powered by electricity, which was generated by steam
N F

Wide chimney, or smokestack
Cow catcher
TAHOE

Burlington Route

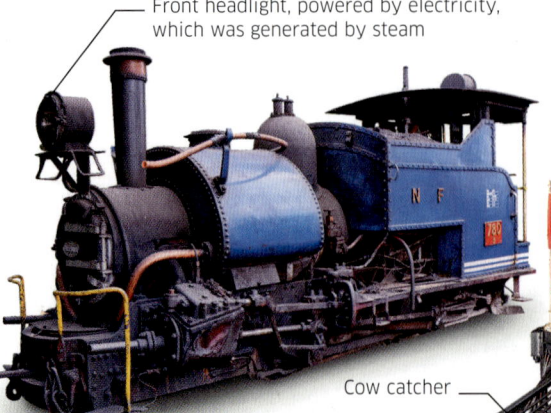

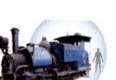

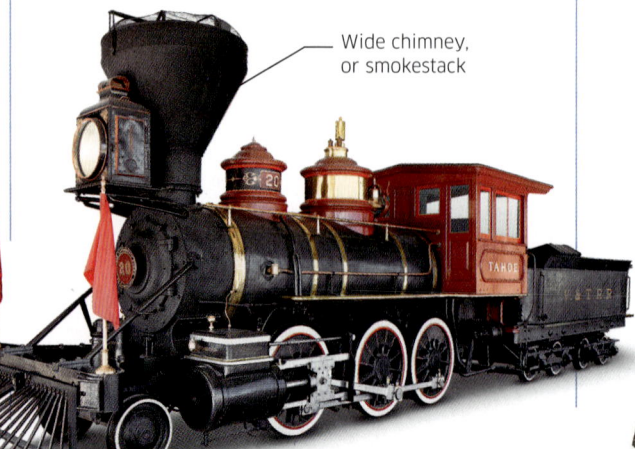

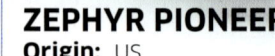

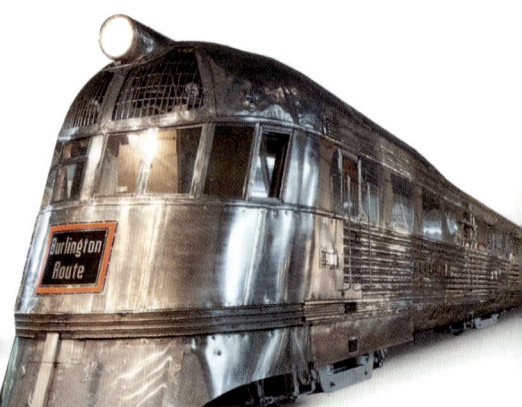

13,452 The number of passenger trains run each day by Indian Railways.

6 days or 5,753 miles (9,259 km)—the longest one-way trip on the Trans-Siberian Railroad.

M-497 BLACK BEETLE

Origin: US
Year: 1966
Length: 85 ft (25.9 m)

This experimental locomotive merged a Budd diesel railcar with two jet engines from a military bomber plane. In tests, it reached 183.7 mph (295.7 kph), still a US rail speed record.

Jet propelled
The two General Electric J47-19 jet engines cost $5,000 and were bought from the US Air Force.

WUPPERTALER SCHWEBEBAHN GTW15

Origin: Germany
Year: 2015–2017
Length: 78 ft, 9 in (24 m)

Around 80,000 commuters a day enjoy a scenic trip along this suspension railway, first opened in 1901. Today, its trains are propelled along the overhead monorail by four electric motors.

SIEMENS AVENIO

Origin: Germany
Year: 2009
Length: 30 ft (9 m) per carriage

Quiet electric motors propel this family of trolleys and light rail trains at speeds up to 50 mph (80 kph). Large double doors allow easy entry and exit for more than 500 passengers across eight carriages.

Light body
Each car is made out of lightweight steel.

SHANGHAI TRANSRAPID

Origin: Germany/China
Year: 2002
Length: 502 ft (153 m)

Hovering above its concrete track due to powerful electromagnets, this maglev (magnetic levitation) train reaches a peak speed of 268 mph (431 kph)—the fastest passenger service in the world.

TGV EURODUPLEX

Origin: France
Year: 2011
Length: 657 ft (200.2 m)

This high-speed electric train pulls eight double-decker passenger cars containing up to 556 seats. The train travels at speeds of up to 199 mph (320 kph) through France, Germany, and Spain.

DUBBELDEKS VIRM

Origin: Netherlands
Year: 1994–2009
Length: 356 ft (108.6 m)

These high-capacity trains feature two decks to pack in passengers. Motors are built into carriages, so no lead locomotive is needed. One set of 4–6 cars is called an "electrical multiple unit," or EMU.

Upper deck
Passengers here enjoy an elevated view but less luggage space.

Train controls
The driver's cab includes controls for the train's electrical system.

498 miles (802 km)—**the length of the longest subway system**, the Shanghai Metro in China.

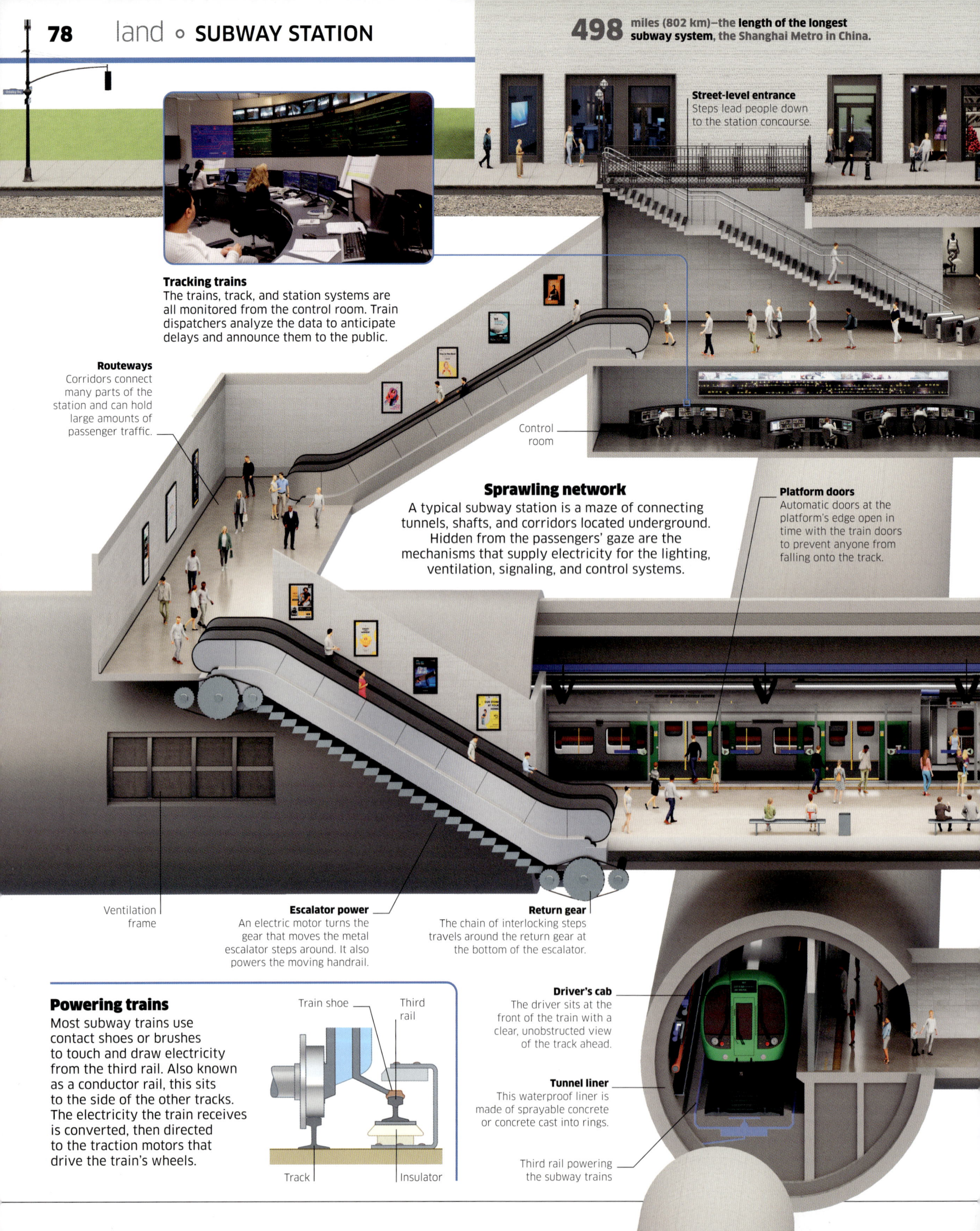

Street-level entrance
Steps lead people down to the station concourse.

Tracking trains
The trains, track, and station systems are all monitored from the control room. Train dispatchers analyze the data to anticipate delays and announce them to the public.

Routeways
Corridors connect many parts of the station and can hold large amounts of passenger traffic.

Control room

Sprawling network
A typical subway station is a maze of connecting tunnels, shafts, and corridors located underground. Hidden from the passengers' gaze are the mechanisms that supply electricity for the lighting, ventilation, signaling, and control systems.

Platform doors
Automatic doors at the platform's edge open in time with the train doors to prevent anyone from falling onto the track.

Ventilation frame

Escalator power
An electric motor turns the gear that moves the metal escalator steps around. It also powers the moving handrail.

Return gear
The chain of interlocking steps travels around the return gear at the bottom of the escalator.

Driver's cab
The driver sits at the front of the train with a clear, unobstructed view of the track ahead.

Tunnel liner
This waterproof liner is made of sprayable concrete or concrete cast into rings.

Powering trains
Most subway trains use contact shoes or brushes to touch and draw electricity from the third rail. Also known as a conductor rail, this sits to the side of the other tracks. The electricity the train receives is converted, then directed to the traction motors that drive the train's wheels.

Train shoe
Third rail
Track
Insulator

Third rail powering the subway trains

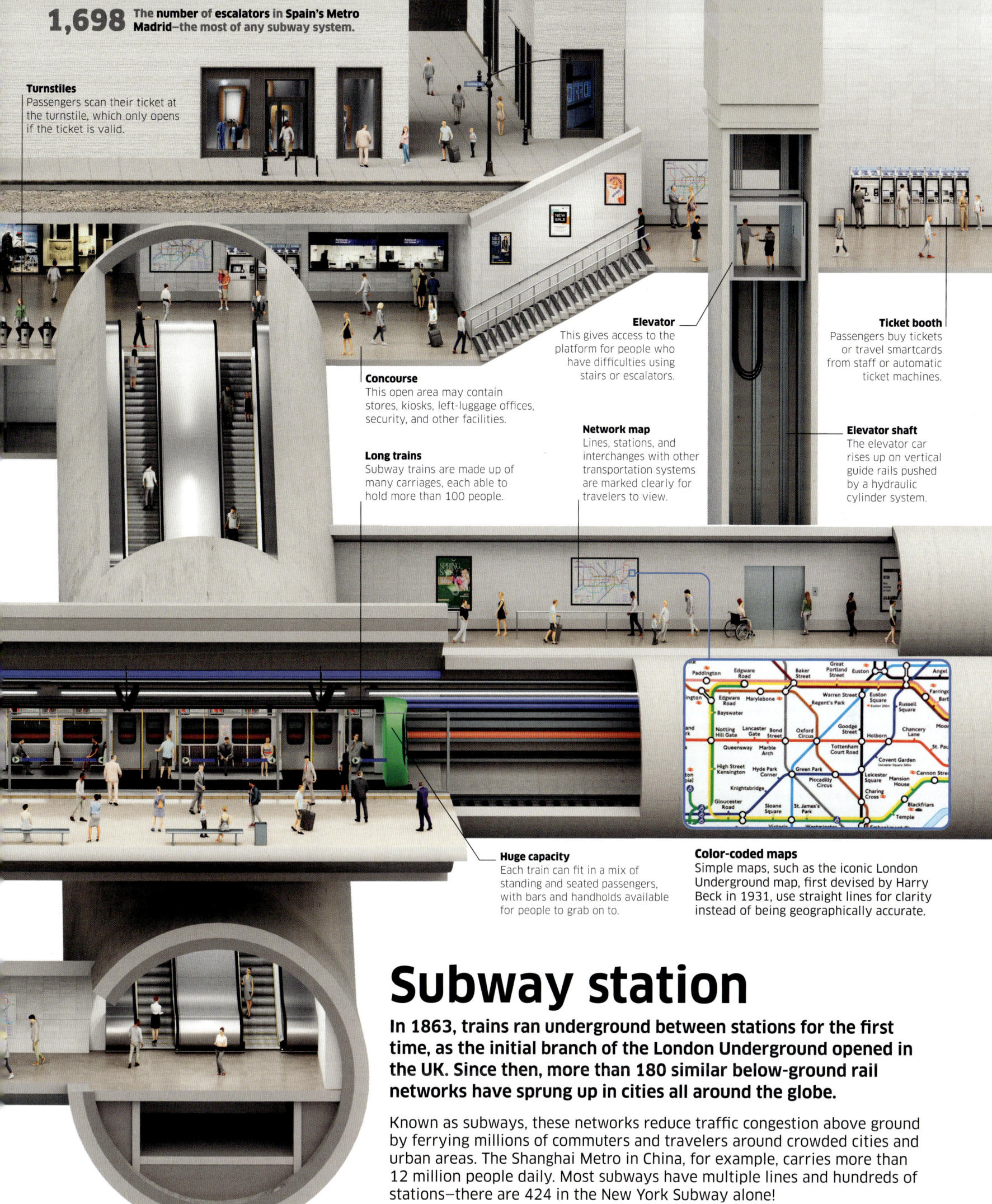

1,698 The **number of escalators** in Spain's Metro Madrid—the most of any subway system.

Turnstiles
Passengers scan their ticket at the turnstile, which only opens if the ticket is valid.

Elevator
This gives access to the platform for people who have difficulties using stairs or escalators.

Ticket booth
Passengers buy tickets or travel smartcards from staff or automatic ticket machines.

Concourse
This open area may contain stores, kiosks, left-luggage offices, security, and other facilities.

Network map
Lines, stations, and interchanges with other transportation systems are marked clearly for travelers to view.

Elevator shaft
The elevator car rises up on vertical guide rails pushed by a hydraulic cylinder system.

Long trains
Subway trains are made up of many carriages, each able to hold more than 100 people.

Huge capacity
Each train can fit in a mix of standing and seated passengers, with bars and handholds available for people to grab on to.

Color-coded maps
Simple maps, such as the iconic London Underground map, first devised by Harry Beck in 1931, use straight lines for clarity instead of being geographically accurate.

Subway station

In 1863, trains ran underground between stations for the first time, as the initial branch of the London Underground opened in the UK. Since then, more than 180 similar below-ground rail networks have sprung up in cities all around the globe.

Known as subways, these networks reduce traffic congestion above ground by ferrying millions of commuters and travelers around crowded cities and urban areas. The Shanghai Metro in China, for example, carries more than 12 million people daily. Most subways have multiple lines and hundreds of stations—there are 424 in the New York Subway alone!

Traveling by Shinkansen produces **92 percent less carbon emissions** than by plane.

High-speed train

With powerful electric motors and streamlined designs, high-speed trains race along tracks at more than 124 mph (200 kph)–slashing journey times around the world.

Following Japan's lead with its first Shinkansen trains in 1964, many countries now have high-speed rail networks, mostly running long distance between major cities. Some early rapid trains ran on steam or diesel, but electric propulsion has proven the most reliable and eco-friendly way of powering rapid intercity rail transportation.

Shinkansen E5

Introduced in 2011, Japan's latest Shinkansen trains can reach speeds of up to 199 mph (320 kph). Each train is made up of eight regular carriages known as cars, plus a power car at both the front and the rear. The entire train can seat 731 passengers.

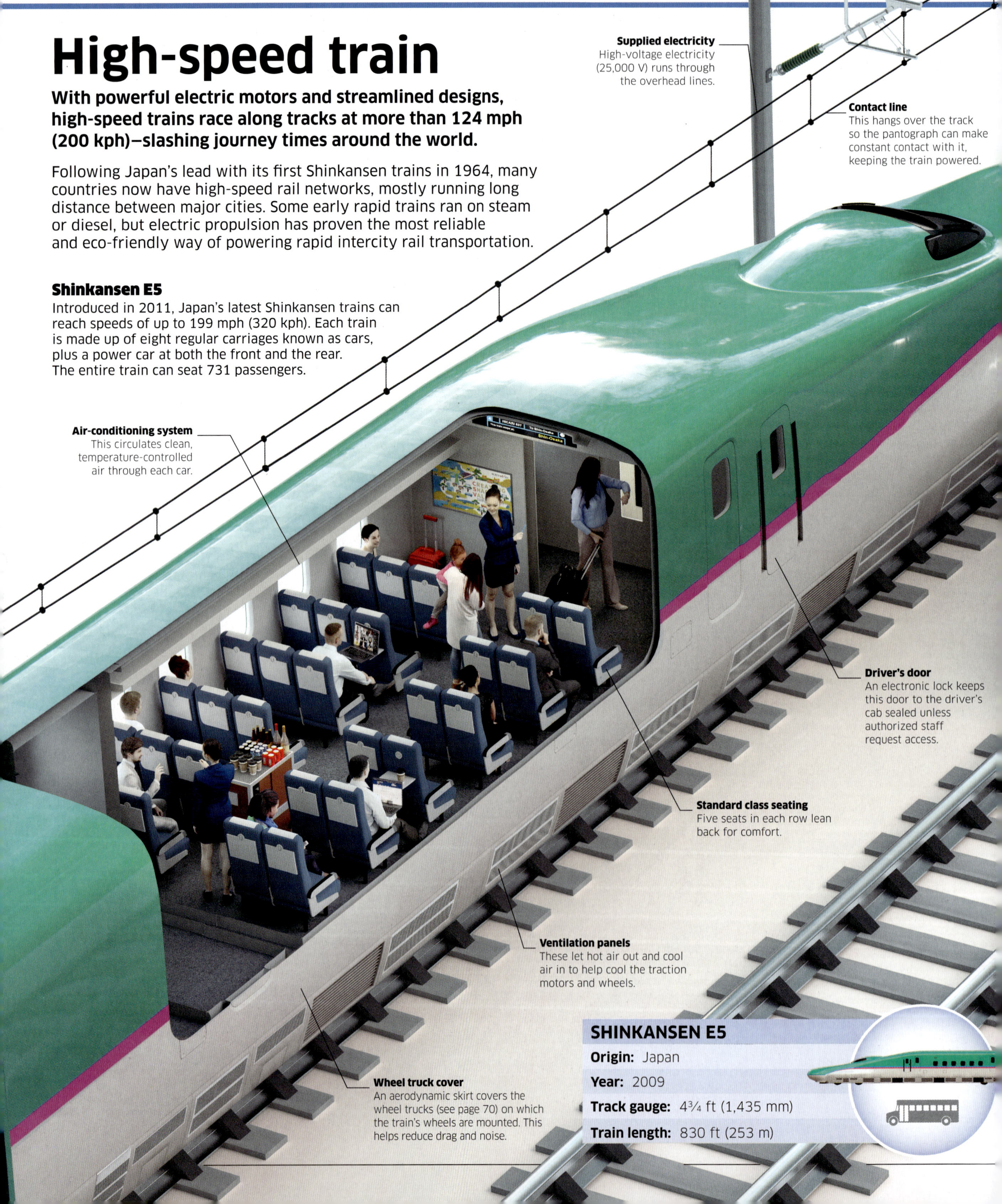

Supplied electricity
High-voltage electricity (25,000 V) runs through the overhead lines.

Contact line
This hangs over the track so the pantograph can make constant contact with it, keeping the train powered.

Air-conditioning system
This circulates clean, temperature-controlled air through each car.

Driver's door
An electronic lock keeps this door to the driver's cab sealed unless authorized staff request access.

Standard class seating
Five seats in each row lean back for comfort.

Ventilation panels
These let hot air out and cool air in to help cool the traction motors and wheels.

Wheel truck cover
An aerodynamic skirt covers the wheel trucks (see page 70) on which the train's wheels are mounted. This helps reduce drag and noise.

SHINKANSEN E5

Origin: Japan

Year: 2009

Track gauge: 4¾ ft (1,435 mm)

Train length: 830 ft (253 m)

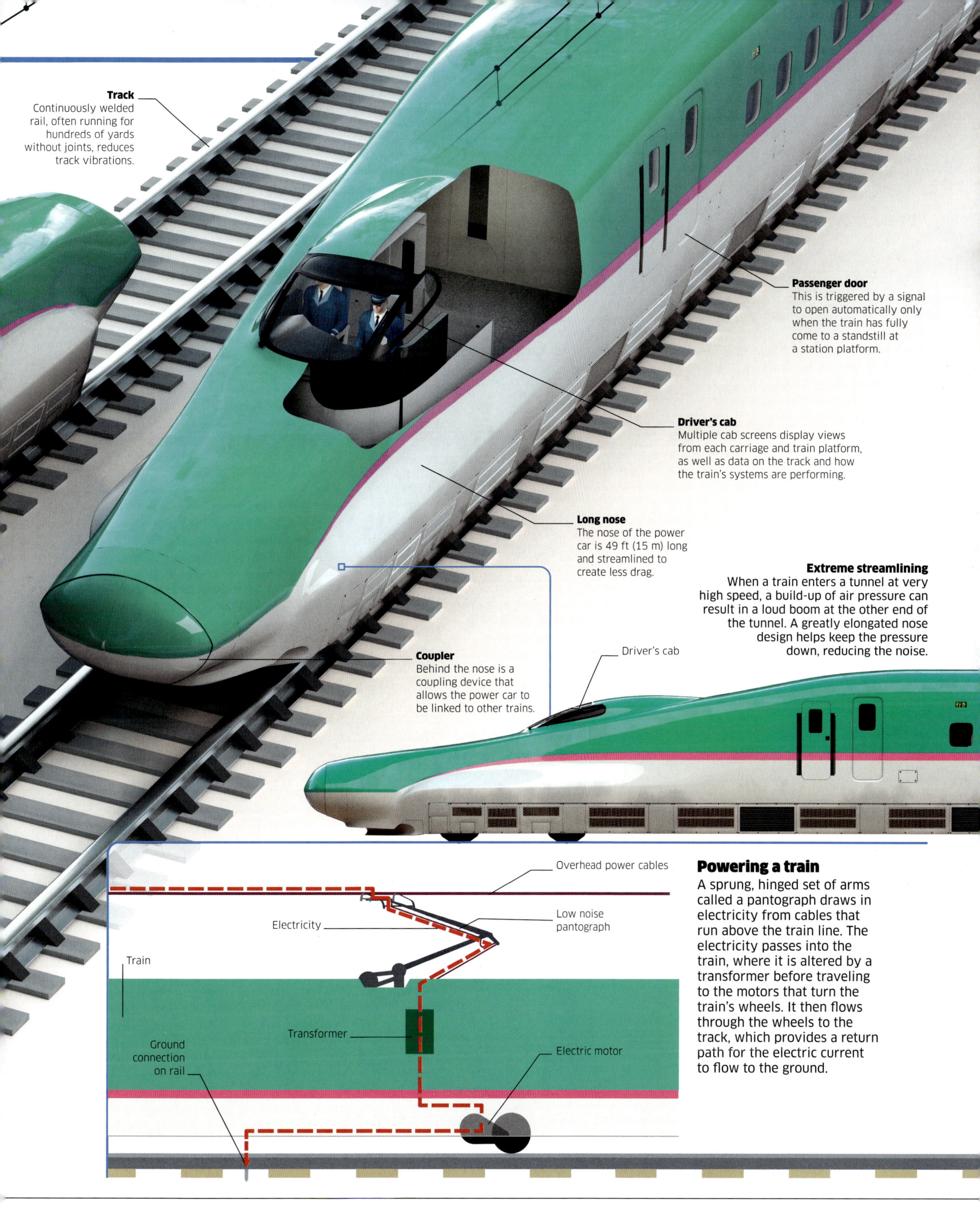

Track
Continuously welded rail, often running for hundreds of yards without joints, reduces track vibrations.

Passenger door
This is triggered by a signal to open automatically only when the train has fully come to a standstill at a station platform.

Driver's cab
Multiple cab screens display views from each carriage and train platform, as well as data on the track and how the train's systems are performing.

Long nose
The nose of the power car is 49 ft (15 m) long and streamlined to create less drag.

Extreme streamlining
When a train enters a tunnel at very high speed, a build-up of air pressure can result in a loud boom at the other end of the tunnel. A greatly elongated nose design helps keep the pressure down, reducing the noise.

Driver's cab

Coupler
Behind the nose is a coupling device that allows the power car to be linked to other trains.

Overhead power cables

Low noise pantograph

Electricity

Train

Transformer

Ground connection on rail

Electric motor

Powering a train
A sprung, hinged set of arms called a pantograph draws in electricity from cables that run above the train line. The electricity passes into the train, where it is altered by a transformer before traveling to the motors that turn the train's wheels. It then flows through the wheels to the track, which provides a return path for the electric current to flow to the ground.

Cliffhanger

Despite being built in 1884, this wooden funicular tramway at Saltburn-by-the-Sea, UK, still runs like clockwork today—traveling up and down a steep 71° gradient.

Cliff-climbing funiculars feature cars at either end of a long cable called the haul rope. This is looped through a pulley and counterweights, so as one car descends the track, the other car climbs up. The Saltburn Cliff Tramway has a 290-gallon (1,100-liter) water tank in each car as a counterweight, repeatedly filled and emptied to keep the cars going.

Armored transport

Equipped with powerful weaponry, clad in heavy armor, and often running on tracks so they can cross difficult terrain, tanks and other military vehicles dominate the battlefield.

Tanks were first used in World War I, when they were such a new idea, people called them "land ships." By World War II, specialized tanks designed for tasks such as mine clearing had been developed. Since then, many more kinds of fighting vehicles have appeared on the battlefield, from engineering machines to mobile hospitals.

MARK IV
Trench crawler
Origin: UK
Year: 1917

The Mark IV was one of the first and most effective tanks of World War I. There were two versions: one equipped with two cannons and three machine guns and the other with five machine guns.

Elevated nose
The tank's diamondlike shape allowed it to pull itself out of a trench if it toppled forward.

Six-pounder gun
This cannon was mounted on a rotating platform, allowing it to move left, right, up, and down.

RENAULT FT-17
Infantry support tank
Origin: France
Year: 1917

The first tank to have a rotating turret, the FT-17's layout was highly influential. It had a rear-mounted engine and crew in the front, which remains the standard tank layout today.

TIGER I
Heavy firepower
Origin: Germany
Year: 1942

The largest tank of World War II, the fearsome Tiger packed a powerful punch with its 88 mm gun, heavy armor, and vast size. However, it was unreliable and was mostly used defensively.

M4A1 SHERMAN
Speedy generalist
Origin: US
Year: 1940

The Sherman was designed to dash through enemy lines after a breakthrough and cause chaos. Almost 50,000 of these were built during World War II and crews liked their rugged reliability.

T-34
Medium tank
Origin: Russia
Year: 1941

Cupola opens for a clear view of the battlefield

Built in vast numbers, the T-34 was highly mobile with wide tracks that could tackle muddy fields. It had sloping armor that antitank weapons were unable to pierce.

SHERMAN V CRAB
Minefield clearer
Origin: US
Year: 1943

This adapted Sherman tank drove slowly over the battlefield, rotating its front-mounted drum to flail a set of chains. When the chains struck a mine, they triggered it at a safe distance from the tank, clearing a path.

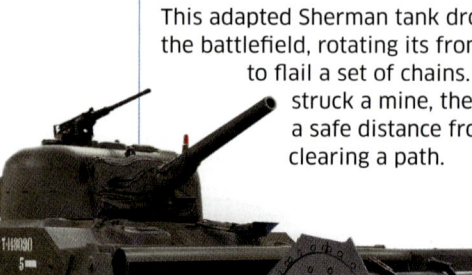

1915 The year the world's first modern tank, "Little Willie," was built.

"Water tanks" was the British codename for the first tanks. The name stuck.

The heaviest tank ever built, the Panzerkampfwagen VIII Maus, weighed more than a blue whale.

85

CRARRV
Repair and recovery vehicle
Origin: UK
Year: 1991

Equipped with winches and an extendable crane, the Challenger Armored Repair and Recovery Vehicle (CRARRV) can tow disabled tanks off the battlefield to stop them from falling into enemy hands.

Reactive armor
This defense system triggers small explosions in reaction to hits from enemy weapons, limiting damage.

HUMVEE
Armored car
Origin: US
Year: 1983

Lightly armed and armored, the Humvee is smaller, faster, and more maneuverable than a tank. It is used for transporting small numbers of troops into places tanks cannot easily go.

TERREX ICV
Amphibious troop carrier
Origin: Singapore
Year: 2009

Fully amphibious, meaning it can function on land and in water, the Terrex ICV can carry 12 soldiers. Cameras are positioned all around the vehicle to give the crew a 360-degree view.

FV104 SAMARITAN
Battlefield ambulance
Origin: UK
Year: 1978

The extra-high roof on this mobile, armored ambulance gave the medics inside plenty of room to work. The Samaritan has space for three casualties seated or on stretchers.

Red cross
This pattern means that this is a medical vehicle and that the enemy should not attack it. The central symbol is the Samaritan's regimental insignia.

Access ramp

LEOPARD C2
Main battle tank
Origin: Germany
Year: 2000

This upgraded version of the Leopard 1 main battle tank (MBT) now serves in Canada. MBTs are top-of-the-line tanks with heavy armor, powerful engines, and strong guns but the speed of a light tank.

Main gun
This 105mm cannon was added by engineers after the tank arrived in Canada.

Road panels
Rubber blocks fitted to the tank's tracks allow it to drive on a road without damaging it.

The Leopard
was so successful that variants of the tank now operate in 15 different countries.

86 land ○ **BATTLE TANK**

9 seconds—the **time it takes** for the M1A2's turret to rotate a full 360 degrees.

Smoke grenade launcher

Gunner's periscope

Thermal imaging viewer

Combat identification panel

Machine gun

Dark detection
Panels on the turret and sides are marked with special low-thermal tape or paint. At night or through thick smoke, allied soldiers can use thermal imagers to check for these markings, identify which side the tank is on, and avoid friendly fire.

Commander
Sitting inside a Kevlar-lined turret basket, the commander can look out through viewing ports in every direction.

Access hatch

Viewpoint
This independent thermal viewer displays areas of heat on the battlefield.

Tough tracks
The broad tracks are made up of more than 150 steel links.

Gunner
This soldier uses a laser rangefinder and other electronic aids to lock the gun's sights onto a target.

M1A2 ABRAMS TANK

Make: General Dynamics Land Systems

Origin: United States

Year: 1992

Length (including gun): 32 ft (9.77 m)

Long-range gun
The rotating cannon can fire shells at targets up to 13,123 ft (4,000 m) away.

Fire!
Many tanks have an automated loader, but in the M1A2, a human loads each explosive shell. At the same time, the gunner locks on to two independent targets. The gun can rapidly and accurately hit both targets.

Battle tank

Heavily armored tanks are some of the most formidable vehicles used in warfare. Their tough exteriors and impressive weaponry make them a key element of armies all around the world.

Advancing across tough terrain, battle tanks can break down walls and clamber over cars, rocks, and any other obstacles and debris blocking their path. Named after US General Creighton W. Abrams, the M1A2 is one of the most sophisticated models in use today, deployed by nations such as the US, Australia, and Egypt. It contains more electronics than an F-16 jet fighter, and much of its exterior is protected by advanced armor, the precise composition of which remains top secret.

Heavy hitter

Weighing up to 73.6 tons (66.8 tonnes), the M1A2 weighs more than 14 monster trucks. Its heavily armored turret weighs 28.7 tons (26 tonnes) alone—two-fifths the entire tank's weight. Rumbling along at a top speed of 42 mph (67 kph), the vast vehicle can only fit a four-person crew inside: a loader, a gunner, a commander, and a driver.

Caterpillar tracks

Made of many individual links, the tracks of a tank provide a large surface area over which to distribute the vehicle's weight. Forming a continuous, flexible belt, they run in a loop between the drive sprocket and front idler wheel. Much of the tracks are covered by armor-plated sideskirt panels that can be opened for maintenance.

Front idler wheel maintains track tension

Gun barrel
Shells exit the barrel at speeds of around 5,249 ft (1,600 m) per second.

Laid-back driver
Half-sitting, half-lying, the driver controls each of the tracks independently.

Drive sprocket
When turned by the engine, this moves the track around, propelling the tank forward.

Roadwheels
These rise upward, enabling the tank to ride up and over obstacles.

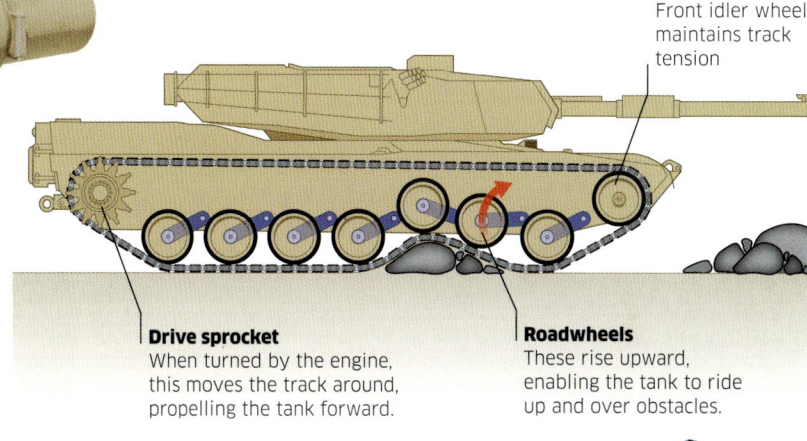

Rubber coating
Each track link has an outer covering of hard rubber to improve grip and traction.

Mine sweeping

Military engineers can add attachments to the M1A2 that allow it to clear deadly land mines buried under the ground. As the tank moves forward, sets of heavy rollers press down in front of it, triggering any mines ahead and leaving the way clear for troops and other vehicles to progress.

WATER

Around 70 percent of our planet is covered by water, so it is not surprising that a lot of transportation is by boat. Oceans, rivers, and canals teem with vessels carrying people, goods, or equipment—be it for business, pleasure, defense, or research.

Windsurfer
Windsurfing, invented in the 1960s, is just one of the many ways in which to compete or just have fun on water.

ONE GRUS, 2019

1950s submarine
Submarines were used in both world wars, but the US launched the first nuclear-powered sub in 1954.

USS NAUTILUS, 1954

HMS DREADNOUGHT, 1905

Dreadnought
A new style of armored warship emerged just before World War I.

1960s ONWARD

Ocean liner
Luxurious steamships carried passengers between continents.

RMS MAURETANIA, 1906–1934

Modern times
Steel replaced wooden hulls, and ships were propelled by steam and, later, diesel engines instead of sails. Ships were still used for transportation and war, but also for pleasure.

1900–PRESENT

1900 ONWARD

Boats through history

From naturally floating tree trunks to ultra-modern warships, humans have continued to come up with new ideas and inventions for traveling across water.

The story of watercraft began many thousands of years ago, when early humans sought ways to cross rivers and lakes, and discovered that some materials and shapes were better suited for floating vessels than others. As science and engineering evolved, boats became larger, faster, or suited for longer journeys, and different types of vessels were built for different purposes.

GEOBUKSEON TURTLE SHIP

Korean warship
The armored "turtle ships" were used for defense and attack from the 15th to the 18th century.

Viking longboat
The Vikings built strong, fast ships that could cross the Atlantic Ocean and navigate far inland along rivers.

700–1100 CE

Timeline of watercraft
This timeline shows how boats have developed from basic craft to sophisticated modern vessels.

BEFORE 500 CE

Early vessels
Around 10,000 years ago, human ancestors who lived near water made rafts and canoes. Thousands of years later, ships sailed around the ancient Mediterranean world and in the Pacific Ocean.

Dugout canoe
Made by carving out a tree trunk, this early form of canoe is still in use in some parts of the world.

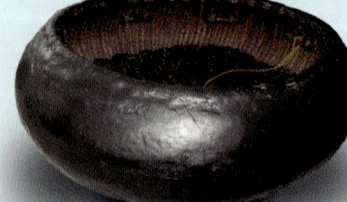

Assyrian quffa
The earliest form of coracle, these large, hide-covered reed baskets were steered with one oar on rivers in Assyria, an ancient civilization in West Asia.

7500 BCE

900 BCE

255 The number of **ships** that took part in the **Chinese** admiral Zheng He's **first expedition** across the Indian Ocean, in **1405**.

Not all boats stay **afloat**—it is estimated that more than **3 million** wrecks, from all periods of history, lie on the **ocean floor**.

91

Container ship
First appearing in the 1950s, container ships today can carry up to 15,000 cargo units.

USS *RONALD REAGAN*, 2001

Aircraft carrier
These gigantic war ships patrol the oceans, carrying fighter planes and helicopters.

RRS *SIR DAVID ATTENBOROUGH*, 2020

Polar research ship
Today, some ships are used for researching and monitoring what is happening to water and wildlife in the oceans.

21ST CENTURY

Steamship
Steamships began to appear in the early 19th century, but *Turbinia* was the first to use a steam turbine.

***TURBINIA*, 1894**

Windjammer
Sailing at high speed, three-masted vessels shipped tea, wool, and passengers between continents.

1880s

HMS *VICTORY*, 1759

1600–1900 ▶

Age of sail
As European countries invaded and colonized other continents, they built ever larger and faster ships for war and for cargo.

Spanish galleon
These armed cargo ships transported silver and gold from Spanish colonies in the Americas.

19th-century warship
Mighty warships, armed with hundreds of guns, battled each other in the many wars of the 18th and 19th centuries.

16TH–18TH CENTURY

Indian Ocean dhow
These merchant ships sailed between East Africa, India, and the Arabian Peninsula.

Trading, raiding, and exploring
As boat-building techniques resulted in better ships and navigation instruments were invented, people set out to sail ever farther away from their home shores.

MEDIEVAL WORLD ◀

Polynesian drua
Polynesian peoples used reed and wooden catamarans to explore the Pacific Ocean from as early as 3,500 years ago. By the 13th century, they had reached Hawaii.

FROM 600 CE

On the Nile
In ancient Egypt, the River Nile was busy with all kinds of boats—from fishing vessels to royal barges.

Greek trireme
The feared warships of ancient Greece dominated the Mediterranean Sea.

3000 BCE

480 BCE

Moving across water

Surrounding continents, mainland coasts, and islands and cutting through countries and cities, water can be an obstacle. But over time, people learned how to journey across it to connect, trade, and travel.

Combining traditional boat building with the latest technology and materials, modern merchant ships are constructed to carry heavy loads, be seaworthy, and (lately) be as environmentally friendly as possible. Boats of all sizes carry passengers, too, or are made for racing or for just having fun on the water. Whatever the size, sailors need to know port from starboard!

BOAT LANGUAGE

Sailors use special terms to talk about left and right at sea and the different parts of their boats. These are some of the most common and useful ones to know.

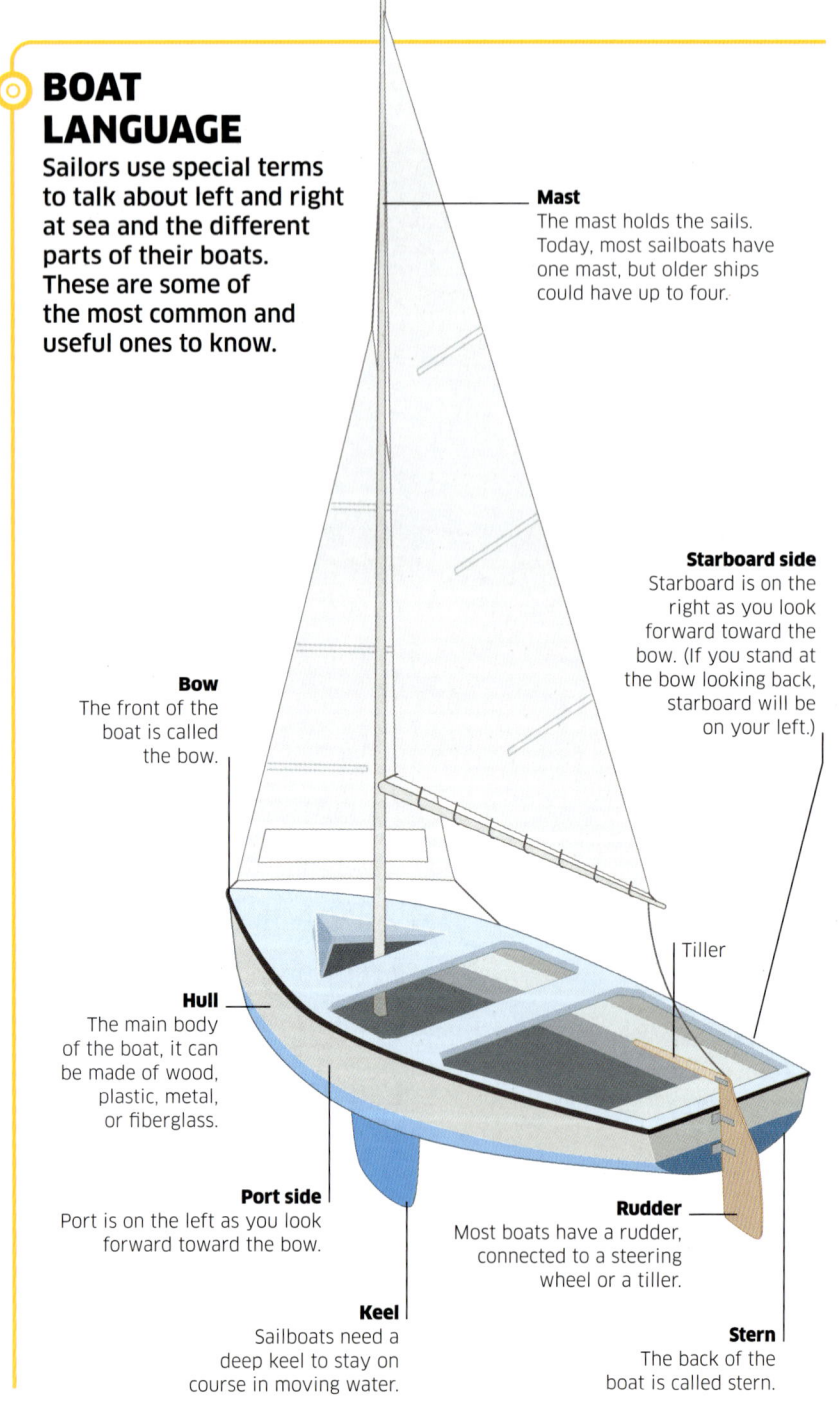

Mast
The mast holds the sails. Today, most sailboats have one mast, but older ships could have up to four.

Starboard side
Starboard is on the right as you look forward toward the bow. (If you stand at the bow looking back, starboard will be on your left.)

Bow
The front of the boat is called the bow.

Tiller

Hull
The main body of the boat, it can be made of wood, plastic, metal, or fiberglass.

Port side
Port is on the left as you look forward toward the bow.

Keel
Sailboats need a deep keel to stay on course in moving water.

Rudder
Most boats have a rudder, connected to a steering wheel or a tiller.

Stern
The back of the boat is called stern.

BOAT SCIENCE

How do heavy boats stay afloat, and what makes them move in an efficient way? Boat builders need to know about physics to create optimal boat designs. The most important is how the boat's density and shape work with the forces (weight, thrust, and drag) that act upon it. This will decide if the boat sinks or swims!

How boats float

When an object is put in water, it displaces (pushes away) some of it. If the object has a higher density (its weight is more than that of the displaced water), it sinks. Boats push aside a lot of water, and the displaced water pushes back up, creating a force called upthrust. If a boat's average density is less than that of the displaced water, the upthrust will be greater, and the boat floats. Because their hulls contain a lot of air, even large, heavy ships can float.

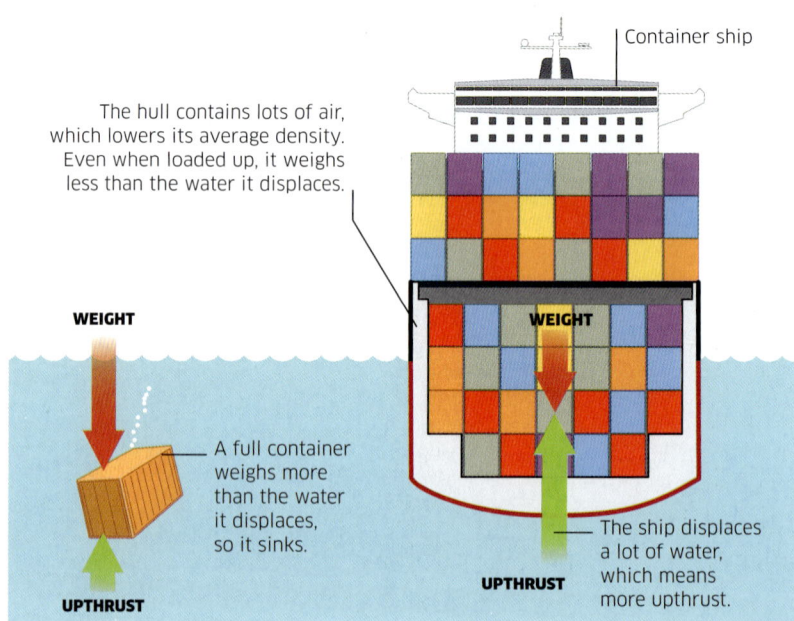

Container ship

The hull contains lots of air, which lowers its average density. Even when loaded up, it weighs less than the water it displaces.

WEIGHT

WEIGHT

A full container weighs more than the water it displaces, so it sinks.

The ship displaces a lot of water, which means more upthrust.

UPTHRUST

UPTHRUST

How boats move

Sailing vessels are propelled by wind, but boats powered by an engine need propellers to move. Propellers rotate to push water backward. This propels the boat forward, creating a force called thrust, which acts against the water resistance (drag). If the thrust is stronger than the drag, the boat moves forward.

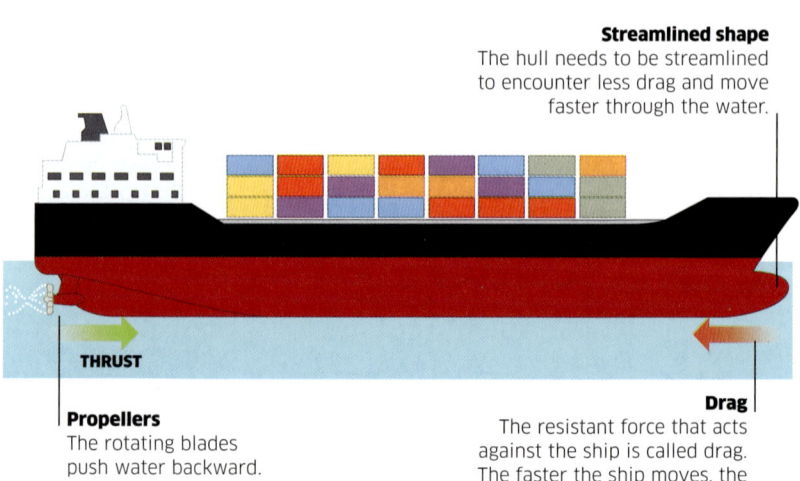

Streamlined shape
The hull needs to be streamlined to encounter less drag and move faster through the water.

THRUST

Propellers
The rotating blades push water backward.

Drag
The resistant force that acts against the ship is called drag. The faster the ship moves, the stronger the drag resistance.

In **China**, the **first human-made canals** were begun as early as the **5th century BCE.**

89 days and 8 hours—the **time** it took *Flying Cloud* to **sail** from **New York to San Francisco** in **1854**, during the *California gold rush.*

93

VITAL WATERWAYS

Before road and rail networks existed, the safest and fastest route between different places was often over water. People traveled on rivers, across lakes, along coasts, and even crossed oceans or sailed around whole continents to reach their destination. Today, long-established routes are still teeming with different types of ships.

Long way around

During the 19th century, the discovery of gold attracted people to set out for North America's west coast. Fortune seekers boarded ships in New York or the English port of Liverpool in Europe. Even though they had to sail all the way around the southern tip of South America, then northward along the other side to reach San Francisco, California, the sea voyage was preferable to the difficult and often dangerous route across the continent on land.

Key to ship routes

🟨 New York to San Francisco
🟪 Liverpool to San Francisco

Clever canals

People build canals to make water routes shorter. The Panama Canal allows ships to cut across a narrow strip of Central America, from the Atlantic to the Pacific, instead of sailing around the dangerous Cape Horn at the tip of South America. It has 12 locks and took 10 years to build, from 1904 to 1914.

School run

Some children living along the Amazon River in Brazil can get to school via boat. This is a much quicker way to travel than making their way over land through the dense rainforest that grows on both sides of the river.

Cargo catastrophe

The Suez Canal connects shipping routes from Europe with those of the Indian Ocean and is one of the world's busiest waterways. But in 2021, the gigantic cargo ship *Ever Given* ran aground, blocking the canal for six days. As hundreds of cargo ships waited to pass, global supply chains were badly affected.

GLOBAL SHIPPING

An enormous range of goods, from oil and car parts to toys, clothes, and food, are carried across water on board cargo ships. Most manufactured goods are shipped in containers between busy container ports (see pages 114–115) from where they continue their journey on land.

Key

The colored lines show how many trips are made along main routes in a year.

🟥 More than 3,000
🟧 1,000–3,000
🟨 Less than 1,000
🔵 World's busiest ports

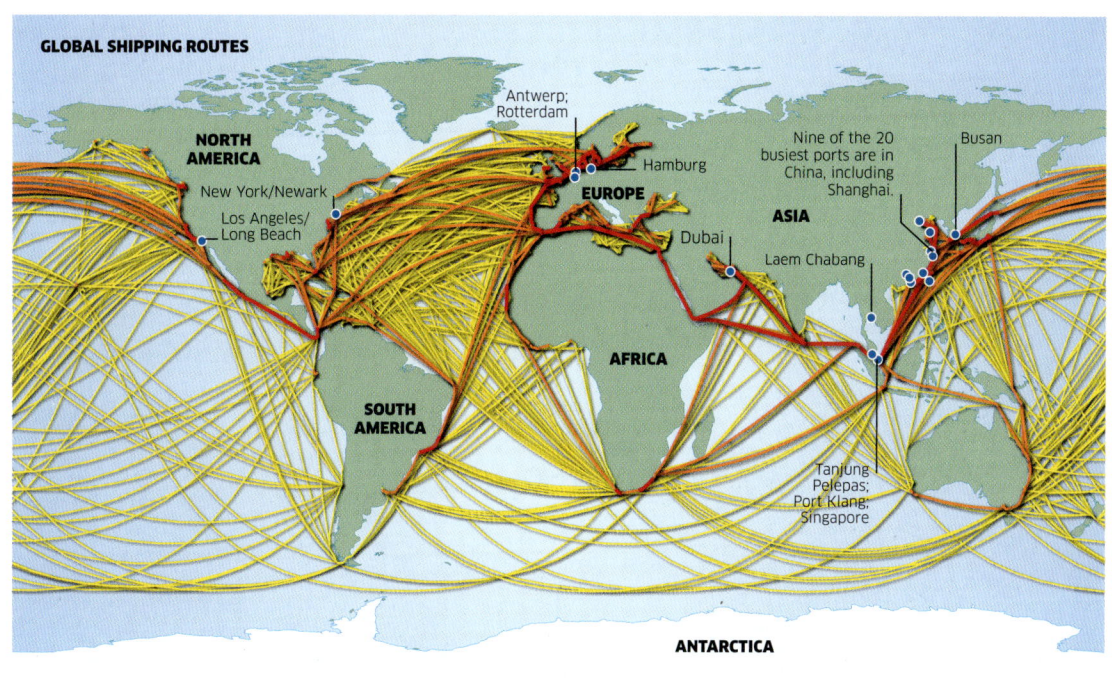

CORACLE
Floating basket
Origin: Assyria
Year: From c. 900 BCE

River traders and fishermen used a single paddle to skillfully propel and steer these baskets, preventing them from just spinning around. Coracles were later used in India, as well as in the British Isles.

Indian coracle
As on all coracles, animal hides cover the outside, making it watertight.

INUIT KAYAK
Hunting craft
Origin: Arctic Circle
Year: From c. 1000 BCE

The Inuit, Yupik, and Aleut people of the Arctic Circle constructed these boats from driftwood, covering them with waterproof seal skin. They used them to hunt whales, seals, and walruses in the icy seas.

Paddle

Cockpit holding one person

SAMPAN
River trading vessel
Origin: East Asia
Year: From c. 200 BCE

This flat-bottomed boat is still used to transport goods up and down rivers in China and Southeast Asia. With one or two oars at the rear, or stern, it has storage space for fish and other goods for sale.

Floating home
This Vietnamese sampan has a sheltered living area.

EGYPTIAN SKIFF
Reed boat
Origin: Nile Valley, Egypt
Year: From c. 3000 BCE

Papyrus reeds are hollow, so when tied together in bundles, they make boats that are very buoyant. Fishing skiffs such as this one navigated the Nile at the time of the pharaohs.

PHOENICIAN GAULOI
Seagoing trader
Origin: Eastern Mediterranean
Year: c. 600 BCE

The Phoenicians, from modern-day Lebanon, were great mariners. Their trading ships sailed between the cities they founded across the Mediterranean, such as Carthage and Tyre.

Oars were used when there was not enough wind to sail.

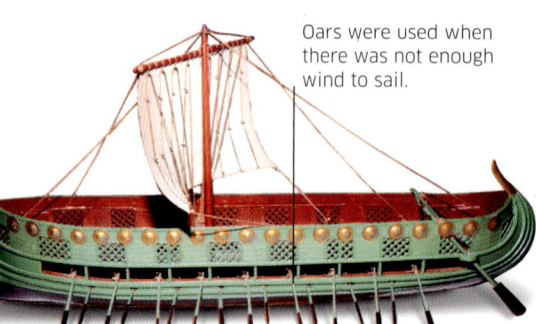

DHOW
Cargo and fishing boat
Origin: Indian Ocean
Year: From c. 600 BCE

With their triangular sails, dhows (or marakabs) are still common in the seas around Asia and East Africa. Their long, thin hulls help them slice quickly through water.

POLYNESIAN CATAMARAN
Ocean explorer
Origin: South Pacific
Year: From c. 1500 BCE

Catamarans, two canoes tied side-by-side—called *drua*, or *waqa tabu* ("sacred boats")—sailed between the 1,000 or so islands of Polynesia. Larger models crossed the Pacific Ocean, reaching Hawaii and New Zealand / Aotearoa.

Platform holding people and cargo

Early boats

The first sailors in ancient civilizations used hollowed-out logs or rafts for fishing. In time, boats became larger and more sophisticated.

The earliest boats were used on rivers and close to shore. Later, the invention of sails allowed mariners to harness the power of the wind, while oars and rudders provided greater control and maneuverability. As societies developed, they created boats for specific uses besides catching fish. There were warships, cargo vessels for trade, and ocean-going ships for exploration. Some models are still in use today.

8000 BCE—the **date of** the **oldest surviving canoe, found in the** Netherlands.

1000 CE—the year **Vikings** first sailed **across the Atlantic** to America.

95

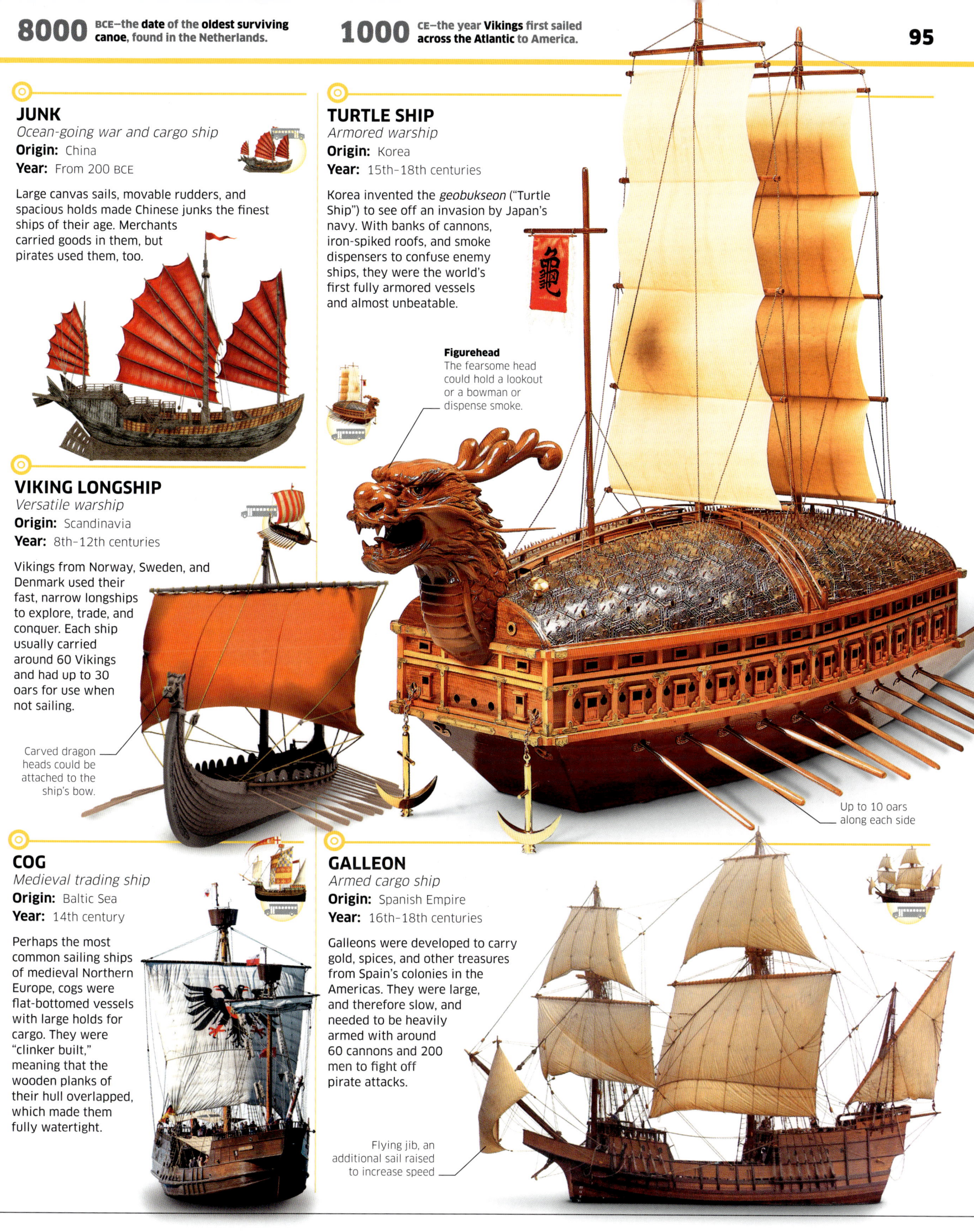

JUNK
Ocean-going war and cargo ship
Origin: China
Year: From 200 BCE

Large canvas sails, movable rudders, and spacious holds made Chinese junks the finest ships of their age. Merchants carried goods in them, but pirates used them, too.

VIKING LONGSHIP
Versatile warship
Origin: Scandinavia
Year: 8th–12th centuries

Vikings from Norway, Sweden, and Denmark used their fast, narrow longships to explore, trade, and conquer. Each ship usually carried around 60 Vikings and had up to 30 oars for use when not sailing.

Carved dragon heads could be attached to the ship's bow.

COG
Medieval trading ship
Origin: Baltic Sea
Year: 14th century

Perhaps the most common sailing ships of medieval Northern Europe, cogs were flat-bottomed vessels with large holds for cargo. They were "clinker built," meaning that the wooden planks of their hull overlapped, which made them fully watertight.

TURTLE SHIP
Armored warship
Origin: Korea
Year: 15th–18th centuries

Korea invented the *geobukseon* ("Turtle Ship") to see off an invasion by Japan's navy. With banks of cannons, iron-spiked roofs, and smoke dispensers to confuse enemy ships, they were the world's first fully armored vessels and almost unbeatable.

Figurehead
The fearsome head could hold a lookout or a bowman or dispense smoke.

Up to 10 oars along each side

GALLEON
Armed cargo ship
Origin: Spanish Empire
Year: 16th–18th centuries

Galleons were developed to carry gold, spices, and other treasures from Spain's colonies in the Americas. They were large, and therefore slow, and needed to be heavily armed with around 60 cannons and 200 men to fight off pirate attacks.

Flying jib, an additional sail raised to increase speed

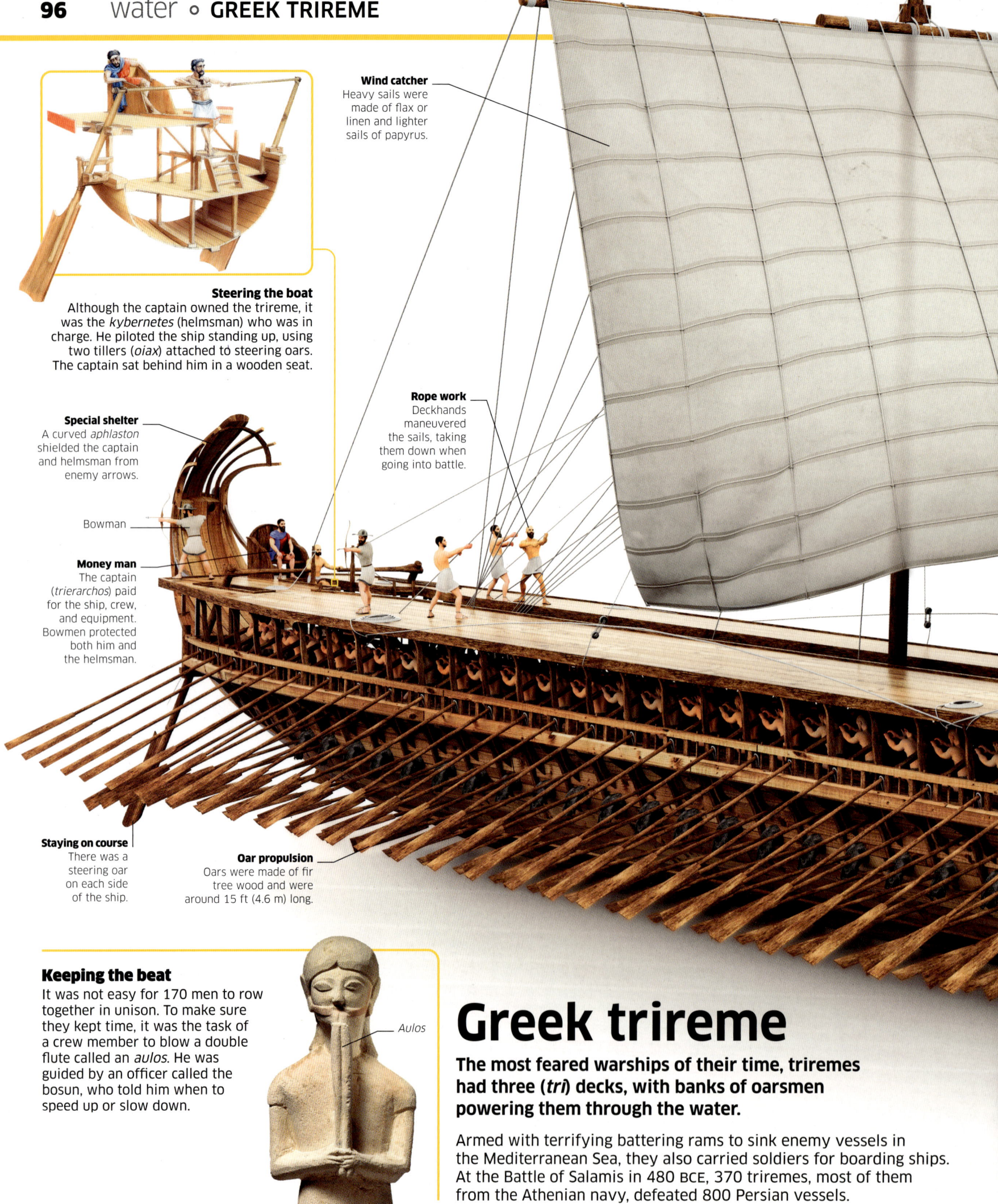

Wind catcher
Heavy sails were made of flax or linen and lighter sails of papyrus.

Steering the boat
Although the captain owned the trireme, it was the *kybernetes* (helmsman) who was in charge. He piloted the ship standing up, using two tillers (*oiax*) attached to steering oars. The captain sat behind him in a wooden seat.

Rope work
Deckhands maneuvered the sails, taking them down when going into battle.

Special shelter
A curved *aphlaston* shielded the captain and helmsman from enemy arrows.

Bowman

Money man
The captain (*trierarchos*) paid for the ship, crew, and equipment. Bowmen protected both him and the helmsman.

Staying on course
There was a steering oar on each side of the ship.

Oar propulsion
Oars were made of fir tree wood and were around 15 ft (4.6 m) long.

Keeping the beat
It was not easy for 170 men to row together in unison. To make sure they kept time, it was the task of a crew member to blow a double flute called an *aulos*. He was guided by an officer called the bosun, who told him when to speed up or slow down.

Aulos

Greek trireme

The most feared warships of their time, triremes had three (*tri*) decks, with banks of oarsmen powering them through the water.

Armed with terrifying battering rams to sink enemy vessels in the Mediterranean Sea, they also carried soldiers for boarding ships. At the Battle of Salamis in 480 BCE, 370 triremes, most of them from the Athenian navy, defeated 800 Persian vessels.

300 The **number of triremes** in the Athenian navy at its height.

50 The **number of strokes per minute** that oarsmen were expected to keep up at attack speed.

97

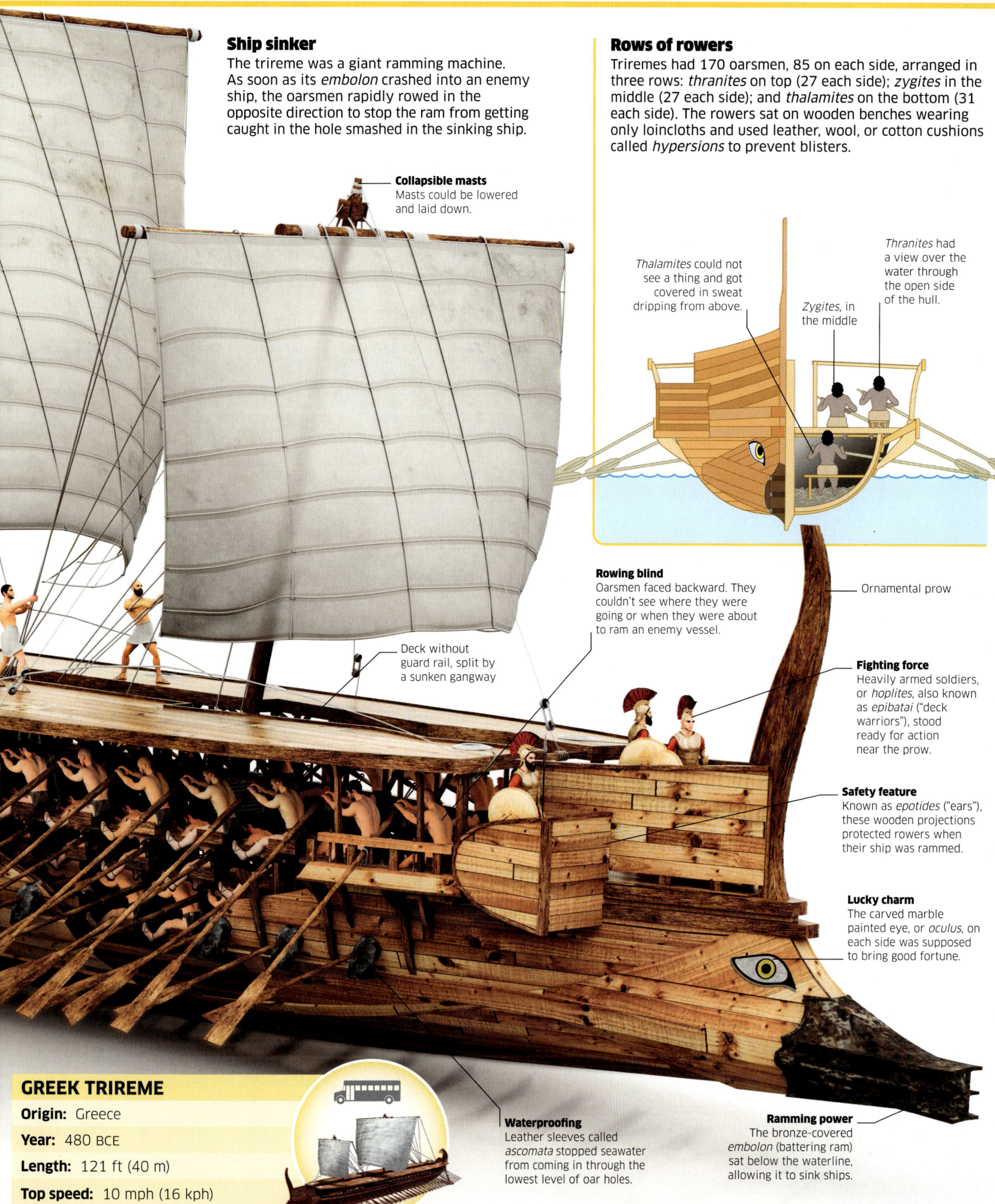

Ship sinker

The trireme was a giant ramming machine. As soon as its *embolon* crashed into an enemy ship, the oarsmen rapidly rowed in the opposite direction to stop the ram from getting caught in the hole smashed in the sinking ship.

Rows of rowers

Triremes had 170 oarsmen, 85 on each side, arranged in three rows: *thranites* on top (27 each side); *zygites* in the middle (27 each side); and *thalamites* on the bottom (31 each side). The rowers sat on wooden benches wearing only loincloths and used leather, wool, or cotton cushions called *hypersions* to prevent blisters.

Collapsible masts
Masts could be lowered and laid down.

Thalamites could not see a thing and got covered in sweat dripping from above.

Thranites had a view over the water through the open side of the hull.

Zygites, in the middle

Rowing blind
Oarsmen faced backward. They couldn't see where they were going or when they were about to ram an enemy vessel.

Ornamental prow

Deck without guard rail, split by a sunken gangway

Fighting force
Heavily armed soldiers, or *hoplites*, also known as *epibatai* ("deck warriors"), stood ready for action near the prow.

Safety feature
Known as *epotides* ("ears"), these wooden projections protected rowers when their ship was rammed.

Lucky charm
The carved marble painted eye, or *oculus*, on each side was supposed to bring good fortune.

GREEK TRIREME

Origin: Greece

Year: 480 BCE

Length: 121 ft (40 m)

Top speed: 10 mph (16 kph)

Waterproofing
Leather sleeves called *ascomata* stopped seawater from coming in through the lowest level of oar holes.

Ramming power
The bronze-covered *embolon* (battering ram) sat below the waterline, allowing it to sink ships.

Dragon boats

Propelled by 20 crew members each, two dragon boats compete in a ferocious race, paddles plunging into the water in time to a driving drum beat.

In China, dragons are closely connected to water, and dragon boat races have been held on the fifth day of the fifth month for more than 2,000 years. Today, people take part in these riotous races all over the world in modern boats made from lightweight fiberglass. The goal is to be the first to pull a flag positioned on the finish line out of the water.

5,500 The **number** of trees used to **build** *Victory*.

First-rate warship

In the 18th and 19th centuries, many nations fought wars at sea. Ships needed to be fast, maneuverable, and have space for lots of guns and sailors.

The finest warship in a navy was known as a first-rate ship of the line. It was among a fleet's largest vessels, with guns arranged over three decks, and would have been the first to sail into battle against a row of enemy craft. One such ship was the British navy's HMS *Victory*.

HMS *Victory* under sail

Victory was a full-rigged ship with three masts. She was square-rigged, meaning most of her sails were arranged in straight, horizontal lines along wooden spars, or yards. Many seamen had to work together to make her sail her best.

Mainmast
The mast in the middle was the ship's tallest mast, at 220 ft (67 m) high.

Topgallant, the uppermost section of a mast

Risky climb
Sailors had to climb the ladderlike part of the rigging, called ratlines, to get up in the masts to work the sails. The ropes were coated in tar, making them slippery when wet.

Working at height
To handle the larger square sails, seamen leaned on the wooden yards while balancing on ropes attached below them. They needed strong arms, a good sense of balance, and to not suffer from vertigo.

Officers on board
The admiral, captain, and other officers stayed in private cabins at the stern (rear) of the ship (see pages 102–103).

The cost of building *Victory* in 1759 was £63,173, which would be more than $60.7 million (£50 million) in today's money.

27 miles (42 km)—the **total length** of **rope** used in *Victory*'s rigging.

101

HMS *VICTORY*

Origin: UK

Year: 1759

Length: 267 ft (69 m)

Weight: 3,556 tons (3,613 tonnes)

Tip-top sails
It took skilled artisans around 1,200 hours to sew together just one single topsail like this one.

Staying in place
Stabilizing ropes called forestays ran between the foremast and the bowsprit.

Jib sail
These triangular, side-on sails helped the ship maneuver in crosswinds.

Flying jib-boom
Adding an extension pole, or flying jib-boom, meant the bowsprit's sails could be raised higher.

Jib-boom

Speedy spritsail
Fitted under the bowsprit, spritsails were raised when the captain wanted to capture the maximum amount of wind to make the ship sail as fast as possible.

Fancy figurehead
Carved wooden sculptures decorated the bow of the ship. They featured all manner of characters, such as lions, mermaids, and women. On *Victory*, two cherubs hold the British coat of arms.

Battle at sea

Sea battles were carefully and strategically planned. Ships lined up at sea, ready to fire broadsides at the enemy line. But wind, weather, and sometimes surprise tactics often meant that things didn't go as planned. And once ships engaged at close quarters, the noise and smoke from guns and burning ships added to the chaos.

Battle of Trafalgar
Victory, commanded by Admiral Horatio Nelson, was the lead ship of the British navy in the Battle of Trafalgar in 1805. After first breaking up the line of Spanish and French ships and then engaging at close range, the British won.

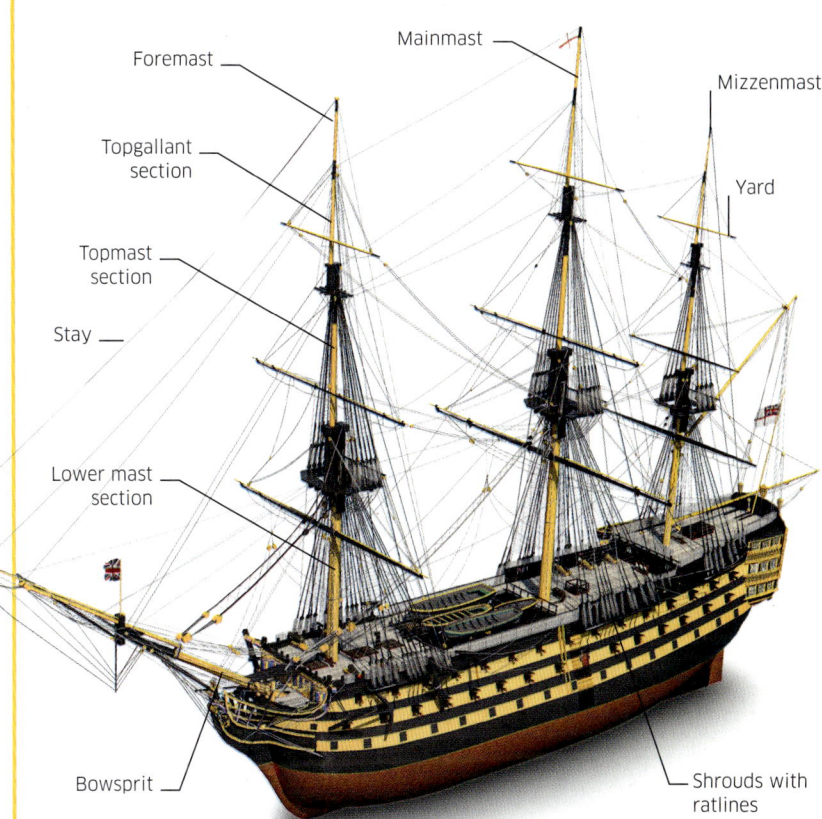

Foremast

Mainmast

Mizzenmast

Topgallant section

Yard

Topmast section

Stay

Lower mast section

Bowsprit

Shrouds with ratlines

Masts and rigging

The three masts, the bowsprit, and the yards were made of wood from pine trees. But no tree would be tall enough for a mast, so masts were made of three sections. Everything was held in place by the strong ropes that form the rigging (stays and shrouds).

102 water ∘ **FIRST-RATE WARSHIP**

820 The number of crew members, of more than 22 nationalities, who were on board *Victory* at the Battle of Trafalgar.

Captain's quarters
Victory's captain lived in relative comfort. He had his own "apartment," with a separate bedroom, an office, and a dining area with elegant furniture.

Deck upon deck
Victory had five full decks. Its core was the upper, middle, and lower guns decks, where 88 of the ship's 104 powerful cannons were arranged. There was also a storage deck, or "hold," and a raised poop deck at the rear.

The British navy's flag, or "colors," known as the White Ensign.

Officers' overview
This officer has a perfect view of everything. He might shout an order to the seamen by the steering wheel, which is just underneath him, on the deck below.

Shrouds with ratlines (see page 100)

Skylight letting light into the captain's cabin below

Lighting up
Each of the three stern lanterns was made from copper, brass, and glass.

Officers wore blue uniforms.

Poop deck

Quarter deck
The rear end of the quarter deck was where the captain had his rooms, directly above the admiral's.

Admiral's quarters
Admiral Nelson resided on the upper gun deck, his day cabin at the very rear.

1. Turning the steering wheel to the right winds up the rope on the left.

Tiller

2. This pulls the tiller to the left.

Rudder

3. This pulls the rudder to the right, which makes the ship turn right.

Copper-plated oak hull

Ship's store
The hold contained enough food and drink for a six-month voyage, plus tons of gravel for ballast to balance the ship.

Steering the ship
The rudder was connected to the ship's wheel by ropes and pulleys. If the wheel or ropes became damaged or broken, the rudder could be controlled manually by two heavy chains.

Nelson's bed
Admiral Nelson slept in a curtained hammock, or "cot," that hung from the ceiling.

Rudder

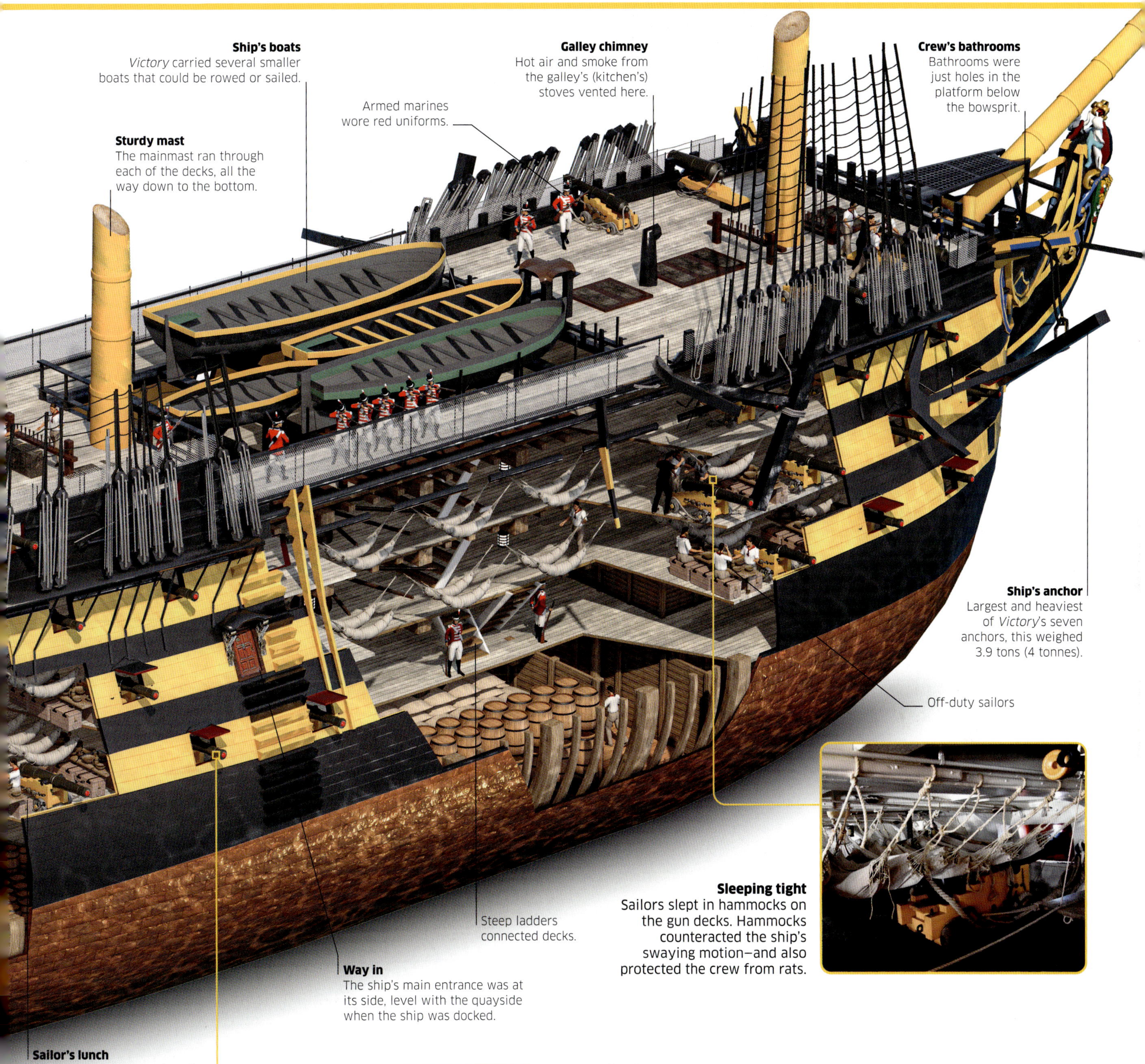

3,923 The number of **copper sheets** covering the part of *Victory*'s hull that sits **underwater**.

47 The **age of Admiral Horatio Nelson** when he **died on board** the *Victory* during the **Battle of Trafalgar**, October 21, 1805.

103

Ship's boats
Victory carried several smaller boats that could be rowed or sailed.

Sturdy mast
The mainmast ran through each of the decks, all the way down to the bottom.

Armed marines wore red uniforms.

Galley chimney
Hot air and smoke from the galley's (kitchen's) stoves vented here.

Crew's bathrooms
Bathrooms were just holes in the platform below the bowsprit.

Ship's anchor
Largest and heaviest of *Victory*'s seven anchors, this weighed 3.9 tons (4 tonnes).

Off-duty sailors

Sleeping tight
Sailors slept in hammocks on the gun decks. Hammocks counteracted the ship's swaying motion—and also protected the crew from rats.

Steep ladders connected decks.

Way in
The ship's main entrance was at its side, level with the quayside when the ship was docked.

Sailor's lunch
The crew ate lots of onions for vitamin C, and tough, chewy, salted pork for protein.

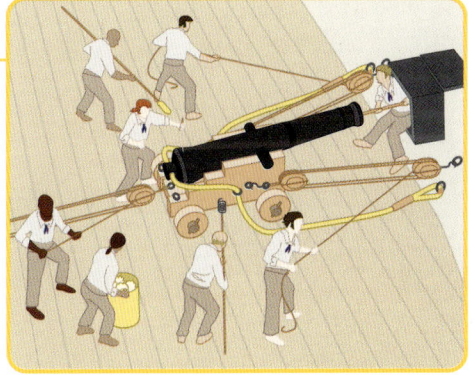

Gun duty
The ship's heavy guns were held in place by strong ropes. To fire them, crew had to quickly move each gun back using pulleys, cool it with a wet sponge on a stick, then load it with gunpowder and a cannon ball before heaving it into position and firing.

Life on board

The living and working conditions for HMS *Victory*'s crew were cramped, dirty, smelly, and often dangerous in daily life as well as in battle.

Handling the ship itself was hard work. The crew took turns (watches), with a few hours of rest in between. There were just six bathrooms, or "heads," for more than 800 crew, and the ship was full of rats. Disease was rife. Nevertheless, sailors considered it a great honor to serve on the British navy's famous flagship.

Full steam ahead

The invention of the steam engine not only made rapid rail transportation possible on land, it also led to a revolution in shipping.

The first steam-powered paddle wheels appeared in the late 18th century, fitted into converted wooden sailing ships. But from the mid-19th century, technological advances meant that ships no longer needed sails. The screw propeller and the steam turbine soon ushered in an age of great passenger ocean liners and powerful cargo vessels.

Trusty three-master
Each mast could carry a full set of sails to help the engine along.

PYROSCAPHE
Revolutionary experiment
Origin: France
Year: 1783

After years of attempts, French engineer Claude de Jouffroy d'Abbans launched the world's first steamboat. A coal-fired boiler turned the two paddle wheels of the *Pyroscaphe* ("heat boat").

Funnel for emitting steam

SAVANNAH
Hybrid steamer
Origin: US
Year: 1818

A converted sailing ship, *Savannah* had two paddle wheels and a coal-fired steam engine. But she didn't have space for all the coal needed for an Atlantic crossing, so she used her sails, too.

The hold carried cargo and overseas mail.

Paddle wheels were powered by the ship's steam engine.

GREAT BRITAIN
Propeller pioneer
Origin: UK
Date: 1843

Designed by Isambard Kingdom Brunel, *Great Britain* was the first great cruise liner and the first iron-hulled vessel powered by a screw propeller instead of a paddle wheel.

INEZ CLARKE
Mail riverboat
Origin: Colombia
Year: 1879

The steel-built *Inez Clarke* transported mail and passengers along Colombia's long Magdalena River. Her front-mounted boiler supplied steam to an engine driving a rear paddle wheel, giving a top speed of 15 mph (24 kph).

The first steamship
to cross the Atlantic Ocean was the *Savannah*, in 1819. It took 24 days.

Room with a view
The boxlike bridge and passenger cabins sat above open-air decks.

Boiler

TURBINIA
Turbine-powered novelty
Origin: UK
Year: 1894

Turbinia, the "Ocean Greyhound," was the world's fastest ship at 40 mph (64 kph). She was the first ship propelled by steam turbines, a breakthrough that would revolutionize maritime engineering.

Lightweight steel hull

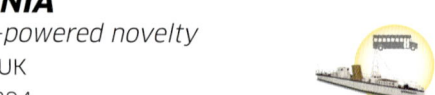

4 days, 10 hours, 51 minutes—*Mauretania*'s speed record for crossing the Atlantic in 1909.

692 ft (211 m)—the length of *Great Eastern* (1858), the world's largest steamship for 41 years.

105

MAURETANIA
Record-breaking cruise liner
Origin: UK
Year: 1906

In her two first years, *Mauretania* set new speed records for crossing the Atlantic—one of which stood until 1929. During World War I, the elegant passenger ship transported troops.

Packing a punch
Mauretania had four funnels, one for each powerful steam turbine.

Color change
The creamy white hull was originally black. During the war (1914–1918), it was given a temporary coat of dazzle paint.

CAP ROIG
Supply ship
Origin: Spain
Year: c.1910

Squat steam-powered vessels like this coal-carrier ferried supplies from dockside to larger ships, while some carried passengers and cargo along the coast.

EARNSLAW
Pleasure steamer
Origin: New Zealand/Aotearoa
Year: 1912

The coal-fired *Earnslaw* originally carried cargo and passengers around Lake Wakatipu. Today, most of her passengers are tourists enjoying the scenery and the nostalgic sound of her steam engines.

SAVARONA
Luxury yacht
Origin: Türkiye
Year: 1931

Built for an American heiress, *Savarona* was the most luxurious private yacht of her day, with a swimming pool and a gold-trimmed staircase. The Turkish government bought her in 1938.

JEREMIAH O'BRIEN
Liberty ship
Origin: US
Year: 1943

The US built 2,751 "Liberty Ships" during World War II to transport arms and supplies to the Allies in Europe. They were simple vessels, each one built from a premade "kit" in just 10 days.

NATCHEZ
Mississippi paddle steamer
Origin: US
Year: 1975

The first of the famous paddle steamers named *Natchez* was built in 1823. Since then, eight more have sailed the Mississippi River between the cities of New Orleans, Louisiana, and Natchez, Mississippi, including this one, launched in 1975.

Wheel of steel
Natchez's mighty oak and steel paddle wheel weighs 28.6 tons (26 tonnes) and is 25 ft (7.6 m) in diameter.

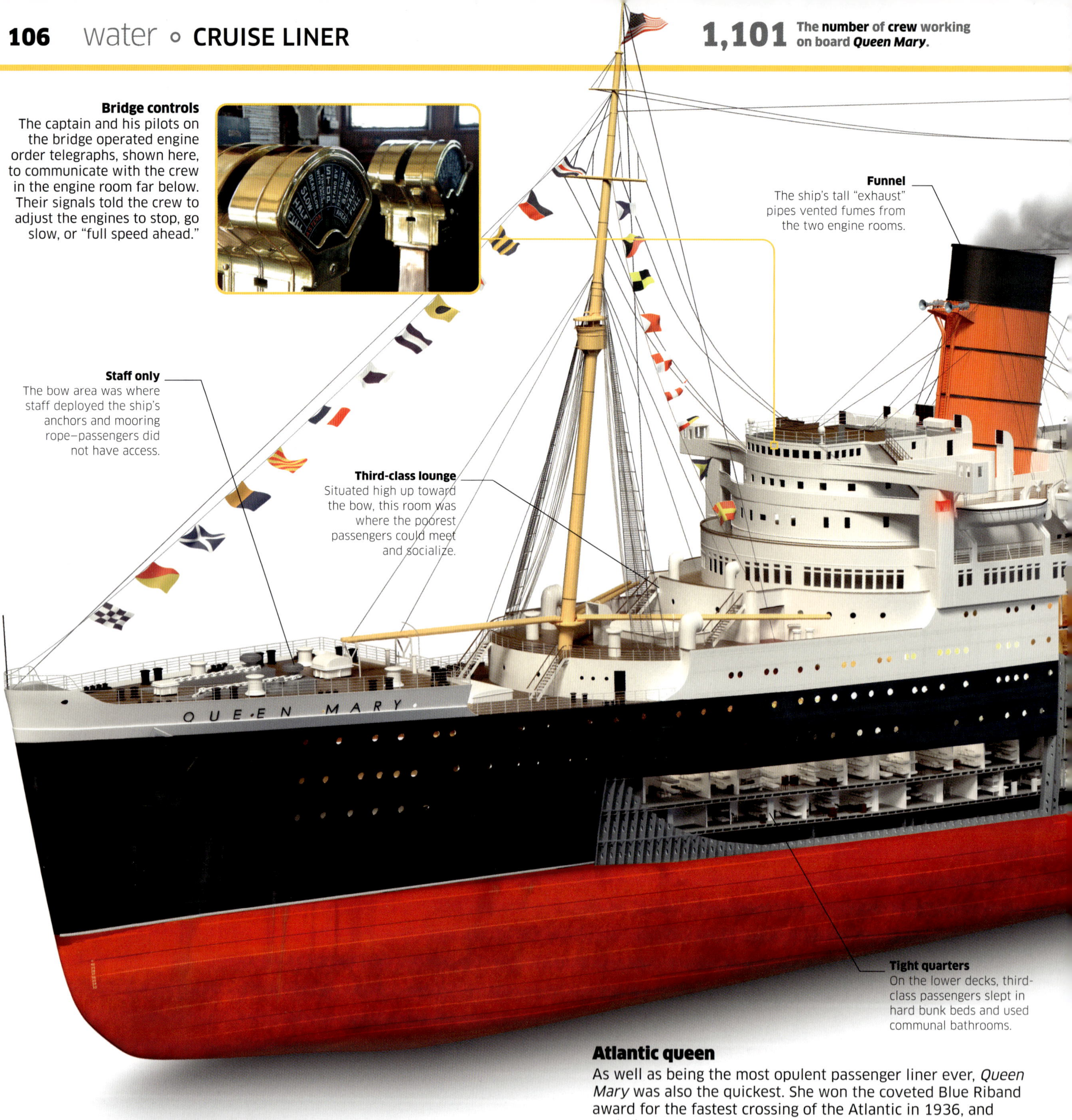

Bridge controls
The captain and his pilots on the bridge operated engine order telegraphs, shown here, to communicate with the crew in the engine room far below. Their signals told the crew to adjust the engines to stop, go slow, or "full speed ahead."

Funnel
The ship's tall "exhaust" pipes vented fumes from the two engine rooms.

Staff only
The bow area was where staff deployed the ship's anchors and mooring rope—passengers did not have access.

Third-class lounge
Situated high up toward the bow, this room was where the poorest passengers could meet and socialize.

QUEEN MARY

Tight quarters
On the lower decks, third-class passengers slept in hard bunk beds and used communal bathrooms.

Atlantic queen

As well as being the most opulent passenger liner ever, *Queen Mary* was also the quickest. She won the coveted Blue Riband award for the fastest crossing of the Atlantic in 1936, and again in 1938, when she completed the voyage in 3 days, 21 hours, and 38 minutes. In World War II, she was painted gray and used to transport up to 15,000 troops.

Cruise liner

During the "Golden Age" of the cruise liner, from 1900 to the 1950s, luxury ships transported travelers across the Atlantic in style and comfort.

One of most famous liners was the steamship *Queen Mary*, whose passengers included Hollywood celebrities, royalty, and the very wealthy. But this huge floating hotel had room on board for those on tighter budgets, too, whether vacationers or emigrants.

QUEEN MARY

Origin: UK

Year: 1934

Number of decks: 12

Passenger capacity: 2,140

1967
The year *Queen Mary* **was** **taken out of service.** She is now a **museum ship** in Long Beach, **California.**

107

Verandah Grill
A fancy restaurant by day, it became the exclusive Starlight Club, where rich and famous passengers mingled by night.

Sundeck
At night, the sun loungers were removed to make an open-air dance floor for the Starlight Club.

Lifeboats
Each of the 24 lifeboats held 145 people and was fitted with a diesel engine.

Windproof promenade
This outside walking and exercise area for first-class passengers was covered to protect people from the elements.

Time for tennis
Three tennis courts on the sports deck kept sporty first-class passengers busy.

Engine room
Four turbines, generating 40,000 hp each, drove the four propellers. Other turbines drove the ship's electricity generators.

Exclusive dining
The vast Grand Salon was a richly decorated, high-class dining room for 815 of the ship's richest passengers.

Swimming pool for first-class passengers

Comfortable cabins
The best first-class accommodation was the 12 private suites. Each had a bedroom, living room, bathroom, and quarters for accompanying servants.

First-class main lounge
Every evening, this spectacular lounge—lined with paintings, tapestries, and carved wooden panels—transformed into a ballroom, where 400 passengers in full evening dress could dance or enjoy concerts performed by a full orchestra.

Safety drill

When *Queen Mary* launched in 1934, the tragic sinking of the *Titanic* 22 years earlier was still fresh in many people's minds. Like on aircraft today before departure, all passengers were taken through a safety drill, where they were shown where to assemble for the lifeboats and how to wear their life jackets.

Separate lives

Passengers had separate—and segregated—eating, sleeping, and recreation areas. The first-class passengers rarely came into contact with second-class passengers and were probably hardly aware of the third-class travelers, whose spaces and facilities were far more basic and mostly out of view.

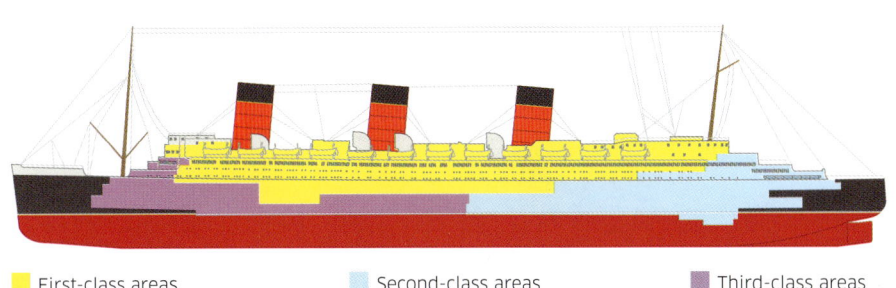

First-class areas   Second-class areas   Third-class areas

Fun on the water

Humans throughout history have taken to the seas for many reasons: to trade, wage war, explore the world, and—just as importantly—to have fun.

Sails and oars powered the boats of the earliest mariners and still propel many modern leisure vessels. But their shapes have evolved over the centuries, and the materials they are made from have changed, too. Today, there are also lots of pleasure crafts driven by pedal power, diesel engines, and gas engines, and there has never been so many ways to enjoy "messing around on the water."

JET SKI
Aquatic motorcycle
Origin: Japan
Year: From 1972

First invented by the Japanese brand Kawasaki, jet skis are now made worldwide. Their gas-powered pump engines create water jets that propel them to speeds of 70 mph (113 kph).

A 1990s
Yamaha model

BEACH CATAMARAN
Speedy double-hull sailer
Origin: US
Year: 1985

Catamarans have two hulls connected by a central platform. They are light, fast, and stable, and often used for racing or leisure sailing. This one is a Supercat 17, so named because it is 17 ft (5.18 m) long.

WINDSURFER
Wind-powered board
Origin: US
Year: From the 1960s

In the late 1960s, US engineer Jim Drake attached a dinghy sail to a surfboard and created a new leisure activity and sport. Windsurfing became an Olympic sport in 1984.

Keeping control
A windsurfer uses the handle, or boom, to steer into and out of the wind.

GIBBS AQUADA
Amphibious vehicle
Origin: New Zealand / Aotearoa
Year: 2003

When this automobile enters the water, its wheels retract and it becomes a boat. The 2.5L V6 engine gives a top speed of 100 mph (161 kph) on land and 35 mph (56 kph) on water.

Car hull
The vehicle has no doors, but a solid "hull" like a boat, so passengers have to climb in.

www.aquada.co.uk

Flexible sail
The sail attaches to a pivot, allowing it to move in any direction.

145 ft (44 m)—hull length of the *Hemisphere*, the world's largest sailing catamaran.

30 million The estimated number of recreational boats in the world today.

109

NARROWBOAT
Canal-based houseboat
Origin: UK
Year: From the 1960s (as pleasure boat)

Once industrial barges, many of these have been converted into diesel-engined boats for people to live and vacation on. They are only around 7 ft (2 m) wide, to fit through narrow canal locks.

Chimney vent

Protective coating
The steel hull is covered in black algae-preventing paint.

PEDALO
Water bicycle
Origin: China
Year: c.960 CE

Said to have first been invented in ancient China as surprise attack boats, pedalos today are plastic and used for more peaceful purposes on lakes and rivers and at the seaside.

Pedal power
Foot pedals operated by passengers drive the underwater paddles.

SPIRIT 46
Modern sailboat
Origin: UK
Year: 2003

The Spirit 46, named for its length of 46 ft (14 m), is just one of thousands of leisure yacht models. Unusually for a modern boat, its hull is made of wood instead of fiberglass.

Windcatcher
This colorful sail is called a spinnaker, used when the wind comes from behind.

Sails made of light but strong woven polyester

ROWBOAT
Oar-propelled craft
Origin: Europe
Year: From c.5800 BCE

The earliest known rowboat, dating from nearly 8,000 years ago, was found in Finland. Today, there are many models; this one is made from overlapping ("clinker-built") wooden boards.

Rowlocks hold the oars in place but allow them to move back and forth in the water.

RIVA AQUARAMA
Luxury motorboat
Origin: Italy
Year: 1960s

Riva's Aquarama motorboats were hand-built and among the most luxurious ever made. Their top speed of around 56 mph (90 kph) earned them the nickname "the Ferrari of the boat world."

Varnished mahogany hull

BANANA BOAT
Towed inflatable
Origin: US
Year: From 1971

Long, air-filled nylon or PVC tubes, Banana Boats have no engine. Instead, they are attached by rope to a motorboat and pulled at speed through the water.

Handles for those who don't want to fall off

Inflated tubes on each side, serving as foot rests

Tight race
Boats compete one to one to qualify as challenger in the final stage of America's Cup, in which the challenger races the winner of the previous cup. In 2021, *Te Rehutai* beat the Italian yacht *Luna Rossa Prada Pirelli*.

Racing yacht

The aim of any racing yacht is to ride the wind and master the water as efficiently as possible. New Zealand/Aotearoa's America's Cup-winner *Te Rehutai* is one of the best.

Yachts designed for the America's Cup are the fastest sailboats in the world. Technologically advanced, these vessels "fly" through the air as much as they slice through the seas at speeds up to 40 knots (46 mph, or 74 kph). Controlling a racing yacht's speed and direction is a specialized job for highly skilled, super-fit crews.

Massive mainsail
The large mainsail is operated by the mainsail trimmer, who uses a device that looks like a computer game controller to fine-tune, raise, and lower it.

Teamwork triumph
A team of 35 designers spent some 90,000 hours planning *Te Rehutai*. Then at least 50 skilled craftspeople, engineers, and boat builders collaborated to put the yacht together over 75,000 working hours. Finally, 11 crew members sailed her to victory.

Camera providing live onboard footage

GPS antenna for tracking the boat

Rudder

Hydraulic controls
The corner of the mainsail contains a hydraulic device that tightens the sail, or moves it along a rack in a controlled way when the boat tacks (changes direction).

Stabilizer
When in the water, the foil acts like a keel, keeping the boat stable.

Rudder foil
The upside-down T shape at the end of the rudder also acts like a foil.

Legendary competition

The first America's Cup race was run in 1851, off the Isle of Wight, UK. It now takes place every three to four years in different parts of the world. Rules regarding the design and measurements of the yachts have changed over the years; these are some of the most famous models.

First winner
The America's Cup was named after the first boat to win—the US schooner *America*.

J-class
In the 1930s, a new type of yacht, with two huge triangular sails, became the standard racer model.

Multihulls
From 1988, two-hulled catamarans and three-hulled trimarans raced alongside traditional monohull yachts.

87 ft (26.5 m)—the **height** of *Te Rehutai*'s **mast**.

111

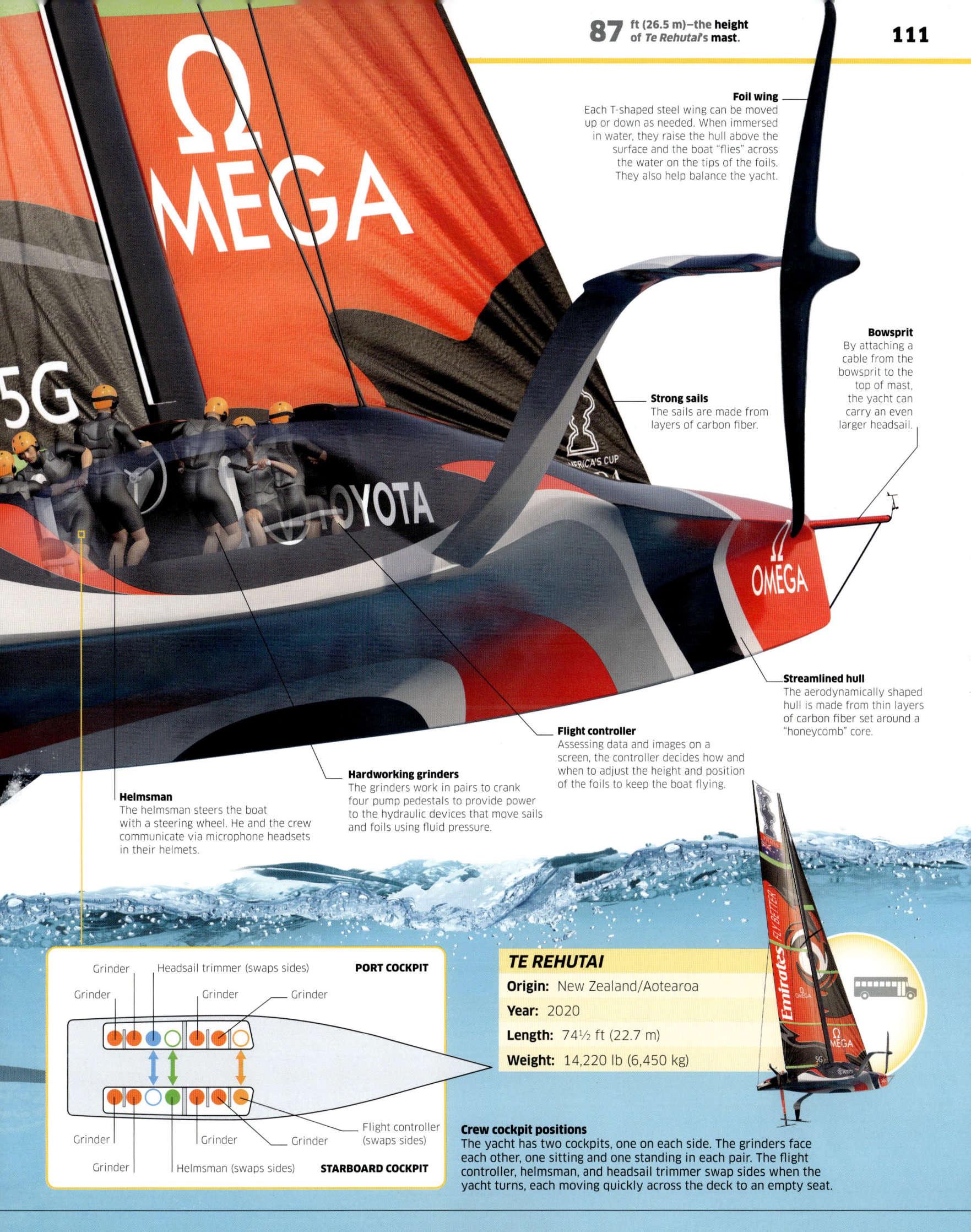

Foil wing
Each T-shaped steel wing can be moved up or down as needed. When immersed in water, they raise the hull above the surface and the boat "flies" across the water on the tips of the foils. They also help balance the yacht.

Bowsprit
By attaching a cable from the bowsprit to the top of mast, the yacht can carry an even larger headsail.

Strong sails
The sails are made from layers of carbon fiber.

Streamlined hull
The aerodynamically shaped hull is made from thin layers of carbon fiber set around a "honeycomb" core.

Flight controller
Assessing data and images on a screen, the controller decides how and when to adjust the height and position of the foils to keep the boat flying.

Helmsman
The helmsman steers the boat with a steering wheel. He and the crew communicate via microphone headsets in their helmets.

Hardworking grinders
The grinders work in pairs to crank four pump pedestals to provide power to the hydraulic devices that move sails and foils using fluid pressure.

TE REHUTAI

Origin: New Zealand/Aotearoa

Year: 2020

Length: 74½ ft (22.7 m)

Weight: 14,220 lb (6,450 kg)

Crew cockpit positions
The yacht has two cockpits, one on each side. The grinders face each other, one sitting and one standing in each pair. The flight controller, helmsman, and headsail trimmer swap sides when the yacht turns, each moving quickly across the deck to an empty seat.

Grinder
Headsail trimmer (swaps sides)
PORT COCKPIT
Grinder
Grinder
Grinder

Grinder
Grinder
Flight controller (swaps sides)
Grinder
Grinder
Grinder
Helmsman (swaps sides)
STARBOARD COCKPIT

112 water ∘ **PRECIOUS CARGO**

73 days–the **fastest trip** made by the speedy **three-master** *Cutty Sark* transporting **wool** from **Sydney** to **London, in 1889.**

FOUR-MASTED WINDJAMMER
Fast-sailing cargo ship
Flag: Germany (as cargo ship)
Year: 1921

Germany's *Magdalene Vinnen II* carried cargo from South America and Australia to Europe. She was given to the Soviet Union in 1945, renamed *Sedov,* and used as a navy training ship. She is still the world's largest sailing ship.

FRUIT JUICE TANKER
Giant juice transporter
Flag: Brazil
Year: 2003

Premium do Brasil carries 37,000 tons (36,000 tonnes) of orange juice from Brazil to the US. The juice is kept in huge refrigerated compartments at a temperature of 32°F (0°C).

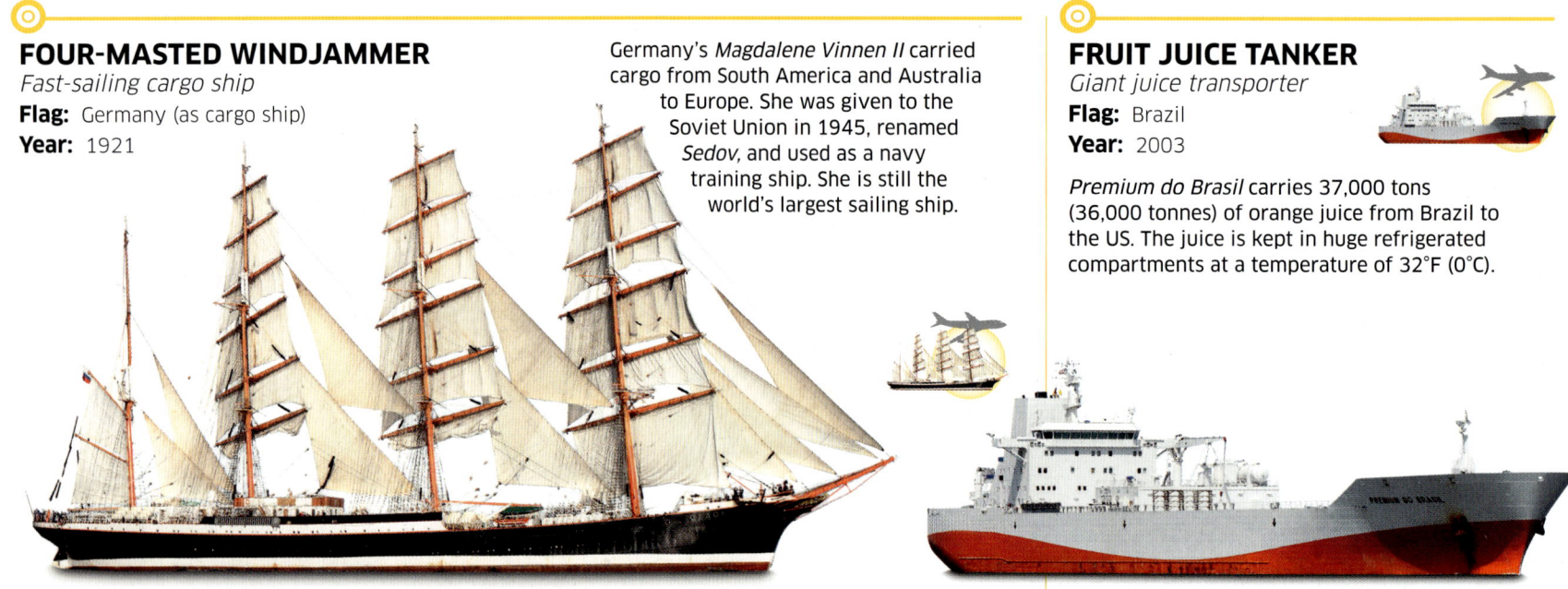

Precious cargo

In our globalized world, we buy, eat, drink, and make things that need to be transported by sea from the other side of the planet.

Merchant ships come in all shapes and sizes and move just about anything imaginable–including passengers on cruise ships. Merchant shipping has begun to come full circle: the earliest cargo vessels were wind-powered, and today a new generation of eco-friendly, partially wind-driven ships are sailing the seas.

CONTAINER SHIP
Super stacker
Flag: Japan
Year: 2019

Around 90 percent of packaged cargo goods are transported in container ships. Vessels such as the *One Grus* below can carry more than 14,000 containers, each one 20 ft (6 m) long.

Box loads
Containers are loaded and off-loaded at vast container ports (see pages 114–115).

OIL TANKER
Fossil fuel carrier
Flag: Liberia
Year: From 1970s

Designed to carry unrefined crude oil, the largest tankers afloat are up to 1,246 ft (380 m) long–the length of four soccer fields–and carry three million barrels of oil.

The largest ever oil spill from a ship occurred when two super tankers collided in the Caribbean in 1979.

Hose holder
A crane is used to lift the heavy hose that pumps oil into the vast hold.

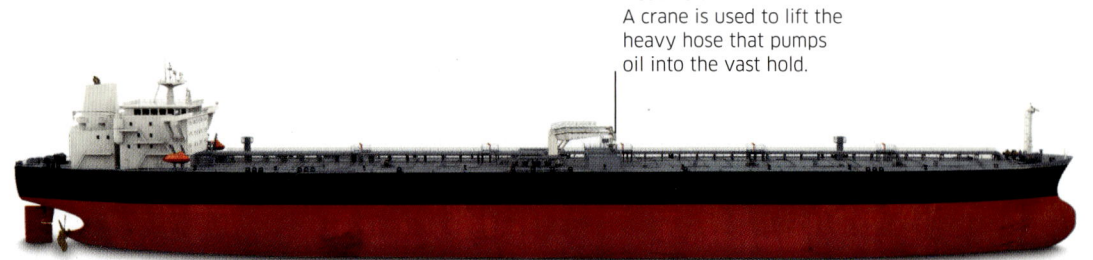

WHALEBACK SHIP
Steam-powered lake transport
Flag: US
Year: Late 19th century

Between 1887 and 1898, 44 whalebacks were built to transport cargo across the Great Lakes of North America. Their gray hulls sat low in the water and looked like surfacing whales.

Smoothly rounded hull

8,883 The **number of** oil tankers navigating the oceans **today**.

$14 trillion (£11.3 trillion)—the **value of all goods shipped on water in a single year**.

113

LIVESTOCK CARRIER
Floating cage

Flag: Panama
Year: 1984

Built as a car transporter, *Ghena* was converted into a livestock carrier in 2010. Pens were installed so that she could ship thousands of live sheep, cattle, and goats across the seas.

Crowded space
Ghena's closed pens can hold up to 85,000 sheep.

MEGA CRUISE SHIP
Luxury floating hotel

Flag: US
Year: 2023

Icon of the Seas is the world's largest cruise ship. Among the many facilities her 5,619 passengers can enjoy are 40 restaurants and bars, an ice rink, and the Thrill Island waterpark.

Protected panorama
The climate-controlled "aqua dome" is made of 673 glass panels.

Multistory living
With 20 decks, the ship is as tall as a tower block.

BULK CARRIER
Merchant fleet workhorse

Flag: Barbados
Year: 2004

Bulk carriers such as *Federal Mackinac* make up one fifth of the world's total merchant fleet. They transport loose, dry items such as coal, concrete, or grain stored in up to five separate holds.

Crane for loading and unloading

LIQUID HYDROGEN TANKER
New fuel carrier

Flag: Japan
Year: 2019

Suiso Frontier is the world's first transporter of liquid hydrogen, an important future energy source. To keep hydrogen liquid, it has to be stored in tanks cooled to -424.4°F (-253°C).

WIND-POWERED CARGO SHIP
Green goods transporter

Flag: Japan
Year: 2023

Pyxis Ocean's innovative steel and fiberglass WindWing sails allow her engines to be turned off in windy conditions. This reduces her fuel consumption by around 30 percent.

123-ft- (37.5-m-) tall sails

PUSHER BOAT
Compact powerhouse

Flag: US
Year: 1950s onward

Pushers are boats whose almost flat bottoms allow them to operate in shallow rivers and lakes. Small and powerful, they push—instead of drag or pull—barges or loaded platforms.

Mississippi barge
Barges loaded with logs, coal, and even vehicles are pushed along many US rivers.

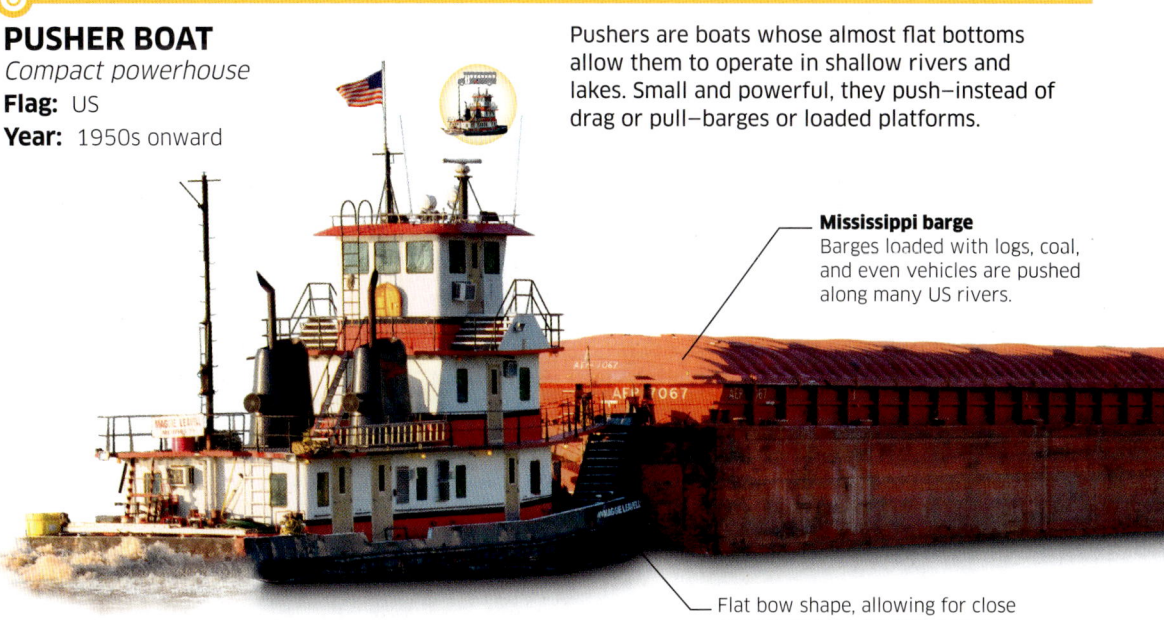

Flat bow shape, allowing for close contact and great pushing power

Giant port
The world's busiest port is Shanghai, China. It handles more than 47 million containers annually and has 191 ship berths, 43 container terminals, and 156 large cranes.

Crane operator's view
Up high, an operator maneuvers the crane. They move the crane's "spreader" along the crane arm until it is positioned over the ship. Then the spreader is lowered, grabs each end of a container, and lifts it off the vessel.

Loading and unloading
In ports around the world, ships arrive to offload containers and replace them with new cargo. Assisted by tugboats, it can still take several hours for a large container ship to dock safely.

Container port

Almost every item manufactured today—from clothing to TVs, toys, computer games, and anything else you can think of—has at some point passed through a container port.

The world's busiest cargo hubs are 24/7 operations, open all day, every day, processing an endless stream of containers. Once, these ports would have been hives of human activity, with dockers and sailors of all nations hard at work. Today, they are increasingly automated places where you are more likely to encounter a robot vehicle or an AI-enabled crane than an actual person.

Unloading cargo
A fully loaded ship has moored in its allocated berth (ship's parking space). The cranes get to work and unloading begins.

Containers
Made of strong steel, these secure boxes are all 8 ft (2.43 m) wide and 8.5 ft (2.59 m) high but come in two standard lengths—either 20 ft (6.06 m) or 40 ft (12.12 m).

Tough tugboat
Small but very powerful tugboats skillfully maneuver the heavy, unwieldy container ships into port, then push, pull, and nudge them into position along the quayside.

Container codes
All shipping containers are marked with the same internationally recognized codes. The ISO code shows the container's size and what it is used for. Other marks show the container's weight and the maximum volume of goods it can carry.

Logo of the owner
RB73521 — ISO code
67,196 LB 30,479 KG — Maximum weight
"Attention!" symbol for workers
8,554 LB 3,880 KG — Empty container weight
Door panel
Locking bar
Corner castings, for stacking

CONTAINER UNIT

Rolling crane
Rubber-tired gantry cranes (RTGCs) roam the dockyard, moving containers onto trucks and trains.

Container freight train
Typically, freight trains can transport at least 50 shipping containers.

Remote control center
This high-tech room is where operators oversee activity remotely. Screens show progress of loading and unloading and the movement of stackers and robo vehicles.

Warehousing
Containers holding high-value items are kept in secure storage facilities.

Container trucks
Containers are lowered onto the flat "skeletal" trailer on the rear of these vehicles.

Reach stacker
Their telescopic arms allow them to stack up to six containers.

Container handlers
These super-sized forklift trucks can pile five or six containers on top of each other.

Gantry cranes
These lift and lower containers from ships directly onto trucks and trains.

Robo vehicle
Driverless transporters use GPS, HD cameras, and infrared sensors to navigate safely around the port.

Quayside moorings
Metal bollards—used to secure the ships' mooring ropes, or hawsers—are spaced along the quay.

Ostfriesland's hull is strong enough to break up the ice that often forms in **winter** along **Germany's North Sea shores.**

Car and passenger ferry

Ferries transport people, cars, trucks, and sometimes even trains across water. Many car ferries are "ro-ro" ferries–cars can roll on at one end and roll off at the other.

Car ferries are designed differently, depending on where they operate and on how many cars and passengers they need to carry. Some larger, seagoing ones are fully enclosed, while others, such as the *Ostfriesland*, have partly open car decks. Many older ferries are now refitted and converted to run on green fuel.

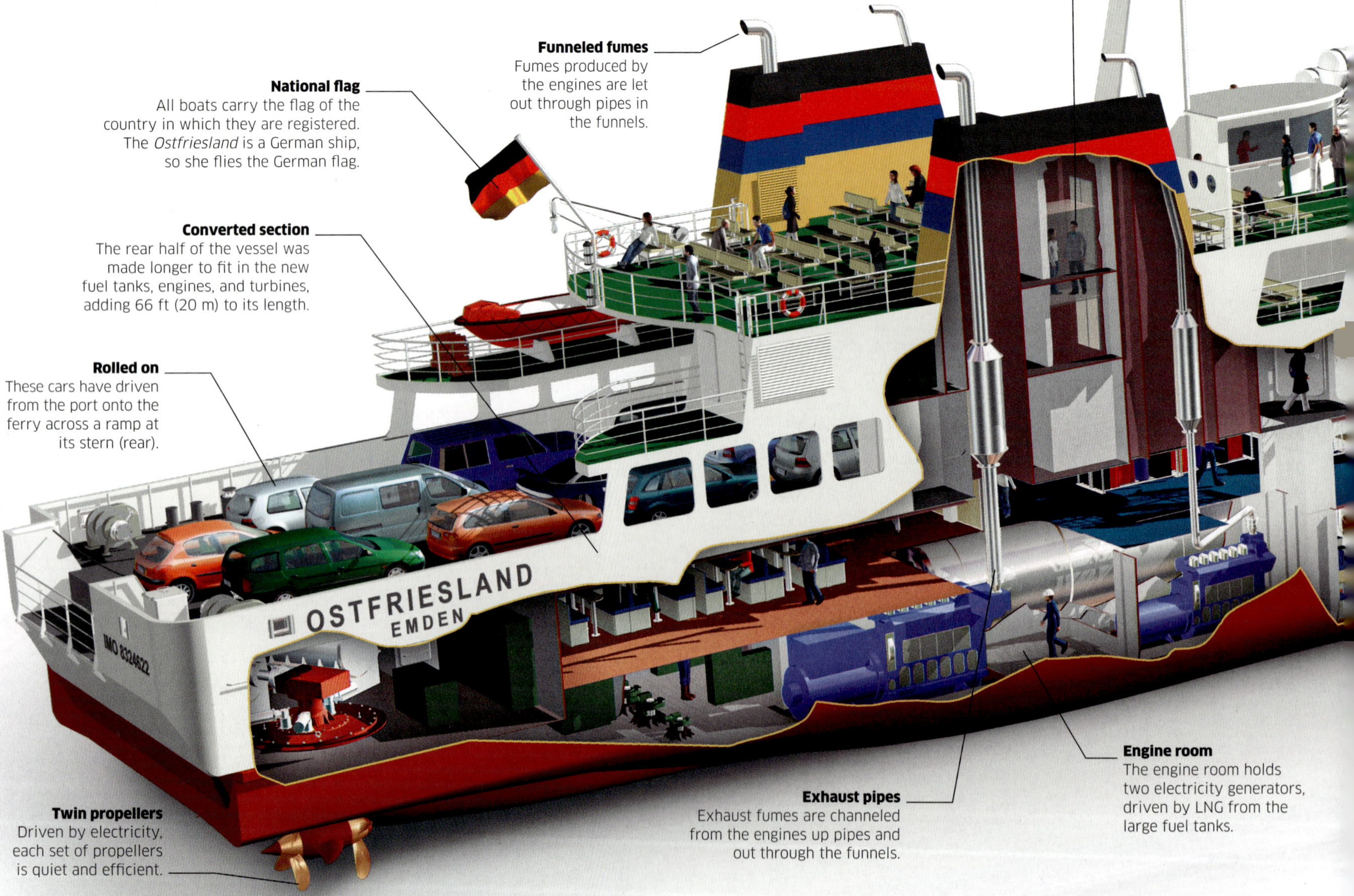

Funnel elevator
Each funnel has an elevator to take passengers between decks.

Funneled fumes
Fumes produced by the engines are let out through pipes in the funnels.

National flag
All boats carry the flag of the country in which they are registered. The *Ostfriesland* is a German ship, so she flies the German flag.

Converted section
The rear half of the vessel was made longer to fit in the new fuel tanks, engines, and turbines, adding 66 ft (20 m) to its length.

Rolled on
These cars have driven from the port onto the ferry across a ramp at its stern (rear).

OSTFRIESLAND
EMDEN

Twin propellers
Driven by electricity, each set of propellers is quiet and efficient.

Exhaust pipes
Exhaust fumes are channeled from the engines up pipes and out through the funnels.

Engine room
The engine room holds two electricity generators, driven by LNG from the large fuel tanks.

Crossing water

People have always needed to travel across water to get to the other side–be it rivers, lakes, or the open sea. Over time, many means of crossing have developed, depending on what these vessels have to carry, ranging from people and animals to cars, trucks, and trains. These are just a few different types of ferries in use today.

River raft ferry
Crossing Mongolia's Tsagaan Nuur river, this man operates a raft by pulling a rope fastened at each bank.

Small island car ferry
Simple roll-on-roll-off ferries ship people, cars, and supplies between islands near Stockholm, Sweden.

Train ferry
Where there is no railway bridge, trains travel by ferry, such as this one between Sicily and mainland Italy.

1,200 The **maximum number of passengers** that can **board** the **ferry** on one **crossing**.

75 The number of **cars** that can **fit** on the **car deck**.

16 knots (18 mph or 30 kph)—the **top speed** of the *Ostfriesland* when **fully loaded**.

117

Night lights
This row of little lanterns lights up when the sky is dark during early-morning and late-night crossings.

Bridge
The captain and other crew work here, navigating and steering the ship through busy ports and waterways.

Fresh-air views
Passengers do not stay in their vehicles, but can take in the views of the Wadden Sea from the upper deck.

Self-Inflating raft
In case of an emergency, 10 inflatable lifeboats automatically expand to take passengers to safety.

Bow ramp
At the front, a ramp is lowered in port to let cars drive on or off.

Car deck
Cars park in lines and stand with engines turned off until it's time to drive off again.

Passenger bathrooms
The crossing takes 50 minutes, so bathroom breaks might be needed.

Environmentally friendly ferry

The *Ostfriesland* is a typical roll-on-roll-off ferry. It carries cars and passengers from Emden on Germany's North Sea coast to the island of Borkum through a protected area known as the Wadden Sea. In 2015, it was rebuilt and converted to run on green fuel to cause as little damage as possible to wildlife and nature.

MS *OSTFRIESLAND*

Country: Germany

Year: 1985, converted 2015

Length: 308 ft (94 m)

Weight: 2,862 tons (2,596 tonnes)

Here I come
Each robot has 3D sensors that help it detect and avoid objects up to 33 ft (10 m) away.

Green fuel

Ferries used to run on diesel, which pollutes the environment. Some modern ships now use low-emission LNG (liquefied natural gas). This is methane gas that has been cooled to -260°F (-162°C) so that it turns into a liquid. Liquefied gas helps protect the environment because emissions such as sulfur dioxide and nitrogen oxides are reduced by up to 80 percent and carbon dioxide by 20 percent. This type of fuel also removes fine dust particles from the exhaust fumes.

Liquid gas tanker
Special tankers are used to transport the liquefied gas, stored in the sphere-shaped containers, between ports.

Restaurant robot
Passengers can get something to eat in the restaurant, where they are served by Bella Bots. These are fully automated battery-operated electric robots that take orders and deliver food and drinks. They are voice activated, programmable, and have heated trays to keep food warm.

TRAWLER
Shrimp-fishing TX 24

Flag: The Netherlands
Year: Early 2000s

Trawlers land a quarter of all fish caught—19 million tons (19.3 million tonnes) per year—and often have onboard processing factories. Smaller ones like *TX 24* fish for shrimp near the coast.

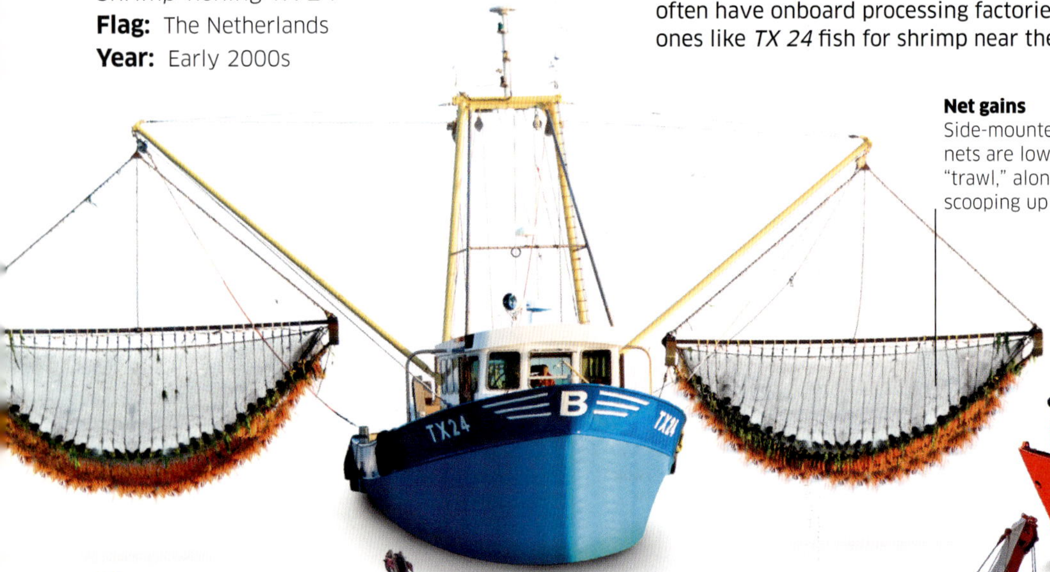

Net gains
Side-mounted outrigger nets are lowered to drag, or "trawl," along the seabed, scooping up shrimp.

LIGHTSHIP
Safety beacon

Flag: UK
Year: 1989

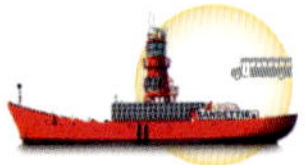

Sandettie's light is visible 17 miles (28 km) away, warning ships of dangerous sandbanks in the seas between France and the UK. Like many lightships, *Sandettie* is also a weather station.

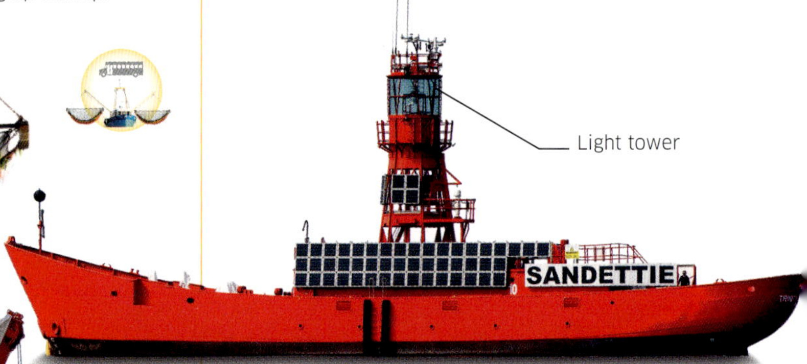

Light tower

DREDGER
Channel-clearing machine

Flag: US
Year: 1980s

Most dredgers use large tubes to suck up sand and silt from clogged-up channels, riverbeds, and estuaries, ensuring they stay navigable. For more precise excavations, clamshell dredgers like this one grab small amounts of material in "claws."

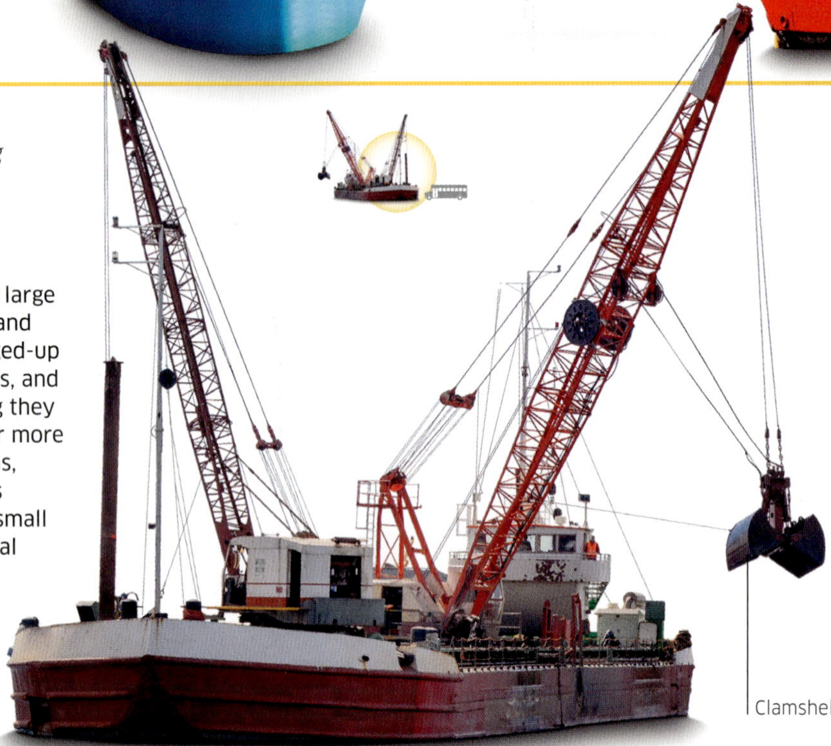

Clamshell claws

Working boats

From catching fish to laying pipes and clearing routes for other ships, there are a huge variety of jobs that require specialized watercraft.

Working boats are designed to perform well on difficult and sometimes dangerous oceans and waterways. These are vessels whose function is much more important than their form. They are always powerful, usually rugged, and they get the job done! Many are put to work across the world, carrying the flag of the country in which they are registered.

SEMI-SUBMERSIBLE CRANE VESSEL (SSCV)
Sea-based multitasker

Flag: Panama
Year: 2019

SSCVs can lay pipes, lift sunken ships, and build underwater foundations for bridges and oil and gas rigs. The gigantic SSCV *Sleipnir* is powered by energy-efficient LNG (liquefied natural gas).

Heavy duty
Cranes can lift weights up to 9,842 tons (10,000 tonnes).

Stable platform
Eight floating, semisubmerged legs, known as pontoons, hold the platform high above water.

1732 The year the **world's first** permanently anchored **lightship** began operating at the Nore Sand Bank in the **River Thames, UK.**

Some **ports**, such as **Hamburg** in Germany and **Sydney** in Australia, put on **shows** where tugboats show off their skills in a **"tug ballet."**

119

TUGBOAT
Hard-working ship mover

Flag: The Netherlands

Year: 2010

Small, slow, and strong, tugboats like Amsterdam's *Athena* are designed to pull larger vessels in and out of port, especially those that are too big to maneuver themselves in small spaces.

PIPE LAYER
Pipeline installer

Flag: Panama

Year: 2015

Sapura Rubi carries, lays, and welds oil and gas pipelines along the ocean floor at depths of 9,843 ft (3,000 m), covering around 5.6 miles (9 km) a day.

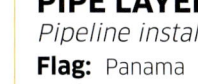

Towering tool
The 168-ft- (51.26-m-) high tower feeds pipe into the laying mechanism via a tilting ramp.

Down the line
Aligner wheels ensure the pipe remains vertical as it is fed through the moon pool (see page 121) into the sea.

ICEBREAKER
Polar pathfinder

Flag: Finland

Year: 1987

Vessels such as *Kontio* keep waterways in cold regions open when they freeze. Their powerful diesel-electric or nuclear engines provide the thrust to break ice up to 16.4 ft (5 m) thick.

The first icebreaker,
known as a *koch*, was invented by Indigenous Arctic peoples as early as the 11th century.

Big bear
The symbol illustrates the ship's name ("bear" in Finnish).

Cozy quarters
Interior crew areas are comfortable and warm, in contrast to the icy outside.

Slide and crush
The bow is shaped so it can glide up on sea ice and crush it by its weight.

WIND FARM SUPPORT VESSEL
Crucial supply craft

Flag: UK

Year: 2014

With a top speed of 30 mph (48 kph), the catamaranlike *Njord Alpha* transports workers, equipment, and spare parts to offshore wind farm turbines that need to be built, repaired, or maintained.

SEMI-SUBMERSIBLE HEAVY-LIFT VESSEL (HLV)
Heavy goods transporter

Flag: Malta

Year: 1999

HLVs such as *Black Marlin* have ballast tanks that allow them to partially sink to take on heavy loads by "floating" them onto their deck. Then they empty the ballast tanks, rise, and set off.

Nine super-heavy barges loaded for transport

36,090 ft (11,000 m)—the **depth** that the ***Sir David Attenborough***'s echo sounders can reach when **mapping the ocean floor.**

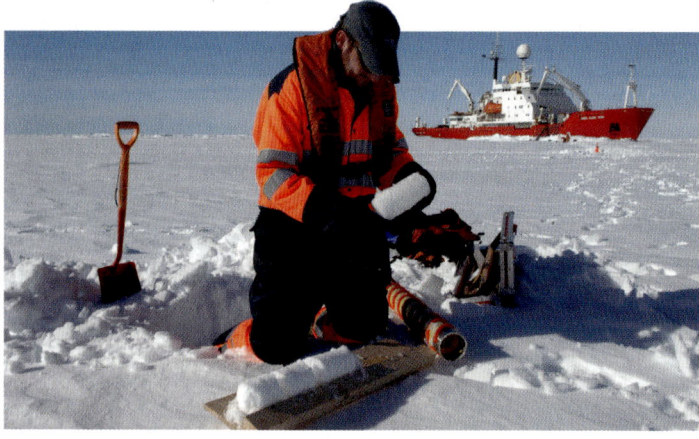

Ice science
By carving out "cores" from the thick sea ice, experts can calculate the level of greenhouse gases in the atmosphere hundreds of years ago.

Research vessel

Polar Research Vessels (PRVs) carry out scientific investigations to monitor our planet's ever-changing environment. One of the world's most advanced PRVs is the *Sir David Attenborough*.

Polar Research Vessels are designed to function in the harsh climate of the Arctic and Antarctic regions. They need to be able to travel safely through ice, waves, and storms, and provide comfortable sleeping and leisure areas for the scientists on board. Scientific instruments, the equipment for handling these, and laboratories for analyzing samples together make up the main part of a PRV.

Polar explorer
The *Sir David Attenborough* conducts research in the coldest, most hostile environments. Its scientists monitor sea ice, sea water, the sea floor, the atmosphere above water, and local ocean biodiversity.

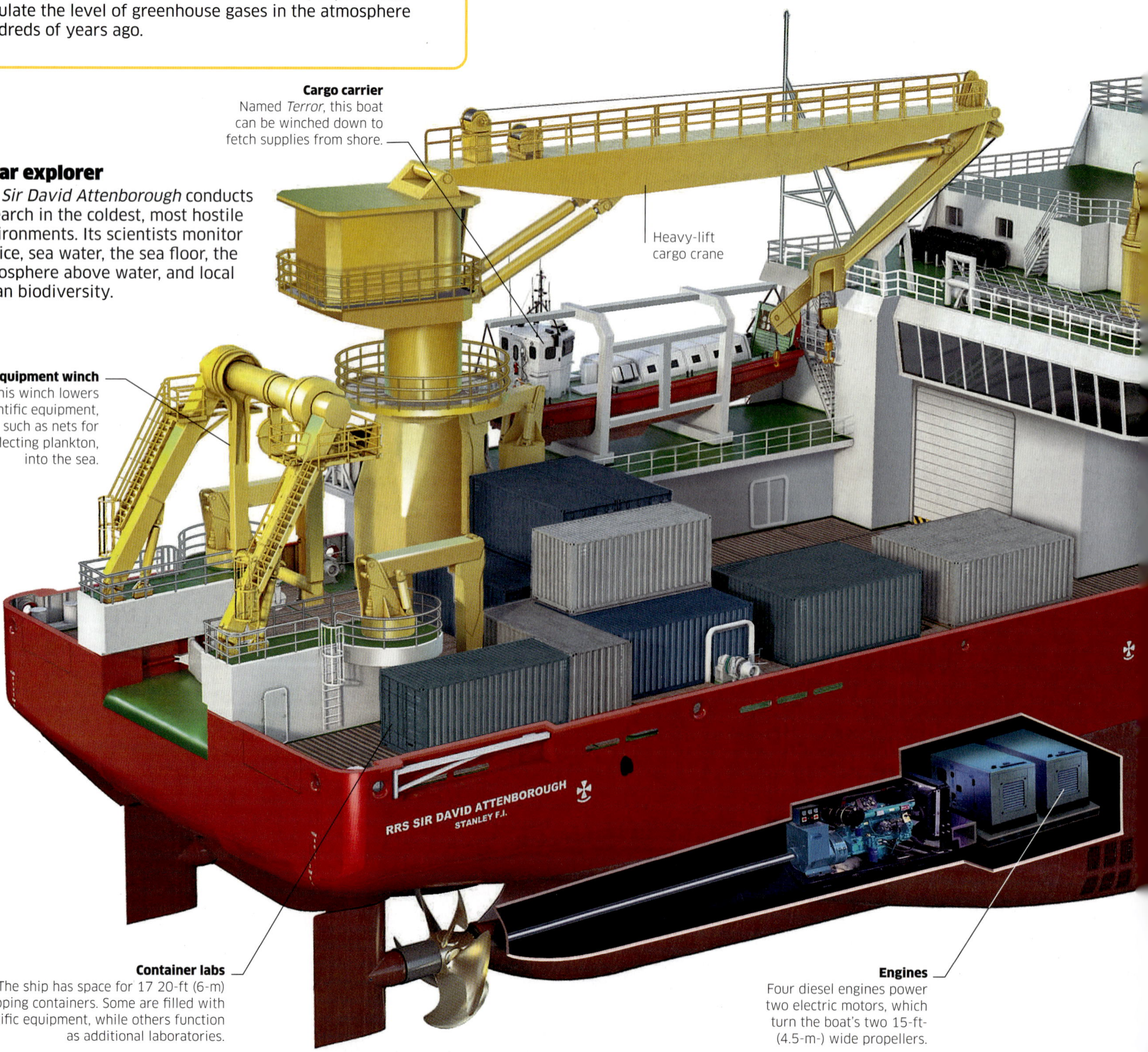
RRS SIR DAVID ATTENBOROUGH
STANLEY F.I.
Heavy-lift cargo crane

Cargo carrier
Named *Terror*, this boat can be winched down to fetch supplies from shore.

Equipment winch
This winch lowers scientific equipment, such as nets for collecting plankton, into the sea.

Container labs
The ship has space for 17 20-ft (6-m) shipping containers. Some are filled with scientific equipment, while others function as additional laboratories.

Engines
Four diesel engines power two electric motors, which turn the boat's two 15-ft- (4.5-m-) wide propellers.

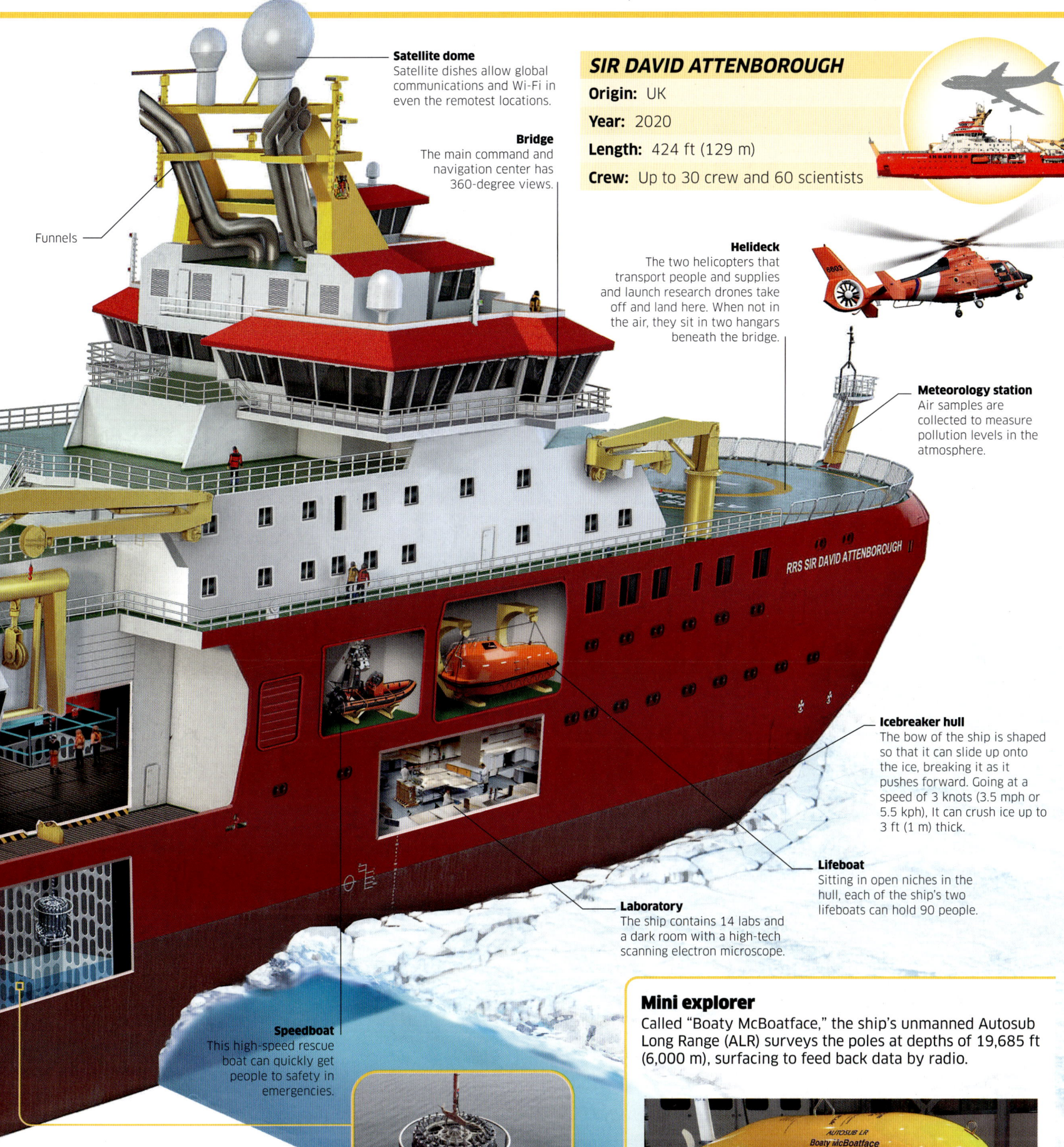

Satellite dome
Satellite dishes allow global communications and Wi-Fi in even the remotest locations.

Funnels

Bridge
The main command and navigation center has 360-degree views.

SIR DAVID ATTENBOROUGH

Origin: UK

Year: 2020

Length: 424 ft (129 m)

Crew: Up to 30 crew and 60 scientists

Helideck
The two helicopters that transport people and supplies and launch research drones take off and land here. When not in the air, they sit in two hangars beneath the bridge.

Meteorology station
Air samples are collected to measure pollution levels in the atmosphere.

RRS SIR DAVID ATTENBOROUGH

Icebreaker hull
The bow of the ship is shaped so that it can slide up onto the ice, breaking it as it pushes forward. Going at a speed of 3 knots (3.5 mph or 5.5 kph), It can crush ice up to 3 ft (1 m) thick.

Lifeboat
Sitting in open niches in the hull, each of the ship's two lifeboats can hold 90 people.

Laboratory
The ship contains 14 labs and a dark room with a high-tech scanning electron microscope.

Speedboat
This high-speed rescue boat can quickly get people to safety in emergencies.

Mini explorer
Called "Boaty McBoatface," the ship's unmanned Autosub Long Range (ALR) surveys the poles at depths of 19,685 ft (6,000 m), surfacing to feed back data by radio.

Moon pool
A vertical hole in the middle of the ship, the moon pool opens directly into the sea. It allows delicate equipment, such as this salt level sampler, to be lowered directly into the ocean depths, avoiding waves or ice.

Giant propeller

Modern ships need huge propellers to power them through the water. The blades have to be kept clean of debris to keep them running smoothly and efficiently.

This is one of three propellers mounted underneath a large modern cruise ship. Each propeller is attached to a pod containing the electric engine that makes it spin to move the ship. The pod and propeller can rotate 360 degrees to allow the ship to maneuver and travel forward, backward, or even sideways.

777 **The number of ships in the modern Chinese navy, the largest in the world.**

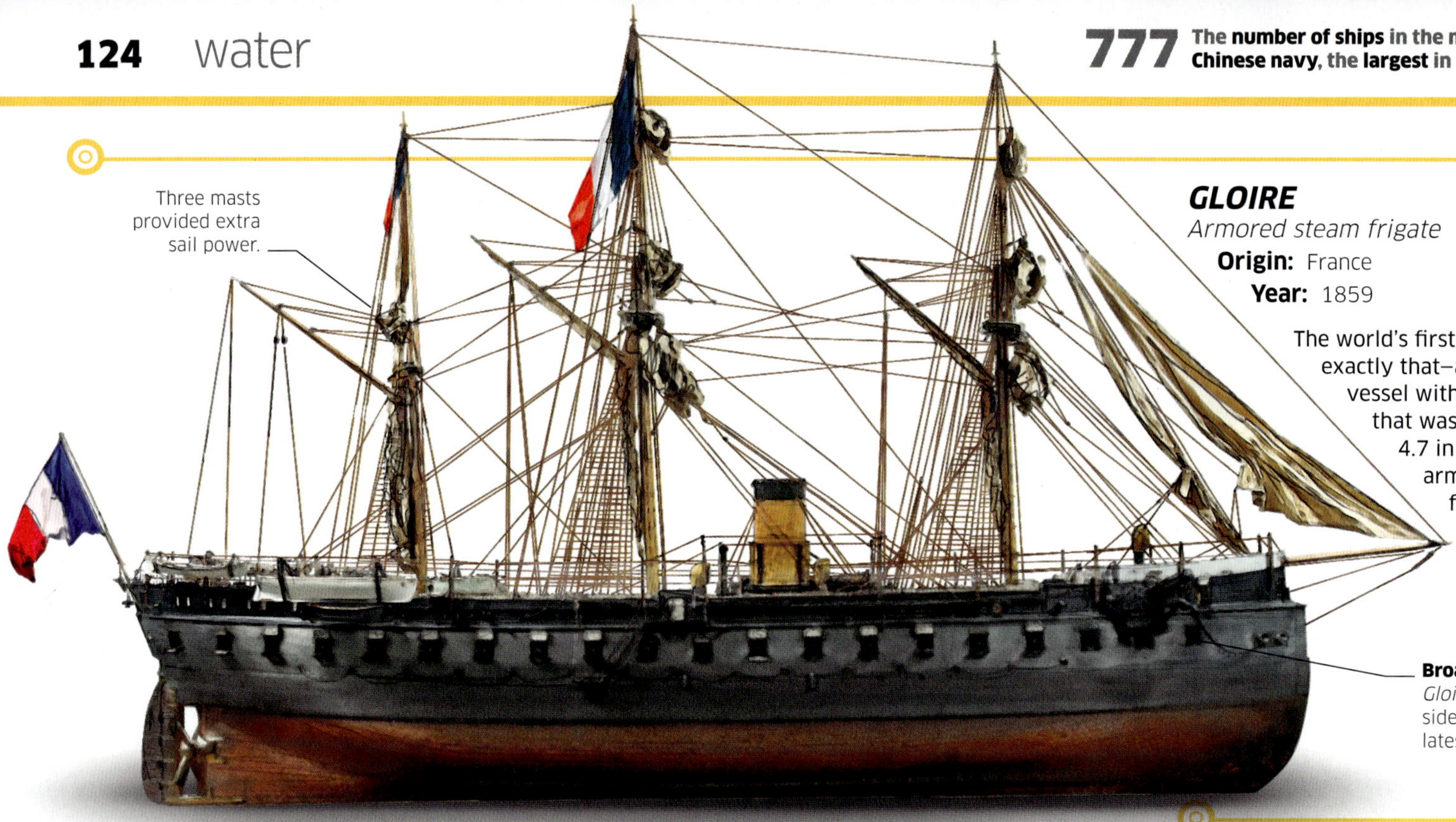

Three masts provided extra sail power.

GLOIRE

Armored steam frigate

Origin: France

Year: 1859

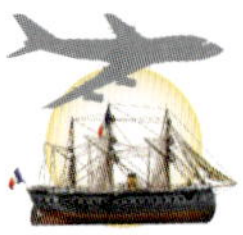

The world's first "ironclad" ship was exactly that—a steam-powered vessel with a hull made of oak that was clad in a layer of iron 4.7 in (12 cm) thick. Her armor could withstand fire from the most powerful guns of the day.

Broadside
Gloire had 18 guns on each side, all 66-pounders of the latest model.

Naval ships

From the middle of the 19th century, naval vessels from across the world began to be constructed from metal—chiefly iron—instead of wood. Different types of craft were developed for different tasks.

Over time, naval ships became increasingly sophisticated. Better engines allowed for higher speeds and thicker armor made them harder to destroy. Their weapons improved, too. Today's ships have sophisticated attack and defense capabilities, backed up by technological developments that make them "invisible" to their enemies.

CSS *VIRGINIA*

US Civil War ironclad

Origin: US

Year: 1862

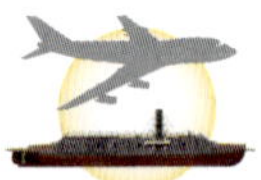

In the war between the southern Confederate states and the northern Union states, the CSS *Virginia* clashed with the USS *Monitor*. This was the first time two ironclad vessels engaged in conflict.

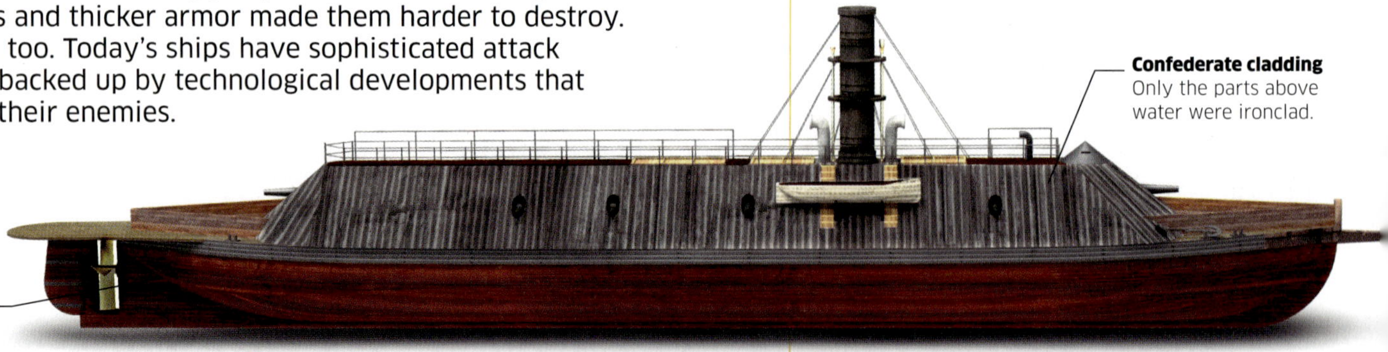

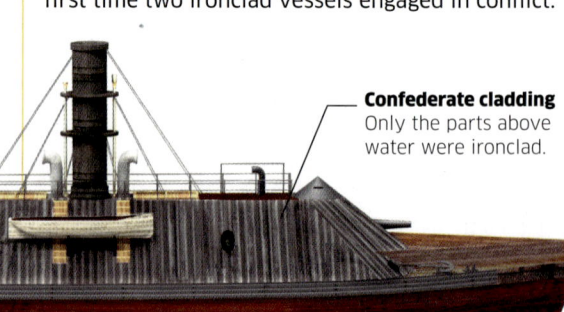

Confederate cladding
Only the parts above water were ironclad.

Recycled hull
The wooden hull was taken from a sailing ship no longer in use.

HMS *DREADNOUGHT*

Spearheading battleship

Origin: UK

Year: 1906

With thick armor, 10 huge 10-in (25.4-cm) guns in five turrets, five machine guns, and 23 quick-firing guns, HMS *Dreadnought* set the new standard for fighting ships until the outbreak of World War II.

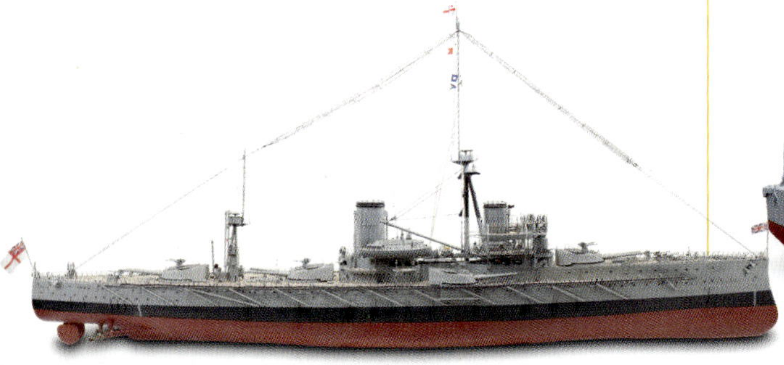

LCM (3)

World War II landing craft

Origin: UK

Year: 1938

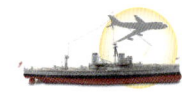

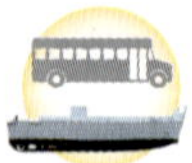

Landing Craft Mechanized 3 (LCM 3) transporters ferried troops, tanks, and other vehicles from ship to shore in World War II, most famously on D-Day. They were 50 ft (15.24 m) long.

Onto the beaches
Once the craft was in shallow water, the ramp lowered to let soldiers off.

MINESWEEPER

Hardworking protector

Origin: Sweden

Year: 1940

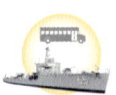

The job of minesweepers like *Bremön* was to detect and disarm mines at sea and to deploy bombs. They also acted as escorts, protecting commercial shipping against enemy attacks in World War II.

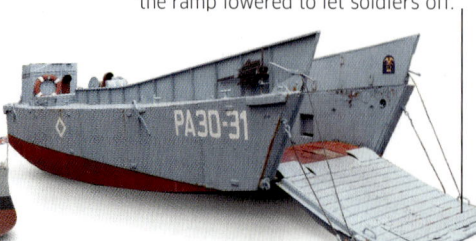

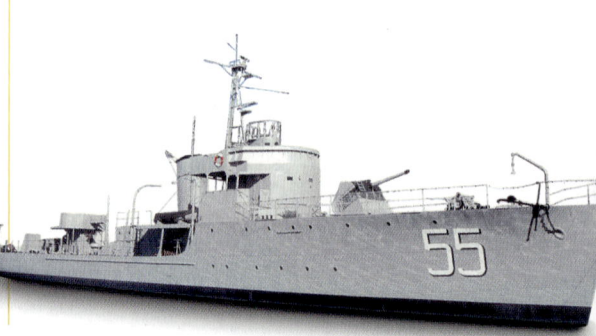

6,939 The number of Allied vessels involved in the 1944 D-Day landings in Normandy, France, during World War II.

1960 The year the first ever nuclear-powered aircraft carrier, the USS Enterprise, was launched.

125

SWIFT BOAT
Naval patrol vessel
Origin: US
Year: 1965

Patrol Craft Fast (PCF), or "Swift Boats," served in the Vietnam War, patrolling the Vietnamese waterways. They were 50 ft (15.24 m) long and armed with two 0.50-caliber machine guns.

HAMINA CLASS MISSILE BOAT
Fast-attack craft
Origin: Finland
Year: 1999

These ships carry missiles that can hit targets 250 miles (400 km) away. They have stealth capabilities, including ducts that vent their infrared-detectable exhaust fumes into the sea.

MISTRAL CLASS BOAT
Amphibious assault ship
Origin: France
Year: 2005

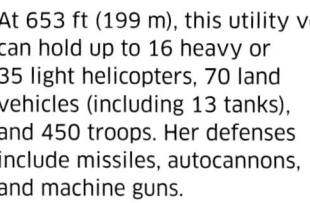

At 653 ft (199 m), this utility vessel can hold up to 16 heavy or 35 light helicopters, 70 land vehicles (including 13 tanks), and 450 troops. Her defenses include missiles, autocannons, and machine guns.

TYPE 45 DARING CLASS
Guided missile-armed destroyer
Origin: UK
Year: 2003

The tall, pyramid-shaped tower contains high-tech air defense systems to track and target enemy missiles and aircraft, which can be shot down with "smart" Sea Viper missiles.

Rear fans, propelling the craft forward

Air-cushion
A rubber "skirt" traps air, pushed underneath the LCAC by a central fan.

LANDING CRAFT AIR-CUSHION
Assault hovercraft
Origin: US
Year: 1986

Hovering above the water, LCACs can "fly" over sea and wetlands at speeds up to 46 mph (74 kph). They carry people, weapons, and heavy vehicles across the sea and straight onto shore.

Air-cushion empties when craft stops.

Ramp lowered onto shore

ZUMWALT-CLASS DESTROYER
Stealth ship
Origin: US
Year: 2013

The low, smooth profile of the Zumwalt class of destroyers is designed so that they can avoid radar detection. Stealth ships like this are the future of naval warfare.

Secret weapons
Guns are hidden inside the turrets.

Command center
The central tower contains sophisticated defense and attack radar systems.

One of *Bismarck*'s most **unusual defenses** was a **smoke generator** that produced **white fog** to **hide** the vessel from sight.

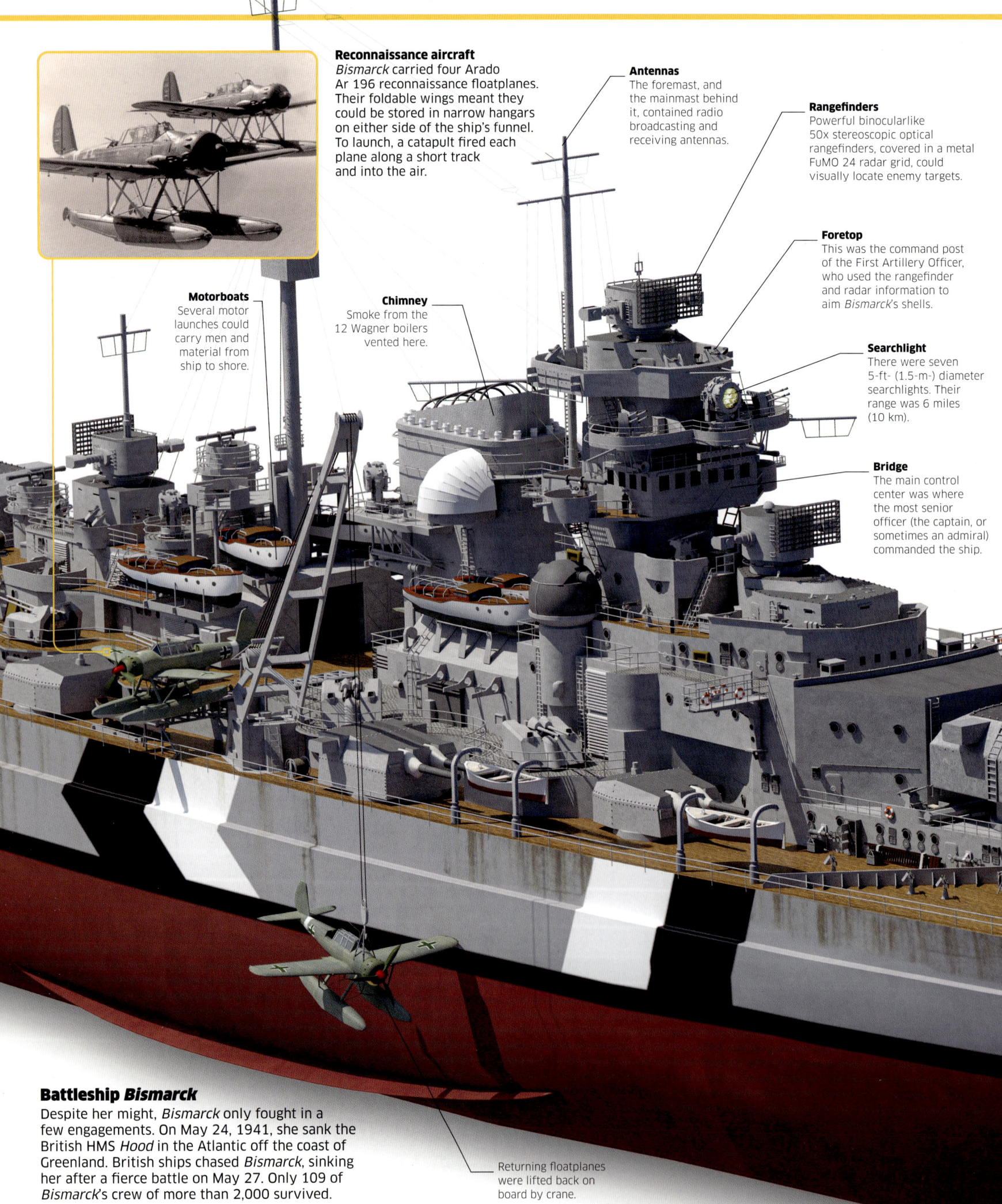

Reconnaissance aircraft
Bismarck carried four Arado Ar 196 reconnaissance floatplanes. Their foldable wings meant they could be stored in narrow hangars on either side of the ship's funnel. To launch, a catapult fired each plane along a short track and into the air.

Antennas
The foremast, and the mainmast behind it, contained radio broadcasting and receiving antennas.

Rangefinders
Powerful binocularlike 50x stereoscopic optical rangefinders, covered in a metal FuMO 24 radar grid, could visually locate enemy targets.

Foretop
This was the command post of the First Artillery Officer, who used the rangefinder and radar information to aim *Bismarck*'s shells.

Motorboats
Several motor launches could carry men and material from ship to shore.

Chimney
Smoke from the 12 Wagner boilers vented here.

Searchlight
There were seven 5-ft- (1.5-m-) diameter searchlights. Their range was 6 miles (10 km).

Bridge
The main control center was where the most senior officer (the captain, or sometimes an admiral) commanded the ship.

Battleship *Bismarck*
Despite her might, *Bismarck* only fought in a few engagements. On May 24, 1941, she sank the British HMS *Hood* in the Atlantic off the coast of Greenland. British ships chased *Bismarck*, sinking her after a fierce battle on May 27. Only 109 of *Bismarck*'s crew of more than 2,000 survived.

Returning floatplanes were lifted back on board by crane.

960 The **number** of shells *Bismarck* stored in her large **magazine** below deck.

12½ in (320 mm)—the **greatest thickness** of *Bismarck*'s hull **armor**.

127

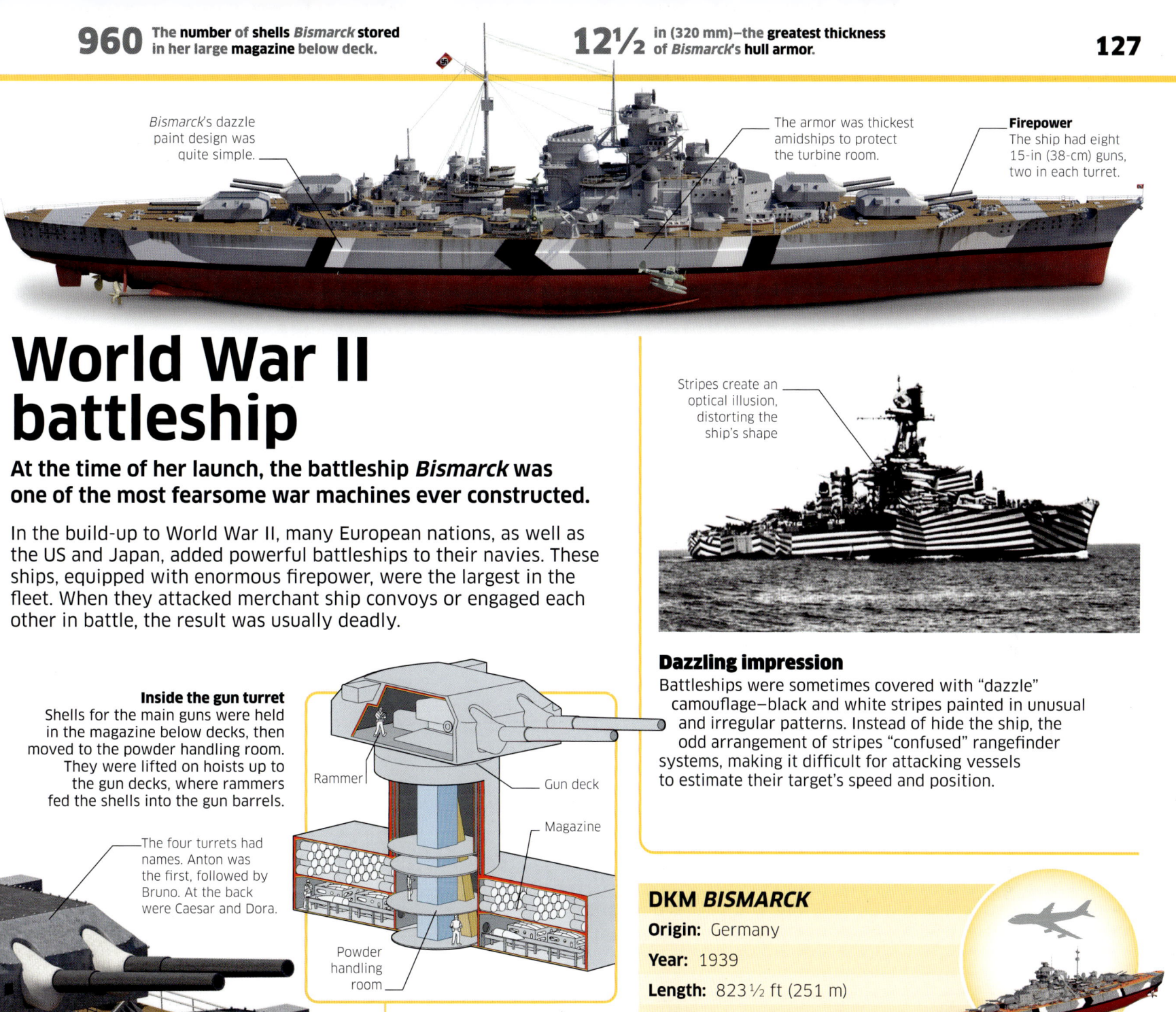

Bismarck's dazzle paint design was quite simple.

The armor was thickest amidships to protect the turbine room.

Firepower
The ship had eight 15-in (38-cm) guns, two in each turret.

World War II battleship

At the time of her launch, the battleship *Bismarck* was one of the most fearsome war machines ever constructed.

In the build-up to World War II, many European nations, as well as the US and Japan, added powerful battleships to their navies. These ships, equipped with enormous firepower, were the largest in the fleet. When they attacked merchant ship convoys or engaged each other in battle, the result was usually deadly.

Stripes create an optical illusion, distorting the ship's shape

Dazzling impression

Battleships were sometimes covered with "dazzle" camouflage—black and white stripes painted in unusual and irregular patterns. Instead of hide the ship, the odd arrangement of stripes "confused" rangefinder systems, making it difficult for attacking vessels to estimate their target's speed and position.

Inside the gun turret
Shells for the main guns were held in the magazine below decks, then moved to the powder handling room. They were lifted on hoists up to the gun decks, where rammers fed the shells into the gun barrels.

Rammer

Gun deck

Magazine

The four turrets had names. Anton was the first, followed by Bruno. At the back were Caesar and Dora.

Powder handling room

DKM *BISMARCK*

Origin: Germany

Year: 1939

Length: 823½ ft (251 m)

Artillery: 64 guns of various size

Capstan
Revolving capstans lowered and raised the large anchors by winding or unwinding their heavy chains.

Anchor
Bismarck had five anchors, two at the back, two at the front, and a spare at the prow.

Spare anchor

DREBBEL'S SUBMARINE
Prototype submarine
Origin: UK
Year: 1620

Dutchman Cornelius van Drebbel made the first working submarine for King James I of England and demonstrated it on the River Thames. Four oarsmen rowed it at a depth of 12 ft (4.5 m).

Leather seals prevent water from entering the oar holes

The DREBBEL

Leather-paddled oars

Under the sea

The world's oceans teem with highly sophisticated military, research, rescue, and exploratory submarines and submersibles.

The idea of navigating vessels safely underwater has been floated since at least the time of the ancient Greeks around 500 BCE. But more than 2,000 years would pass before the first submarine was built, and it could only stay submerged for a few minutes. After centuries of experimentation, the submarine age truly began at the onset of World War I. Today, more advanced models descend ever deeper.

TURTLE
Pedal-powered minelayer
Origin: US
Year: 1775

A 7-ft- (2.13-m-) tall wooden ball coated in waterproof tar, *Turtle* could stay underwater for 30 minutes. Made during the Revolutionary War, it was meant to surface next to British ships so that the operator could attach mines to them.

Detachable mine

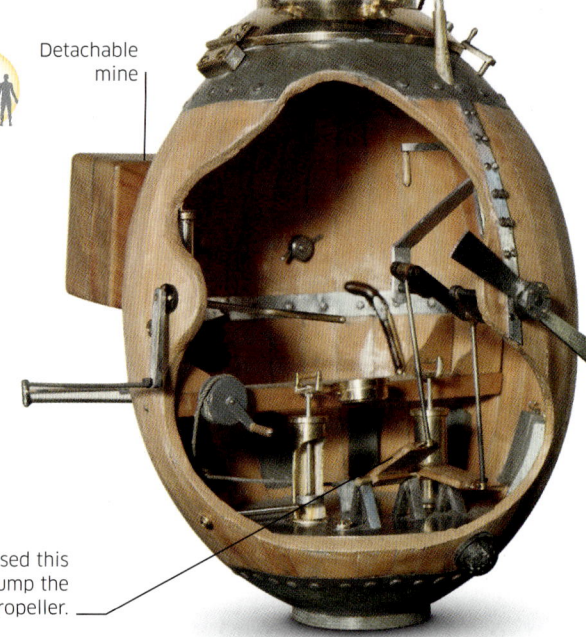

Wooden hull lined with pigskin on the inside

The pilot used this pedal to pump the outside propeller.

ICTINEO II
Steam-powered pioneer
Origin: Spain
Year: 1862

Spanish engineer Narcis Monturiol built the first-ever steam-powered submarine. It was fitted with a device for eliminating carbon dioxide and replenishing oxygen to allow for longer dives.

Rudder for steering

SM U-9
World War I submarine
Origin: Germany
Year: 1910

Funnel releasing smoke from the gas engines

Torpedo hatch
The *U-9* fired torpedoes from hatches, two at the front and two at the back.

Germany's U-boats (*Unterseeboot*, or "undersea" boat) were developed just before World War I. They terrorized shipping–on September 22, 1914, a *U-9* sank three British cruisers on the same day.

I-400 SUBMARINE
Aircraft-carrying sub
Origin: Japan
Year: 1944

Japan's I-400 class were the largest submarines of World War II. They were built specifically to attack Allied shipping in the Panama Canal and carried three bomber floatplanes for aerial assault.

Crane for lifting aboard planes after landing

Hangar with space for three floatplanes during dives

Small floatplanes could take off from the deck but would land on water.

3,000 The number of Allied **ships sunk** by German **submarines** in World War II.

574 ft (175 m)—the overall **length** of a Soviet Typhoon-class submarine, the **largest submarine** ever built.

129

TRIESTE
Deep diver
Origin: Italy
Year: 1953

Designed in Switzerland and built in Italy, *Trieste* was a bathyscaphe, an early type of submersible. On January 23, 1960, she descended 35,797 ft (10,911 m) into the Challenger Deep in the Pacific Ocean, the lowest known point on Earth.

Entrance with tunnel leading to the observation gondola

Propellers sit on top of the hull.

Observation gondola

USS NAUTILUS
Nuclear submarine
Origin: US
Year: 1955

The world's first operational nuclear-powered submarine, *Nautilus* was the first vessel to sail under the North Pole, in 1958. She clocked more than 500,000 miles (800,000 km) in service.

Tower
This contains sonar and radar masts plus search and attack periscopes.

NAUTILE
Research vessel
Origin: France
Year: 1984

Able to dive down to 3.7 miles (6 km), *Nautile* has explored the *Titanic* wreck. She has robot arms, cameras, and strong spotlights. She is carried to her dive sites on a ship, then lowered into the water.

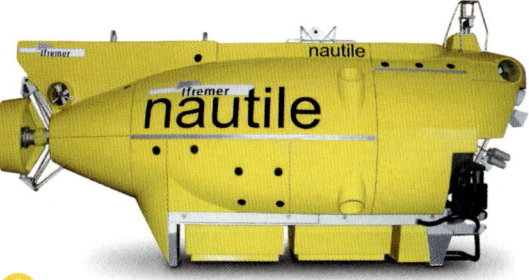

NSRS
Rescue submersible
Origin: France/Norway/Britain
Year: 2008

The NATO Submarine Rescue System (NSRS) helps submerged submarines in distress. It attaches to a downed submarine's hatch, allowing crew members to escape through a decompression chamber.

This part attaches to the stricken submarine's hatch.

DEAPSEA CHALLENGER
Submersible sampler
Origin: Australia
Year: 2012

This bullet-shaped craft was designed to dive in the Mariana Trench, the deepest point on the Pacific Ocean floor. It reached 38,787 ft (10,908 m), where it collected samples for scientists.

Floodlights

Directional spotlight

A pilot operates the arms.

Camera

Tool for picking up samples

DEEP FLIGHT AVIATOR
Undersea tourist craft
Origin: US
Year: 2003

Designed to resemble an aircraft, *Deep Flight Aviator* takes passengers on undersea tours to depths of 1,500 ft (458 m) to explore coral reefs and wildlife. It "flies" through water using wings and thrusters to dive and rise instead of water ballast tanks like normal subs (see page 131).

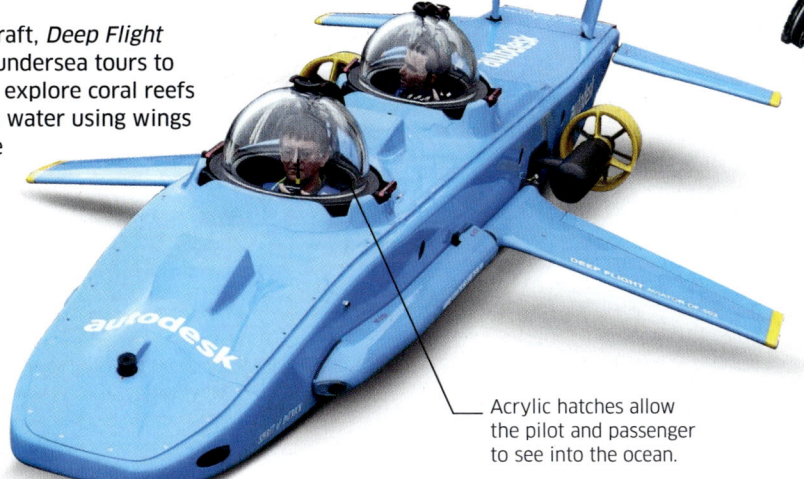

Acrylic hatches allow the pilot and passenger to see into the ocean.

28⁴/₅ ft (8.8 m)—the diameter of the *Suffren* at her widest point.

Bunking down

The men and women serving on the *Suffren* live and work in cramped spaces. When not on duty, they can relax in the canteen or in their cabins. Each bunk has its own light, power outlet, and USB port, so submariners can access recorded movies and TV shows.

Nuclear-powered submarine

Almost undetectable while gliding silently and swiftly beneath the oceans, the most cutting-edge subs are powered by onboard nuclear reactors.

Heat from nuclear fission in the reactors turns water into steam to power the submarine's turbines and electricity generators. Only six nations operate nuclear-powered submarines: the US, Russia, China, India, the UK, and France. The French *Suffren* is one of the newest and most sophisticated examples of these underwater vessels.

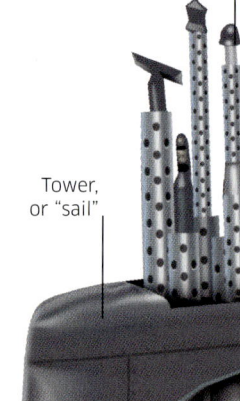

Infrared, HDTV, and digital "optronic masts" and range finders and thermal imagers that can be retracted into the tower.

Tower, or "sail"

SUFFREN

Origin: France

Year: 2019

Submarine class: Barracuda

Length: 326 ft ½ in (99.5 m)

Sleeping quarters
Each cabin contains two sets of three-tier bunk beds.

Access hatch
Crew accesses the ship via a watertight hatch. There is another hatch at the vessel's stern.

Torpedo and missile hatches
These four metal shutters open and close automatically when torpedoes or missiles are fired.

Weapons bay
Suffren's supply of Exocet and Storm Shadow missiles and A3SM sea-to-air missiles are kept in this storage area.

Ballast tank
These take on or release seawater to help the sub stay underwater or rise to the surface.

MOAS sonar
A specialized Mine and Obstacle Avoidance Sonar (MOAS) array is housed in the sub's nose.

Cylindrical main sonar
This large, drum-shaped sonar array can detect objects up to 1⅕ miles (1.9 km) away.

Torpedo tubes
Four 21-in- (553-mm-) diameter tubes guide F21 Artemis torpedoes through to the four hatches at the nose.

Superior sub

Suffren is the first of six Barracuda-class nuclear submarines. It can go 10 years without refueling, has an unlimited range, and can fire smart and guided missiles at distances of up to 620 miles (1,000 km).

12²/₅ miles (20 km)—the **length** of all the **pipework** running through *Suffren*.

70 The number of days *Suffren* can stay at sea submerged before food supplies run out.

131

Underwater dining
Fresh meals are cooked everyday for a crew of 65. Officers have their own dining area (above), while the rest of the crew have theirs on the floor below.

Special forces hangar
A Dry-Dock Shelter (DDS) attached to the rear hatch stores submersibles used by *Suffren*'s special forces operatives. Divers enter and exit via an airlock connecting to the floor below.

Mini sub
The Special Warfare Underwater Vehicle (SWUV) is used to transport attack divers and commandos to military targets.

Water power
The pump-jet propulsion system sucks in water and forces it out at high speed to drive the propellers.

Airlock

X rudders
Four rudders arranged in an X shape make the sub easier to steer and maneuver.

Reactor compartment
A uranium-fueled K15 nuclear reactor drives the turbines and creates electricity and power.

Batteries included
Excess electricity generated by the reactor is stored in lithium-ion batteries to provide back-up power.

Flank array sonar
Sonars on the port and starboard flanks detect objects and threats coming from either side.

Control room
All combat, navigation, and maintenance and monitoring activities are controlled from the high-tech operations room. It has 10 multifunction digital screens and a high-tech and interactive "tactical table."

How subs dive and rise
Submarines have ballast tanks at the front, middle, and rear. Filling and emptying these of water controls the vessel's rise and fall.

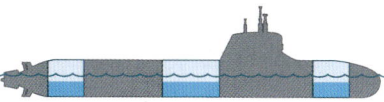

Horizontal position
An equal amount of water in each tank keeps the sub even. The more water in the tanks, the deeper the sub sinks, in a controlled, level way.

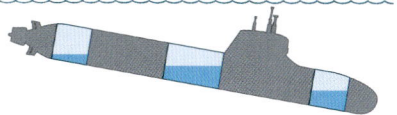

Diving
Emptying only the rear tank forces the nose downward, and the sub dives at an angle.

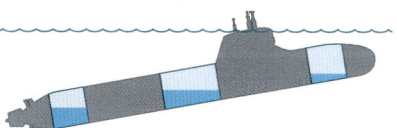

Rising
Emptying only the forward tank pushes the nose up. As water is let out and air replaces it, the sub surfaces.

How sonar works
Suffren has "passive" and "active" sonar systems. Passive sonar devices, or "arrays," work by listening for sounds made by external objects. Active sonar arrays operate by emitting pulses of noise and calculating the size and distance away of any object they bounce off.

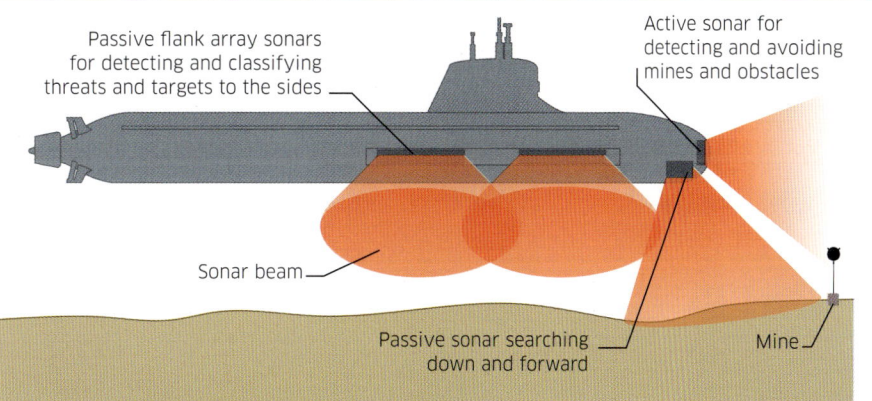

Passive flank array sonars for detecting and classifying threats and targets to the sides

Active sonar for detecting and avoiding mines and obstacles

Sonar beam

Passive sonar searching down and forward

Mine

$4.5 billion The **total cost** of building the USS *Ronald Reagan*.

USS *RONALD REAGAN*

Origin: US

Year: 2001 (launched)

Length: 1,092 ft (332.8 m)

Speed: 35 mph (56.3 kph)

Aboard the USS *Ronald Reagan*
Named after the 40th president of the United States and active since 2003, this ship is a hive of constant activity. Aircraft must be serviced and combat-ready at all times, the vessel itself must be maintained, and the crew of around 6,000 people need to be fed and housed.

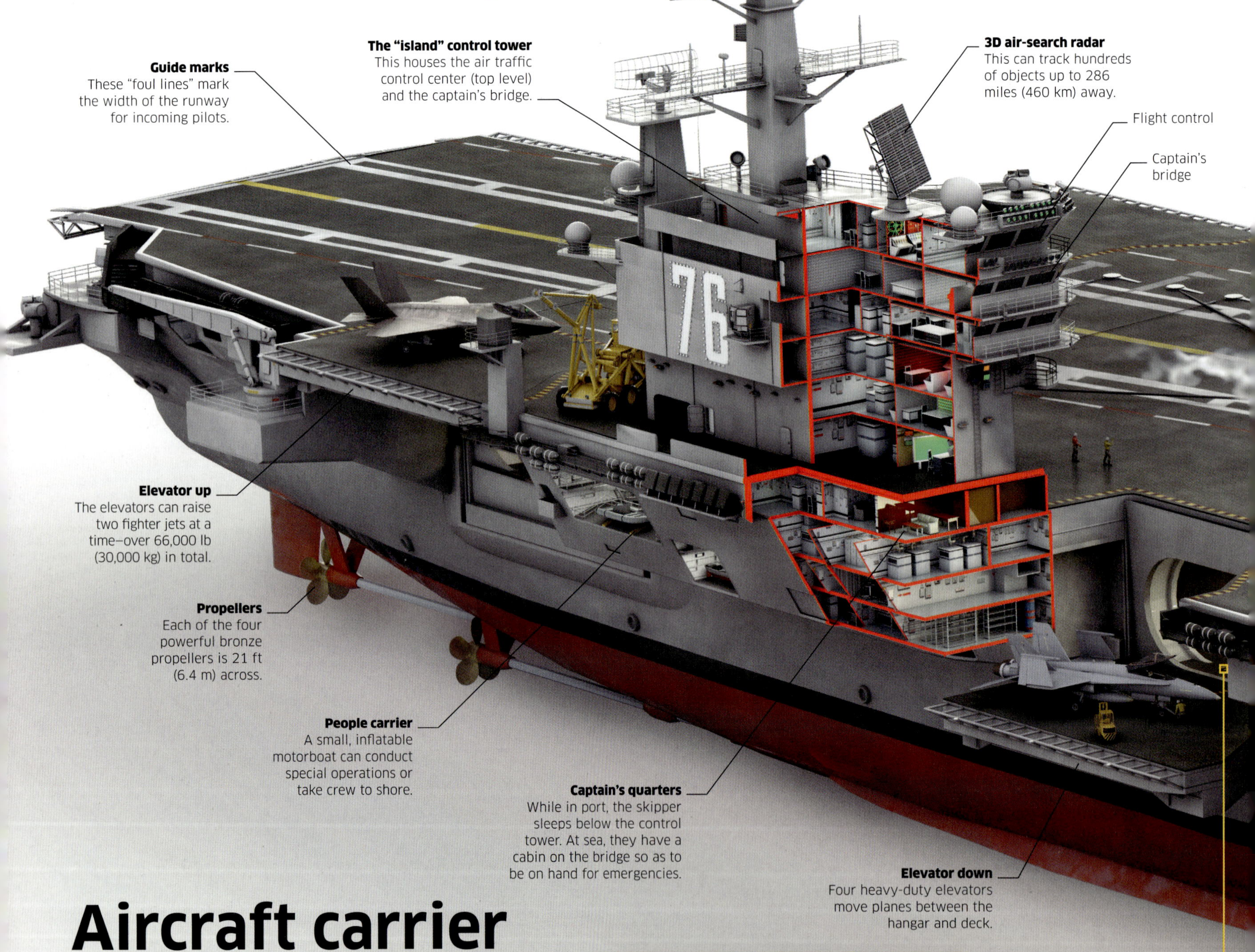

Guide marks
These "foul lines" mark the width of the runway for incoming pilots.

The "island" control tower
This houses the air traffic control center (top level) and the captain's bridge.

3D air-search radar
This can track hundreds of objects up to 286 miles (460 km) away.

Flight control

Captain's bridge

Elevator up
The elevators can raise two fighter jets at a time—over 66,000 lb (30,000 kg) in total.

Propellers
Each of the four powerful bronze propellers is 21 ft (6.4 m) across.

People carrier
A small, inflatable motorboat can conduct special operations or take crew to shore.

Captain's quarters
While in port, the skipper sleeps below the control tower. At sea, they have a cabin on the bridge so as to be on hand for emergencies.

Elevator down
Four heavy-duty elevators move planes between the hangar and deck.

Aircraft carrier

Nuclear-powered Nimitz-class "supercarriers" are among the most sophisticated and powerful vessels ever to take to the seas.

Patrolling the oceans as part of a battlegroup of up to 12 ships, and with 75 aircraft on board, these mobile air bases patrol oceans around the world. They can travel 800 miles (1,300 km) a day and strike a target from over 500 miles (800 km) away. With 10 Nimitz-class carriers and one larger Ford-class, the US has double the landing deck-space of every other navy combined.

Belly of the beast
Two levels below the flight deck, and running two-thirds of the ship's length, the hangar is three decks (26 ft or 8 m) high and holds up to 60 aircraft, plus stores and weapons. The vast space is split into three sections, separated by massive metal doors that can slide shut in case of a fire.

244 ft (74 m)—the ship's **total height** from keel to **mast** is the same as a **24-story building!**

As a **plane touches down, the pilot accelerates** in case they have **missed the arresting cables** and need to **take off again!**

133

Precision parking

Landing on the moving deck of an aircraft carrier takes nerves of steel. To help the incoming pilot, lights seen through special "Fresnel" lenses and color filters indicate the correct "glidepath." As the plane's approach angle changes, an amber bar of light in the center appears to move up or down, while green lights on either side remain fixed. The pilot must line up the amber light with the green to descend safely.

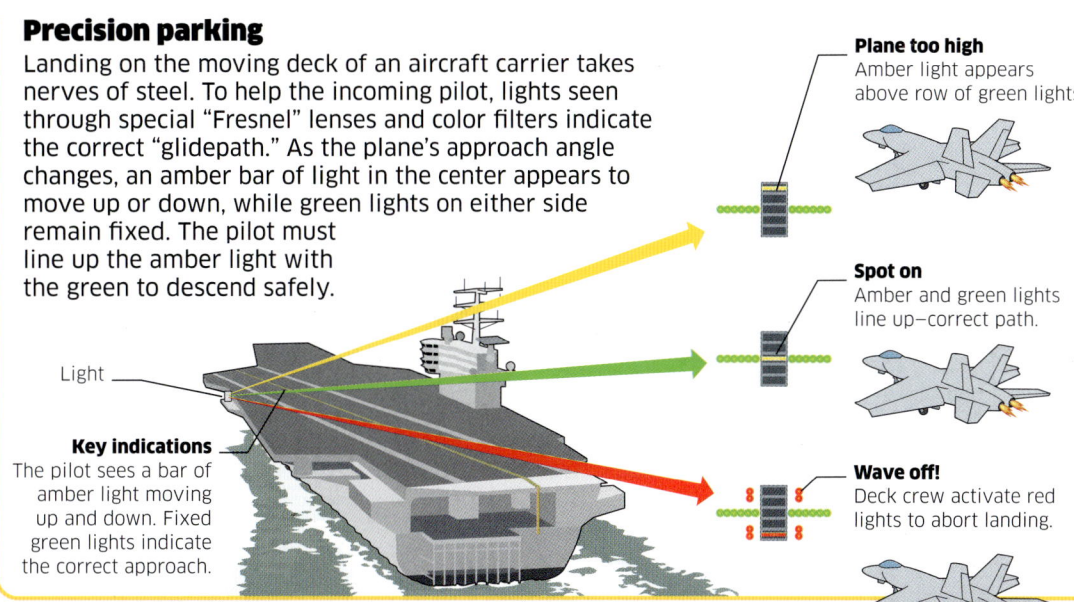

Plane too high
Amber light appears above row of green lights.

Spot on
Amber and green lights line up—correct path.

Wave off!
Deck crew activate red lights to abort landing.

Light

Key indications
The pilot sees a bar of amber light moving up and down. Fixed green lights indicate the correct approach.

Snappy landings

To land, the pilot lowers a tailhook to catch one of four cables strung across the deck, which brings the jet to a halt. Gears, pulleys, and hydraulics control the cable's tension—too tight and the aircraft will flip as it lands; too loose and the plane may shoot off the edge of the ship!

Hornet's wings
Many planes designed for aircraft carriers have foldable wings to save space.

Temporary parking
Planes are cleared from the runway before landing operations begin.

Double capacity
An angled rear runway allows planes to land at the same time as jets take off from the straight forward deck.

Firepower
The ship carries a formidable arsenal of weapons, which are brought up in special bomb elevators.

Engines
Four steam turbine engines rotate the ship's propellers, generating 260,000 hp.

Nuclear reactor
Two Westinghouse A4W uranium reactors superheat water to drive the steam turbines.

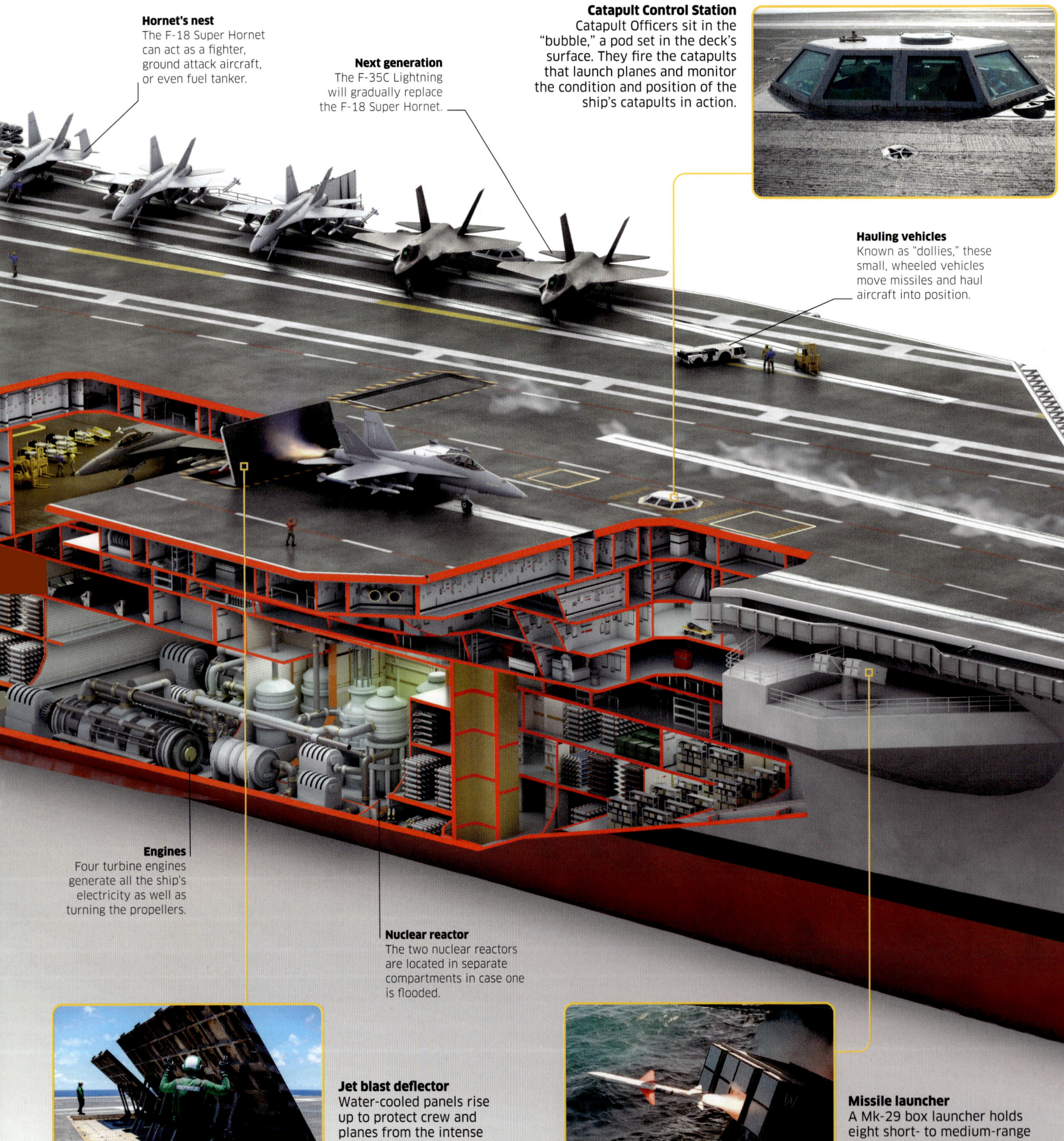

Hornet's nest
The F-18 Super Hornet can act as a fighter, ground attack aircraft, or even fuel tanker.

Next generation
The F-35C Lightning will gradually replace the F-18 Super Hornet.

Catapult Control Station
Catapult Officers sit in the "bubble," a pod set in the deck's surface. They fire the catapults that launch planes and monitor the condition and position of the ship's catapults in action.

Hauling vehicles
Known as "dollies," these small, wheeled vehicles move missiles and haul aircraft into position.

Engines
Four turbine engines generate all the ship's electricity as well as turning the propellers.

Nuclear reactor
The two nuclear reactors are located in separate compartments in case one is flooded.

Jet blast deflector
Water-cooled panels rise up to protect crew and planes from the intense heat and flames of a jet's exhaust. They are raised for take-off, then quickly lowered so the next plane can move into position.

Missile launcher
A Mk-29 box launcher holds eight short- to medium-range "Sea Sparrow" radar-guided missiles. Four of these launchers defend the carrier against enemy aircraft or antiship cruise missiles.

5 dentists and 5 doctors are stationed on board.

20 years—the life of the ship's nuclear reactors before they need refueling.

4 The number of **launch catapults** – each helps planes reach 170 mph (274 kph) in 2 seconds.

135

Color-coded uniforms

All deck crew have specific jobs. Those in yellow (the "Yellow Shirts") guide planes around the deck. Red Shirts load weapons, while Purple Ships fuel aircraft. Blue Shirts operate elevators and drive dollies, Brown Shirts service and maintain planes, while Green Shirts look after the catapult and arresting gear. White Shirts perform final checks before take-off.

Take-off!

With the roar of its jet engine and the help of the aircraft carrier's high-powered catapult, a fighter jet launches into the air and soars over the ocean.

As the jet is catapulted from the deck, it accelerates from 0 to 165 mph (265 kph) in 2 seconds flat—fast enough to begin climbing. Another plane can launch from the same track just 30 seconds later.

Stealth tape
Special tape covering hatch edges scatters enemy radar signals.

Range extender
The top fuel hatch is designed for air-to-air refueling.

Catapult track
A steam-powered "arm" runs along this channel, flinging planes forward for take-off.

Identifying markings
All ships have hull identification numbers (HINs). The higher the number, the newer the ship.

Secondary bridge
If the tower is damaged, the ship can be run from a back-up bridge behind these portholes.

Into the air

The F-35C Lightning is the first stealth fighter developed specifically for use on aircraft carriers, with larger wings for extra lift and foldable wingtips for compact storage. A fully loaded F-35C takes off with 20,000 lb (9,000 kg) of fuel, one Gatling gun, and four missiles. It has a range of 1,381 miles (2,222 km) or more with air-to-air refueling.

Bulbous bow
The bulb changes how water flows around the hull to reduce drag, making the ship faster and more efficient.

Bow anchor
The anchor and chain weigh 152.9 tons (138.7 tonnes). There is a second anchor at the stern.

On the deck

First, there's a deafening roar, then a bone-rattling "WHOOSH!" as a jet fighter blasts off. Propelled by a catapult, with its engines on full power, this EA-18G Growler hits 170 mph (275 kph) in just 2 seconds.

It's fast and furious work on the flight deck of an aircraft carrier, with planes taking off and landing every 37 seconds. Disaster could strike at any moment. Making sure it doesn't is the small army of highly trained ground crew. Each team member has a specific job to do to keep the intricate machinery running like clockwork.

AIR AND SPACE

Since humans first achieved powered flight in 1903, aviation has advanced rapidly. Every day, around 100,000 flights carry at least 10 million passengers globally. People have even walked on the Moon and landed spacecraft on Mars.

Moon landing
The US took a clear lead in the space race when Apollo 11 landed on the surface of the Moon in 1969.

APOLLO 11, 1969

SPACE SHUTTLE, 1981

Space Shuttle
Five shuttles were built in total, carrying hundreds of astronauts into space and back between 1981 and 2011.

MIL MI-26, 1983

Heavy-lift helicopter
The world's most powerful helicopter can carry anything from armored vehicles to a frozen woolly mammoth.

Supersonic airliner
Concorde, first tested in 1969, broke a new frontier, carrying passengers at twice the speed of sound.

CONCORDE, 1969

Human spaceflight
In 1961, Soviet cosmonaut Yuri Gagarin became the first human in space. He orbited Earth in Vostok 1 several times before reentering the atmosphere.

VOSTOK 1, 1961

Reaching for the sky

Until the dawn of modern science, practical human flight was all but impossible. But once scientists began to understand the properties of air and to develop engines, progress was swift.

The first craft to take to the air were balloons, filled with hot air or light gas. Then experimental gliders and early gas engines led to the airplane. Two world wars accelerated research, and the jet age made air travel affordable. Meanwhile, the Cold War brought intense competition between the US and the Soviet Union. Just a few decades after the Wright Flyer first took off, humans soared beyond air and into space. One day this century, a human may even set foot on Mars.

Helicopter
Igor Sikorsky invented the modern helicopter, using a small tail rotor to stop the craft from spinning out of control.

SIKORSKY R-4, 1942

HEINKEL HE 178, 1939

Jet plane
The first ever jet aircraft took off in 1939. The invention would revolutionize air travel after World War II.

BOEING 314 CLIPPER, 1938

Long-range flying boat
The Boeing Clipper competed with airships for long-distance travel, landing on water to refuel and take on new passengers.

Timeline of air and space travel
This timeline shows a selection of craft that have propelled humans into the air and beyond, from the first balloons and aerofoil wings to jet engines and rockets.

1750–1900

Leaving the ground
After centuries of dreaming, humans began to make real progress toward sustained, powered, controllable flight.

Hot air balloon
The Montgolfier brothers harnessed the energy of rising hot air to take to the skies in 1783–the first sustained human flight.

MONTGOLFIER BALLOON, 1783

Airship
The Giffard airship was the first to mount an engine and propeller below a gas balloon to control direction.

GIFFARD AIRSHIP, 1852

Biggest jetliner
In 2005, the enormous double-decker A380 became the world's largest airliner, with space for up to 853 passengers.

AIRBUS A380, 2005

DJI PHANTOM 4, 2016

Camera drone
The DJI Phantom 4 was the first commercial quadcopter drone to offer high-resolution video and target tracking.

Starship
The most powerful launch vehicle ever made, Starship aims to take humans to the Moon, Mars, and beyond.

STARSHIP, 2023

Satellite
The Soviet Union launched a satellite into orbit around the Earth in 1957, starting a race into space with the US.

SPUTNIK 1, 1957

U-2, 1955

B-17 BOMBER, 1943

Spy plane
The U-2 spy plane took crewed flight to an altitude of 70,000 ft (21,000 m) in 1955–higher than ever before.

Interceptor
The Spitfire Mk IX proved nimble and deadly in the battle for the skies over Britain during World War II.

SPITFIRE MK IX, 1943

Heavy bomber
The B-17 "Flying Fortress" bomber flew vast distances with heavy loads and formidable defensive weaponry.

1945 TO PRESENT ▶

Jets and rockets
After World War II, the US and the Soviet Union raced to send craft into space. Alongside the space race, jet airliners brought faster, cheaper travel to the masses.

HINDENBURG, 1936

Transatlantic airship
Giant airships were the first aircraft to carry passengers nonstop across the Atlantic, far faster than a ship could manage.

AUTOGIRO, 1923

Autogyro
Combining rotors with a propeller, the autogyro was a key step toward working helicopters.

FOKKER EINDECKER, 1915

Sleek fighter plane
The Eindecker allowed the pilot to fire bullets through the propeller's arc for the first time.

CAYLEY GLIDER, 1853

Steerable glider
George Cayley's heavier-than-air glider achieved sustained flight with increasing control but no power.

1900–1945 ▶

Experiments in flight
Designers and pilots pushed the boundaries of aviation with bigger aircraft, better engines, and longer flights. People began to realize the full potential of flying machines, in war and peace.

First airplane
The Wright brothers combined the latest in glider technology and gas engines to fly the first airplane.

WRIGHT FLYER, 1903

IN A FLAP

Many early pioneers sought to mimic nature by building ornithopters—machines that flapped their wings like birds. Neither humans nor engines could match birds' lightweight hollow bones, powerful chest muscles, or complex hinged wings. Generating lift by propelling fixed, aerofoil-shaped wings through the air has proven more efficient.

Wings become aerofoils on upstroke.

Wings push air down.

Natural flyer
Birds produce lift on the downstroke by pushing air downward, but on the upstroke, they swivel their curved wings to act as aerofoils, generating more lift.

Edward Frost's ornithopter of 1904 managed just a tiny hop off the ground The gas engine flapped wings made of willow, silk, and real bird feathers.

THE WRIGHT BROTHERS

Inspired by German glider pioneer Otto Lilienthal, American brothers Orville and Wilbur Wright made a systematic study of flight principles. They tested many wing designs and even constructed their own wind tunnel. On December 17, 1903, their Wright Flyer biplane flew 118 ft (36 m) on the first controlled, powered, heavier-than-air flight. The Flyer lacked ailerons, instead using wires that caused the wingtips to twist, a system known as wing warping.

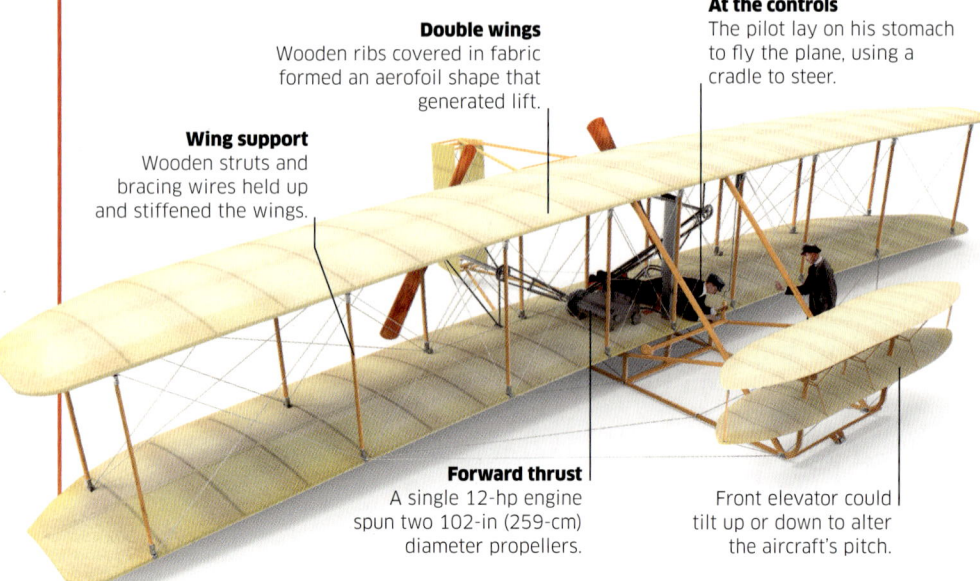

Double wings
Wooden ribs covered in fabric formed an aerofoil shape that generated lift.

At the controls
The pilot lay on his stomach to fly the plane, using a cradle to steer.

Wing support
Wooden struts and bracing wires held up and stiffened the wings.

Forward thrust
A single 12-hp engine spun two 102-in (259-cm) diameter propellers.

Front elevator could tilt up or down to alter the aircraft's pitch.

How to fly

For many centuries, people have looked up and longed to fly like birds. Daredevil attempts to flap artificial, arm-mounted wings never achieved lift-off, and it was only after significant advances in science that we could finally take flight.

Humans would never have taken to the skies without understanding the principles of flight. Pioneers had to find ways to produce lift and thrust effectively to overcome the force of gravity, which pulls any weighty object earthward. They also needed methods to steer and control their craft. First, vessels lighter than air began to make dreams reality—early balloons full of hot air or hydrogen. Another 120 years of testing and ingenuity led to the miracle of heavier-than-air craft.

FOUR FORCES OF FLIGHT

Four forces govern objects that fly: lift, weight, thrust, and drag. For an aircraft to get off the ground, it must generate more lift than the force of gravity acting on the plane's weight and pulling it back toward the ground. And for an aircraft to fly forward, its force of thrust must be greater than the drag or air resistance it encounters.

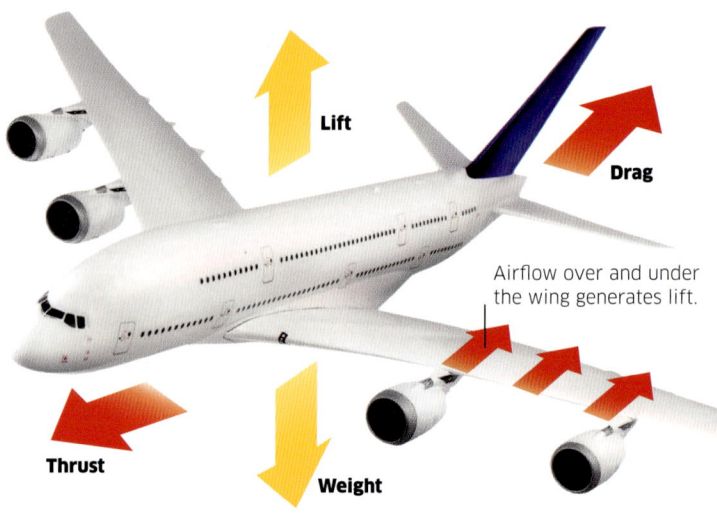

Lift

Drag

Airflow over and under the wing generates lift.

Thrust

Weight

LIGHTER THAN AIR

Filling large bags, or envelopes, with lighter-than-air gases made a craft lighter than the surrounding atmosphere so that it could rise. The balloon created by the Montgolfier brothers in 1783 was filled with heated air, which is lighter than the atmosphere's cooler air. Other early balloons were filled with even lighter gases, such as hydrogen or coal gas.

Bunch of balloons
These modern hot air balloons fly over a dramatic landscape in Türkiye. They are equipped with burners that heat the air inside the balloon's envelope.

9,000 The estimated number of aircraft in the air at any one moment.

385 ft (117 m)–the wingspan of the Stratolaunch Roc, the largest of any plane ever flown.

143

AEROFOILS

An aerofoil is an object shaped so that one surface is more curved than the other. Passing through air with the curved side on top, it produces lift at right angles to the airstream. The faster the air flows, the more lift is generated, which is why aircraft accelerate along a runway to take off. Aircraft wings are aerofoil shaped in order to generate large amounts of lift.

Over and under the wing

The airflow is split by the wing's leading edge so that it travels both over and under the wing. The air travels faster over the curved top surface, creating lower air pressure above the wing and higher air pressure below it. This difference in pressure pushes the wing up, producing lift.

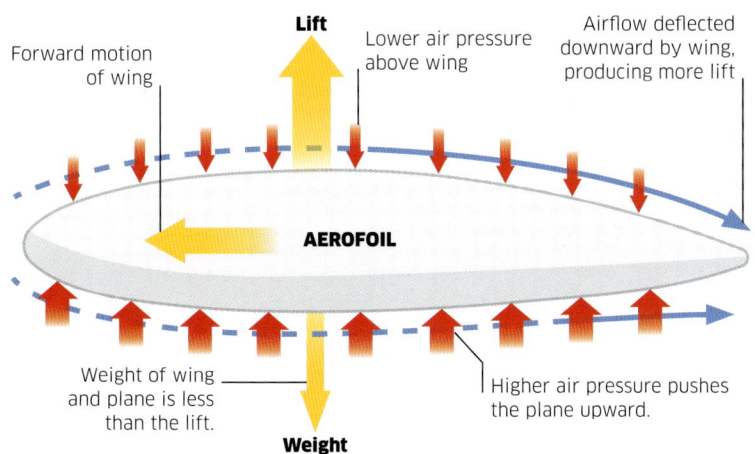

Lift

Forward motion of wing

Lower air pressure above wing

Airflow deflected downward by wing, producing more lift

AEROFOIL

Weight of wing and plane is less than the lift.

Higher air pressure pushes the plane upward.

Weight

Angle of attack

Much of the lift comes from the angle of attack–the angle between the oncoming air and the wing. Air is pushed downward, creating an upward force that adds to the difference in pressure above and below the wing. However, if the angle is too steep, the plane will stall.

1 LOW LIFT
With the wing almost level relative to the airflow, air is only deflected downward to a limited extent, and a moderate amount of lift is produced.

2 MAXIMUM LIFT
When the airflow meets the wing at a steeper angle of attack, the wing deflects more air, generating higher pressure beneath it and boosting lift.

3 STALLING
An extreme angle of attack causes the airflow to separate from the wing. Lift decreases sharply and the plane stalls and begins to fall.

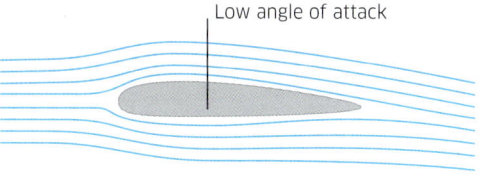

Low angle of attack

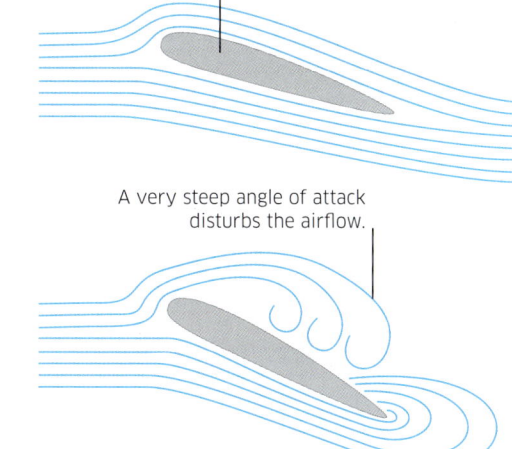

High angle of attack

A very steep angle of attack disturbs the airflow.

CONTROL SURFACES

To steer a plane in midair involves deflecting the airflow around parts of the aircraft to push the plane in different directions. Control surfaces are hinged panels in the wing (ailerons), tail (rudder), and tailplanes (elevators), which alter the aircraft's orientation along one of its three axes. Ailerons and elevators are operated by the pilot's control column, while the rudder is moved by rudder pedals in the cockpit.

Rolling with it

Three aircraft from the Royal Jordanian Falcons display team perform a spectacular aerobatic move. Two of the airplanes are using their ailerons to produce roll—one going from level flight into a quarter roll to fly on one side, and one doing a half roll to fly upside down.

Changing direction

An aircraft uses its control surfaces to alter its movement in three dimensions. These movements are known as pitch (climbing and diving), yaw (turning), and roll (tilting to one side). Altering both yaw and roll at the same time allows aircraft to make smooth, banked turns.

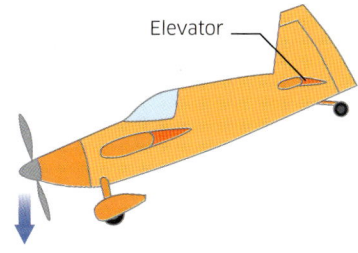

Elevator

Pitch
Lowering the elevators causes the tail to rise and the nose to point downward so that the plane descends.

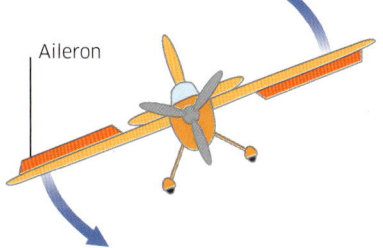

Aileron

Roll
Tilting one wing's aileron up and the other down causes the plane to roll, with one wing traveling upward and the other downward.

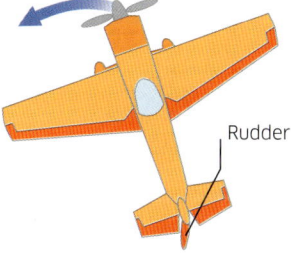

Rudder

Yaw
Tilting the rudder to one side causes the deflected airflow to move the tail in the opposite direction. This pushes the nose sideways so that the plane turns.

400,000 The number of **spectators** who **watched** the **Charlière balloon launch** in Paris.

German pioneer **Otto Lilienthal** made over **2,000 glider flights** in his lifetime.

Taking to the air

People have long looked to the skies and dreamed of flying. Many disastrous attempts to mimic birds occurred before the principles of flight were discovered and exploited.

The first craft to become airborne were balloons filled with gases lighter than the surrounding air. However, these were at the mercy of winds until powered, steerable airships were developed. Scientific advances in the 19th century led to heavier-than-air gliders using aerofoil-shaped wings to generate lift. The next step was to find a light yet powerful engine, with inventors testing steam, electricity, and even pedal power. Following the Wright Brothers' pioneering 1903 Wright Flyer, early aircraft with internal combustion engines became reality and aviation really took off!

Passenger gondola

GIFFARD AIRSHIP

First powered, steerable airship

Year: 1852

Longest flight: 16¾ miles (27 km)

Suspended beneath this cigar-shaped hydrogen balloon, a steam engine spun a 23-ft (7-m) wooden propeller. The French ship flew at 6 mph (10 kph)—not fast enough to overcome headwinds.

MONTGOLFIER BROTHERS' BALLOON

The first powered flight

Year: 1783

Longest flight: 5½ miles (9 km)

After a successful test carrying a duck, rooster, and sheep, the brothers built a 49¼-ft- (15-m-) wide balloon filled with hot air from a fire. It carried two pilots 2,950 ft (900 m) above Paris.

LA CHARLIÈRE BALLOON

Harnessing hydrogen

Year: 1783

Longest flight: 22½ miles (36 km)

Launched in Paris just 10 days after the Montgolfier balloon, this was the first craft to use hydrogen gas to produce lift. The inflated silk balloon flew for two hours, five minutes.

ADER ÉOLE

Almost an airplane

Year: 1890

Longest flight: 164 ft (50 m)

Clément Ader's craft had a 46-ft (14-m) wingspan and a steam engine burning alcohol as fuel. This French invention flew just 8 in (20 cm) above the ground on a short hop—just as well, with no steering controls!

Bamboo propeller

The batlike wings did not flap.

CAYLEY GLIDER

Pioneering glider

Year: 1853

Longest flight: 820 ft (250 m)

After many experiments, Englishman Sir George Cayley built this glider, towed downhill by a galloping horse for take-off. A boat-shaped body held the intrepid pilot—Cayley's coach driver.

36 minutes, 30 seconds—the **duration** of Blériot's first cross-Channel flight in 1909.

1912 The year Harriet Quimby became the first woman to fly across the English Channel, in a Blériot XI.

145

LILIENTHAL NORMALSEGELAPPARAT

Influential German glider

Year: 1894
Longest flight: 820 ft (250 m)

Launched from hillsides, Lilienthal's "Standard soaring apparatus" could glide to a gentle landing, steered by the pilot shifting their bodyweight. Nine of these willow and cotton gliders were built.

LEBAUDY NO.1

"The Yellow One"

Year: 1902
Longest flight: 61 miles (250 m)

Slung from hemp netting below this French airship were a steel keel and a gondola holding a gas engine. This 40-hp engine spun two propellers to give the 185⅜-ft- (56.5-m-) long craft a cruising speed of up to 22 mph (35 kph).

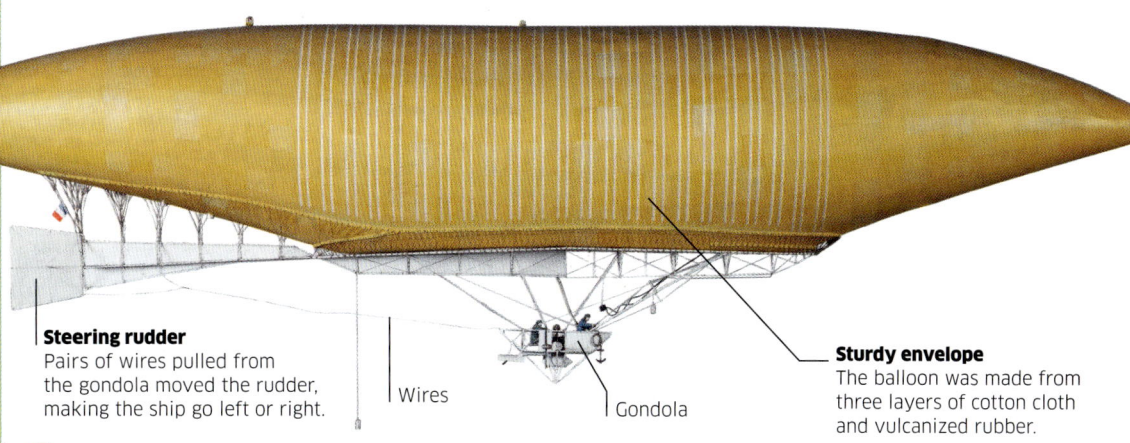

Steering rudder
Pairs of wires pulled from the gondola moved the rudder, making the ship go left or right.

Wires

Gondola

Sturdy envelope
The balloon was made from three layers of cotton cloth and vulcanized rubber.

VOISIN-FARMAN I

Europe's first successful aircraft

Year: 1907
Longest flight: 17 miles (27 km)

The first aircraft to fly a 0.62-mile (1-km) circular route, this French biplane was pushed forward by a 50-hp gas engine spinning a two-bladed propeller. It lacked ailerons to control rolls.

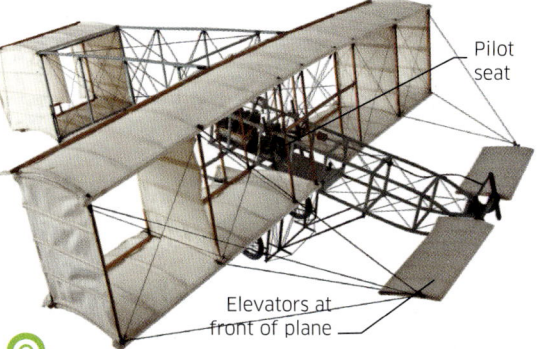

Pilot seat

Elevators at front of plane

SANTOS-DUMONT DEMOISELLE NO.20

Public domain plane

Year: 1908
Longest flight: 11 miles (18 km)

This ultralight monoplane weighed just 242 lb (110 kg), with a tiny 18-ft (5.5-m) wingspan. The Brazilian inventor released the plans to the public for free, and around 50 were built.

Elevators on the tail

BLÉRIOT TYPE XI

International flight

Year: 1909
Longest flight: 75 miles (120 km)

This simple but robust French design used an ash wood box-frame fuselage, part-covered in waterproof fabric. When Blériot flew across the English Channel in 1909, the publicity sparked 103 orders.

FABRE HYDRAVION

Early seaplane

Year: 1910
Longest flight: 3.5 miles (5.6 km)

This ungainly French seaplane was the first to take off from water under its own power, with a top speed of 55 mph (89 kph). Once in the air, the aerofoil-shaped floats produced extra lift.

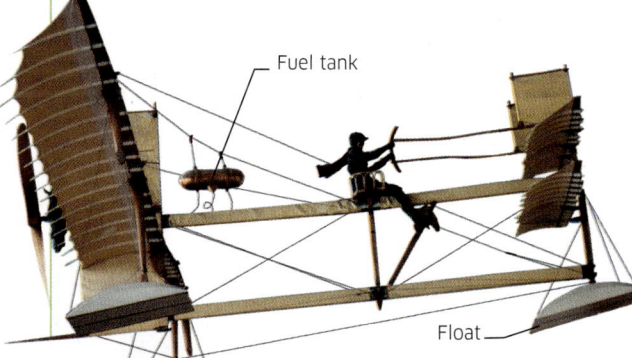

Fuel tank

Float

Shock absorbers
The wheels were braced by bungee cords in case of heavy landings.

Warping wires
Like the Wright Flyer, the XI used wires to control pitch by bending the wings.

Aerial experiments

If a biplane with two pairs of wings can fly, would more wings make an aircraft fly even better? Aviation pioneers attempted to test questions such as this in the 1900s.

French naval engineer the Marquis d'Ecquevilly devised a seven-winged multiplane in 1907. It had a 10-hp engine and balanced on four bicycle wheels. There were no control surfaces, and the pilot had to stand upright in the central cage. It failed to fly and met its end when it was damaged in a fire in 1908.

The *Hindenburg*

The largest rigid-frame airship ever built, the *Hindenburg* flew more than 186,400 miles (300,000 km) in total. It carried a crew of around 50 and sleeping berths for up to 72 passengers. A giant envelope, the length of two-and-a-half soccer fields, was filled with hydrogen gas, giving the airship unparalleled lifting power of 256 tons (232 tonnes).

Luxury travel

Passengers could enjoy panoramic views, gourmet dining, a cocktail bar, and a reading area. Private sleeping cabins were on the main deck, with bathrooms a level below. There was also a smoking room, pressurized to keep out the highly flammable hydrogen!

Axial corridor

The crew used a metal walkway to pass through the airship from nose to tail. The ladders were up to 62 ft (19 m) high!

Triple skin

Two layers of cotton canvas sandwiched a layer of rubberized canvas.

Cool cover

Aluminum powder in the coating reflected sunlight to stop the gas bags from overheating.

Mooring point

A steel cable tied the airship to a tall docking mast.

Panoramic windows

Dining room

25 private cabins for two

Lower deck held bathrooms, bar, and smoking room

Steps folded down for boarding

Furniture was made of aluminum alloy to save weight—even the piano!

Crew quarters

Some crew slept near the control car, and others just behind the passenger area.

Navigation area held charts, clocks, and an altimeter (height gauge)

Observation area housing the airship's magnetic compass

Control area with steering, ballast, and gas valve controls

Control car

The airship was flown from this gondola, divided into observation, navigation, and control sections. Two ship's wheels controlled the tail's rudders and elevators to steer.

Passenger quarters

On either side of the private sleeping cabins were grand public areas, including a lounge, writing room, and promenades with seats offering glorious views and fresh air from the windows.

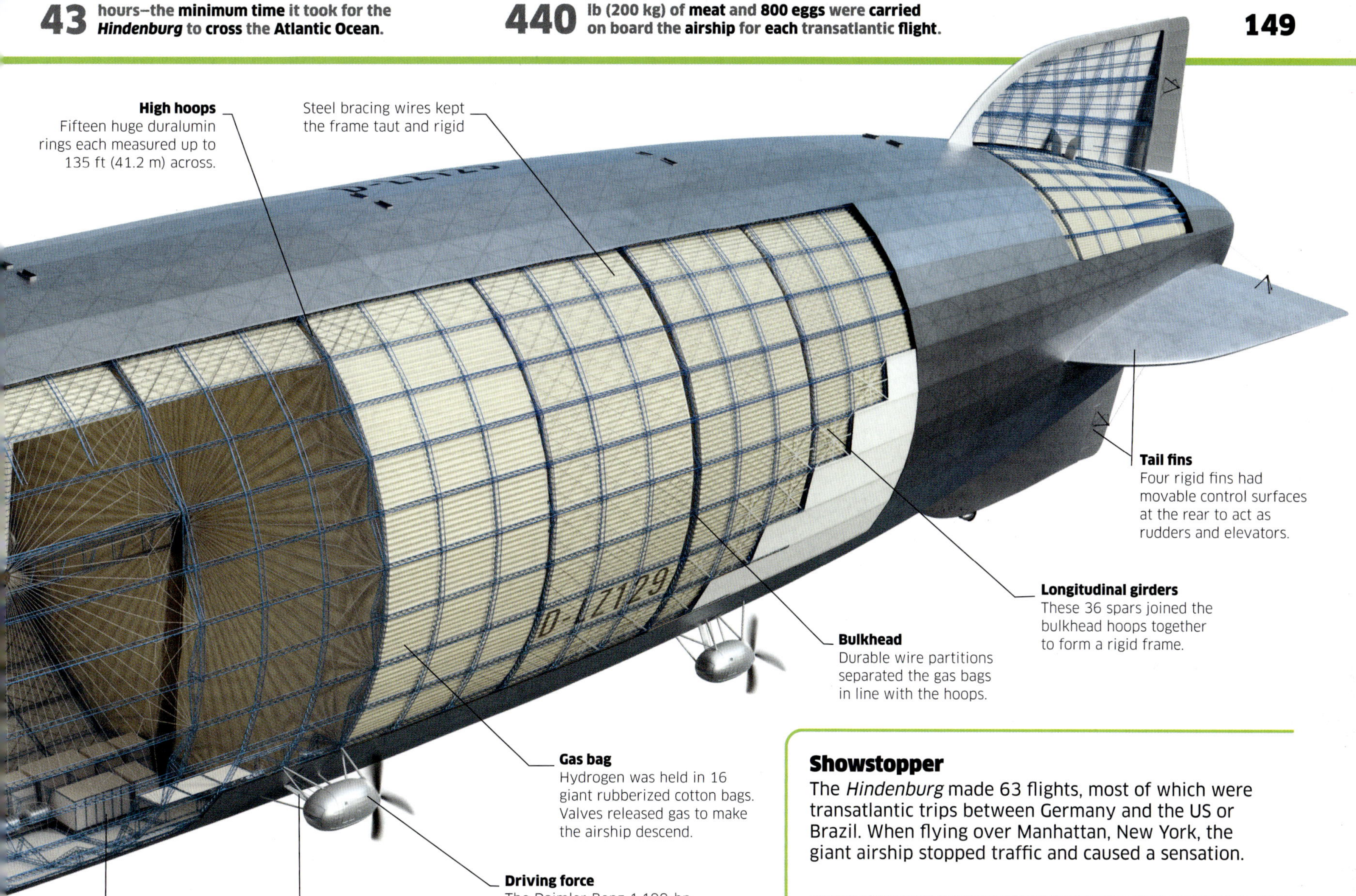

43 hours–the minimum time it took for the *Hindenburg* to cross the Atlantic Ocean.

440 lb (200 kg) of meat and 800 eggs were carried on board the airship for each transatlantic flight.

149

High hoops
Fifteen huge duralumin rings each measured up to 135 ft (41.2 m) across.

Steel bracing wires kept the frame taut and rigid

Tail fins
Four rigid fins had movable control surfaces at the rear to act as rudders and elevators.

Longitudinal girders
These 36 spars joined the bulkhead hoops together to form a rigid frame.

Bulkhead
Durable wire partitions separated the gas bags in line with the hoops.

Gas bag
Hydrogen was held in 16 giant rubberized cotton bags. Valves released gas to make the airship descend.

Driving force
The Daimler-Benz 1,190-hp diesel engine spun a four-bladed propeller. A mechanic in each of the four engine cars kept them running smoothly.

Don't look down!
Crew reached the engine car via an open-air walkway.

Cargo stores
Priority post and cargo between Europe and the Americas was sent by airship for speed.

LZ 129 *HINDENBURG*

Origin: Germany

Year: 1936

Length: 803¾ ft (245 m)

Top speed: 85 mph (135 kph)

Showstopper

The *Hindenburg* made 63 flights, most of which were transatlantic trips between Germany and the US or Brazil. When flying over Manhattan, New York, the giant airship stopped traffic and caused a sensation.

Fiery end

Following a 1937 flight from Germany to New Jersey, the *Hindenburg* burst into flames, killing 36 people. Film and radio reports of the disaster shocked the world–the golden age of passenger airships was over.

Airship

Airships are self-propelled aircraft filled with gases such as hydrogen or helium, which make them lighter than air. They may have a rigid frame or be frameless blimps.

Large, passenger-carrying airships were at their peak in the late 1920s and 1930s. While airplanes of the time had to fly in short hops with frequent pauses to refuel, airships could stay aloft for days at a time, enabling nonstop travel over oceans at a speed far faster than ships. They were also the comfortable option, avoiding rough seas.

150 air and space • WORLD WAR I AIRCRAFT

50 The number of **support crew** needed to handle a giant Staaken R. VI on the ground.

CAUDRON G.3
Reconnaissance biplane
Top speed: 66 mph (106 kph)
Length: 21 ft (5.4 m)

This basic French plane initially lacked ailerons, instead using wing-warping like the Wright Brothers' pioneering planes. It was stable in flight but slow and vulnerable to attack.

LUFT-VERKEHRS-GESELLSCHAFT (LVG) C. VI
Artillery-spotting two-seater
Top speed: 103 mph (165 kph)
Length: 24½ ft (7.5 m)

This German aircraft carried front and rear machine guns and an early radio transmitter, which was used to send back details of enemy positions using Morse code. Some 1,100 were built.

BRÉGUET BR. M5
Escort fighter and night-bomber
Top speed: 83 mph (136 kph)
Length: 32½ ft (9.9 m)

This French two-seater was armed with both a 1.5-in (37-mm) cannon and a machine gun. It could be equipped with a searchlight and pressed into service on perilous nighttime bombing missions.

World War I aircraft

World War I (1914–1918) triggered advances in aviation. Aircraft went from fragile curiosities to more reliable, practical craft with many roles.

Early wartime planes were mostly used for observation—their only weapon was a pistol carried by the pilot! As engines increased in power and airframes in strength, bigger aircraft able to travel longer distances could drop bombs on enemy territory and soldiers. Faster, more nimble fighter planes were developed to combat these and to battle each other for control of the skies.

ALBATROS D.VA
Streamlined fighter plane
Top speed: 116 mph (186 kph)
Length: 24 ft (7.3 m)

Introduced in 1917, this slimline biplane's 180-hp engine and advanced aerodynamics had the Allies scrambling to catch up. It was flown with great success by many of Germany's top aces.

German insignia— the "Iron Cross"

Eyes front
The pilot sat in a wooden tub cockpit at the nose with a clear view ahead.

Tail skid—less weight and drag than a rear wheel

AIRCO D.H.2
"Pusher" fighter plane
Top speed: 93 mph (150 kph)
Length: 25¼ ft (7.7 m)

To counter the Fokker Eindecker, this British single-seater was designed with the engine and propeller behind the pilot to let them aim and fire straight ahead without striking the blades. Angled guns were hard to aim during dogfights.

Wheels down
Fixed wheels caused drag, but retractable landing gear was rare until the 1930s.

Some **early pilots** used grappling hooks, grenades, or **pistols to attack** enemy aircraft!

The **Sopwith Pup** fighter plane weighed less than a grand piano!

67,987 The number of aircraft produced by France during the war, more than any other nation.

151

ZEPPELIN LZ 96
High-altitude bomber and spy
Top speed: 66 mph (107 kph)
Length: 644⅜ ft (196.4 m)

This U-class German airship was made lightweight enough to fly higher than enemy aircraft. It carried 5 engines, 19 crew, and more than 4,400 lb (2 tonnes) of explosives and could stay aloft for up to 100 hours.

Zeppelins held hydrogen in huge gas bags made from **cow intestines**. A single airship required **250,000 cows!**

The outer skin of an airship was called the envelope

Control gondola
The pilot and crew operated the Zeppelin from the main cabin toward the front.

Rear engine gondola
Each gondola held an engine and a propeller—fuel was supplied from tanks above.

Rudders
Steering was operated via cables from the control gondola 500 ft (150 m) away.

ROYAL AIRCRAFT FACTORY S.E.5A
Speedy fighter
Top speed: 138 mph (222 kph)
Length: 21 ft (6.4 m)

Fast, stable, and capable at higher altitudes, the British S.E.5a was designed to counter the German Albatross. Its narrow fuselage ended in aluminum panels to protect the engine, and all four wood and cotton wings had ailerons for swift maneuvering.

Aileron hinge
Around one third of the wing surface was movable.

FOKKER EINDECKER
Straight shooter
Top speed: 87 mph (140 kph)
Length: 23⅝ ft (7.2 m)

This German fighter dominated early aerial combat thanks to its interrupter gear, which enabled gunfire straight ahead—through the propeller arc. Until then, pilots often shot their own blades off. Faster planes superseded it by late 1916.

SOPWITH BABY
Naval scout and airship hunter
Top speed: 100 mph (160 kph)
Length: 23 ft (7 m)

Carried on ships, this British floatplane scouted the waters ahead for German sea vessels and airships. Some were armed with rocket-powered explosive darts and small bombs to attack the airships.

Hollow wooden floats for landing on water

CAPRONI CA.3
Italian heavy bomber
Top speed: 85 mph (137 kph)
Length: 36¼ ft (11.1 m)

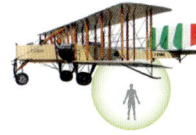

Two wings and three engines—one rear-facing—carried a crew of four and up to 1,764 lb (800 kg) of bombs. The rear gunner doubled as the plane's mechanic, seated near the spinning rear propeller.

ZEPPELIN STAAKEN R. VI
Giant heavy bomber
Top speed: 84 mph (135 kph)
Length: 72½ ft (22.1 m)

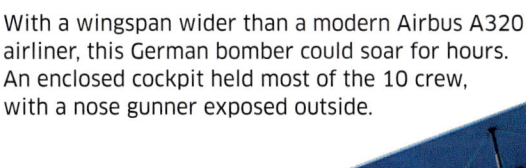

With a wingspan wider than a modern Airbus A320 airliner, this German bomber could soar for hours. An enclosed cockpit held most of the 10 crew, with a nose gunner exposed outside.

138½-ft (42.2-m) wingspan

Rear gunner position

Pilot and copilot at front

Triple tail helped control the heavy bomber

Pilot sat in wicker chair directly in front of fuel tanks

Cables stopped frame bending during turns

Guns fired between propeller blades

Air intake supplied clean, cool air to engine

Fuel tanks

Synchronization gear

A gearing system, driven by the engine, prevented the machine guns from firing when the propeller passed in front of the gun barrels. This stopped the pilot from shooting their own plane's propeller off. Before this invention, some pilots taped the blades to limit splintering and hoped for the best!

Twin Vickers machine guns fired 0.303" rounds

Rotary engine consumed 12 gallons (45 liters) of fuel per hour

Propeller made of laminated layers of wood

Cartridge case ejection chute

Spokes

Tire inflation valve

Sopwith Camel

Unstable and difficult to fly but deadly in the hands of skilled pilots, the Camel was the Allied forces' most successful fighter plane. Camel pilots claimed to have downed 1,294 enemy aircraft in less than 18 months of action, and more than 5,000 Camels were produced in total.

Fabric covers protected wheel spokes from catching on debris.

British insignia to reduce risk of "friendly fire"

SOPWITH CAMEL

Top speed: 115 mph (185 kph)

Wingspan: 28 ft (8.53 m)

Empty weight: 929 lb (422 kg)

Crew: 1

Clerget 9b rotary engine

The cylinders of a rotary engine are set in a circle around a crankshaft and spin along with the propeller. The spinning engine pulled the Camel rightward so that it could turn sharply to the right but only slowly to the left, a trait that both pilots and enemies learned to exploit.

The **Sopwith Camel** was **named** for the **humped fairing** over its **machine guns**.

80 The number of aircraft **downed** by the Red Baron.

153

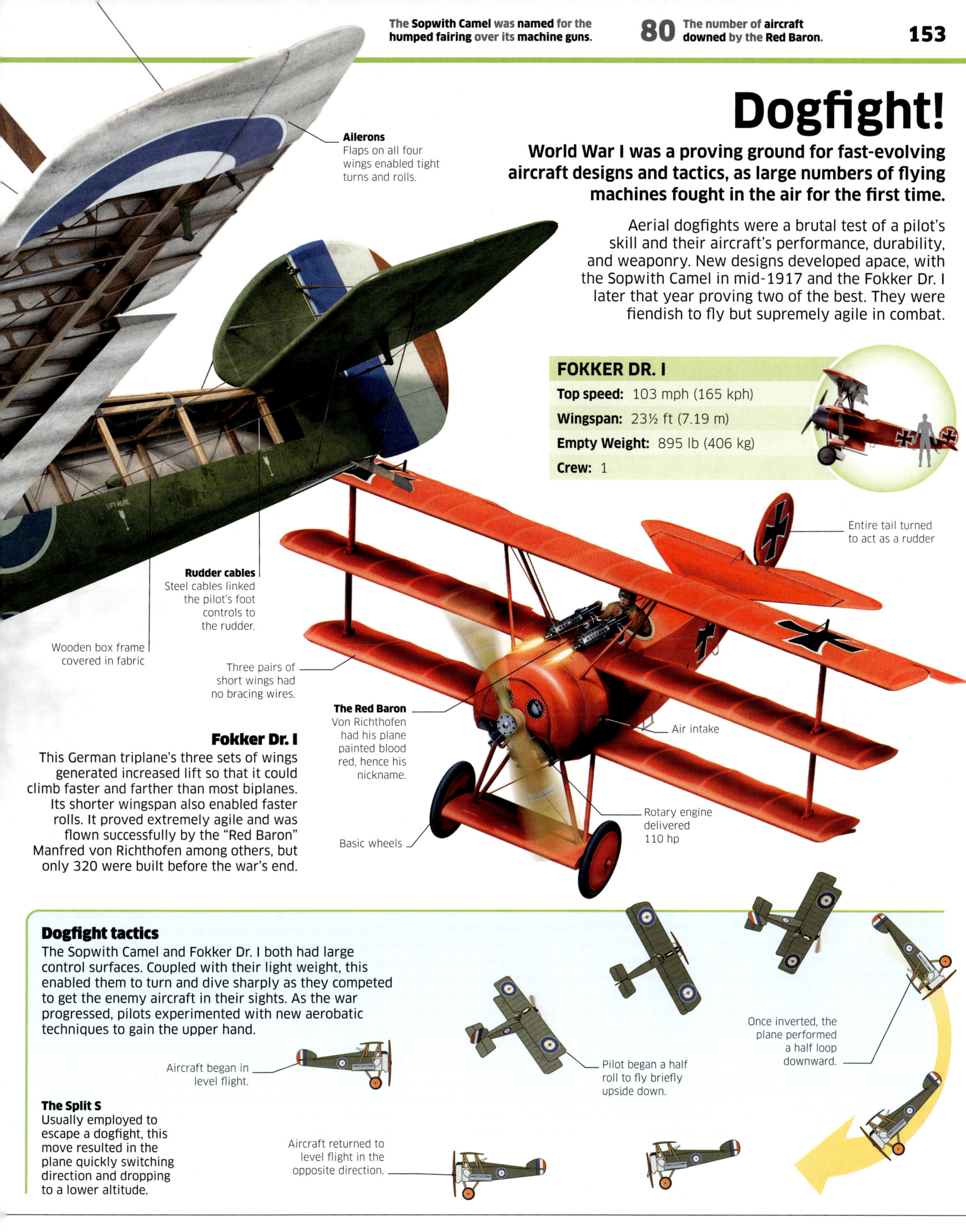

Dogfight!

World War I was a proving ground for fast-evolving aircraft designs and tactics, as large numbers of flying machines fought in the air for the first time.

Aerial dogfights were a brutal test of a pilot's skill and their aircraft's performance, durability, and weaponry. New designs developed apace, with the Sopwith Camel in mid-1917 and the Fokker Dr. I later that year proving two of the best. They were fiendish to fly but supremely agile in combat.

FOKKER DR. I

Top speed: 103 mph (165 kph)

Wingspan: 23½ ft (7.19 m)

Empty Weight: 895 lb (406 kg)

Crew: 1

Ailerons
Flaps on all four wings enabled tight turns and rolls.

Rudder cables
Steel cables linked the pilot's foot controls to the rudder.

Wooden box frame covered in fabric

Three pairs of short wings had no bracing wires.

The Red Baron
Von Richthofen had his plane painted blood red, hence his nickname.

Entire tail turned to act as a rudder

Air intake

Rotary engine delivered 110 hp

Basic wheels

Fokker Dr. I

This German triplane's three sets of wings generated increased lift so that it could climb faster and farther than most biplanes. Its shorter wingspan also enabled faster rolls. It proved extremely agile and was flown successfully by the "Red Baron" Manfred von Richthofen among others, but only 320 were built before the war's end.

Dogfight tactics

The Sopwith Camel and Fokker Dr. I both had large control surfaces. Coupled with their light weight, this enabled them to turn and dive sharply as they competed to get the enemy aircraft in their sights. As the war progressed, pilots experimented with new aerobatic techniques to gain the upper hand.

The Split S
Usually employed to escape a dogfight, this move resulted in the plane quickly switching direction and dropping to a lower altitude.

Aircraft began in level flight.

Pilot began a half roll to fly briefly upside down.

Once inverted, the plane performed a half loop downward.

Aircraft returned to level flight in the opposite direction.

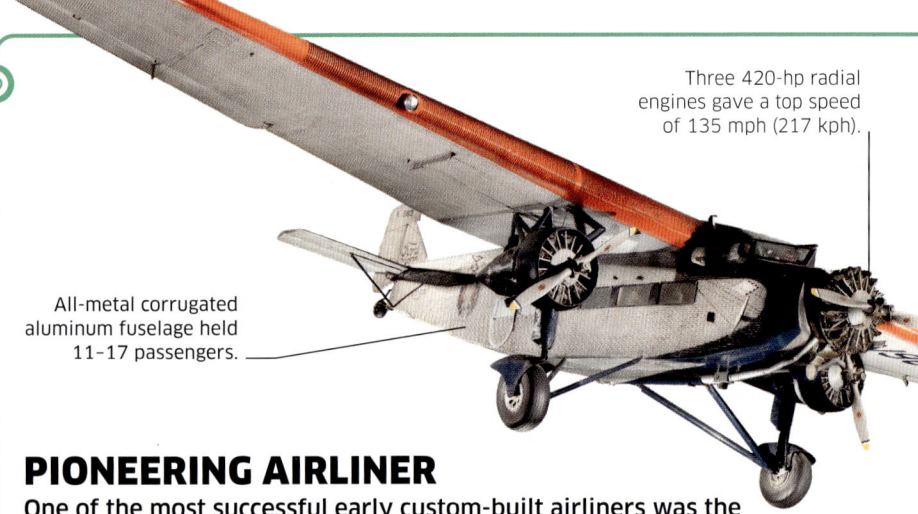

Three 420-hp radial engines gave a top speed of 135 mph (217 kph).

All-metal corrugated aluminum fuselage held 11–17 passengers.

Growth of civil aviation

A mere handful of people had flown as passengers before World War I. Advances in airplane design and technology during the war led to a boom in civil aviation.

A surplus of wartime military pilots and planes helped propel new, peacetime uses for aircraft as airmail carriers, passenger transporters, and cropdusters for spraying farm fields with pesticides. Daring feats, barnstorming displays, and historic flight firsts captured the public imagination and helped boost demand for civil air transportation. The first passenger airlines formed and new, larger civilian aircraft were built in numbers for the first time.

PIONEERING AIRLINER

One of the most successful early custom-built airliners was the Ford 5-AT Trimotor. Simple to repair, robust, and reliable (for the time), the "Tin Goose" first flew in 1926, and some of the 199 models built saw more than 50 years of service.

WAYFINDING BEACONS

After World War I, the US developed a far-reaching airmail service across the country. Planes flew only during the day, handing their cargo over to trains at night, until a beacon network was established in the mid-1920s. The Transcontinental Airway System featured powerful lights 10 miles (16 km) apart, guiding night fliers along the route. It was the world's first ground-based air navigation system.

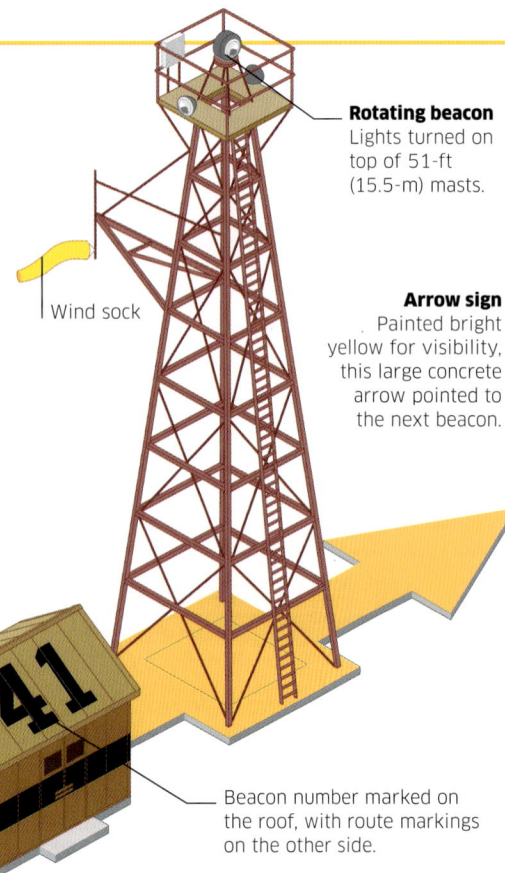

Rotating beacon
Lights turned on top of 51-ft (15.5-m) masts.

Wind sock

Arrow sign
Painted bright yellow for visibility, this large concrete arrow pointed to the next beacon.

Electrical generator
Next to each beacon was a shed housing a generator. An outside tank held 515 gallons (1,950 liters) of fuel to power it.

Beacon number marked on the roof, with route markings on the other side.

ACROSS THE ATLANTIC

The first nonstop flight across the Atlantic Ocean was made in a Vickers Vimy bomber in 1919. As aircraft advanced, so transatlantic passenger flights became possible, first by airship in 1928, then by flying boats in the 1930s, with stops along the way. The arrival of jet airliners—with longer ranges and no need to refuel en route—sped up flight times greatly, culminating in supersonic transatlantic travel.

Graf Zeppelin
The first nonstop transatlantic service carried 24 passengers on a 111-hour journey from Germany to the United States aboard this giant airship.

Boeing Clipper
This large flying boat (see pages 156–157) ferried 22 passengers from New York to Marseille, France. The trip took 42 hours, including 30 hours of flying time.

Douglas DC-4
This piston-engined aircraft for 44 passengers brought the westward flight time across the Atlantic down to a scheduled 17 hours, 40 minutes.

Boeing 707
Powered by four jet engines, this airliner flew the first scheduled jet service across the Atlantic in 1958. Paris to New York flights took under seven hours.

Concorde
The supersonic airliner flew from London to New York in just 2 hours, 52 minutes, and 59 seconds, an average speed of 1,249 mph (2,010 kph).

BARNSTORMERS

In the 1920s, pilots thrilled crowds across the US with "barnstorming" flying displays, stunts, and races. Daredevil performers walked on wings or even jumped between aircraft in midair. The shows boosted interest in aviation by showing off the increasing agility and sturdiness of planes, while thousands of people paid a small fee for a "joyride" to experience their first, exhilarating taste of flight.

Anyone for tennis?
Wing walkers Gladys Roy and Ivan Unger play tennis on the top wing of a Curtiss JN-4 Jenny biplane in 1925.

120 decibels—the **loudness of a Ford Tri-Motor** during **take-off**, loud enough to cause some **hearing damage**.

12½ days—the journey time of the route from England to Australia, run by Imperial Airways and Qantas Empire Airways in 1935.

155

AIRFIELDS TO AIRPORTS

Early air travel saw planes take off from strips of mown grass. As aircraft became larger and heavier, and passengers more numerous, concrete and tarmac runways were constructed, as were buildings, to handle the flow of freight and passengers. Control towers to guide air traffic, maintenance hangars, and other infrastructure arrived, turning basic airfields into busy airports.

Croydon Airport, UK
Just south of London, this early airport featured the first air traffic control tower in the world in 1922. It also had the first dedicated air terminal in the UK, handling 26,000 passengers in the year it opened (1928). It closed in 1959.

Changi Airport, Singapore
The world's tallest indoor waterfall flows at Singapore's ever-expanding international airport. From its four terminals, 99 airlines operate more than 6,300 flights carrying over 1 million passengers each week.

AMELIA EARHART

This pioneering US pilot was the first woman to fly the Atlantic solo and the US coast-to-coast (both in 1932) and held many other aviation records. Seen here in the cockpit of a Stearman Hammond Y-1 aircraft, she was an international celebrity who inspired women to fly. She went missing in the Pacific in 1937 on an attempted around-the-world flight.

ROUTE-FINDING

Early airlines such as KLM and Imperial Airways sent experienced pilots to map out potential air routes before commencing commercial services. With the limited range of early passenger aircraft, longer flights could only be completed as a series of shorter "hops" across a continent. Flights were desperately slow compared to modern standards, but far faster than the many weeks or months it took to make the same journey over land or by sea.

Global travel

Early airlines publicized their routes with posters marketing the adventure of travel to places few customers had been before. Posters showcased the speed of travel and the modernity of their aircraft.

KLM poster, 1930s
KLM emphasized their fast planes and extensive networks.

Pan Am poster, 1938
A plane flies over the ancient ruins of Machu Picchu in Peru.

Flying long distance

In the 1930s, air passengers could reach faraway places, as shown on this map of the longest scheduled air routes of the time. KLM's itinerary from Amsterdam in the Netherlands to Jakarta in Indonesia, for example, took 10 days and involved 19 separate flights with a total of 81 hours in the air. A return ticket cost as much as a new car at the time.

Key to 1930s air routes

- Aéropostale: Toulouse–Punta Arenas (1930)
- Imperial Airways: African Route (c.1934)
- QANTAS Empire Airways: Australian Route (c.1934)
- KLM: Amsterdam–Jakarta (1935)
- Pan American: Transatlantic Route (1939)

3,685 miles (5,930 km)—the **Clipper's** normal **range**, enough to **cross the Atlantic in a single flight.**

Making a splash

A Clipper glides to a halt on the water. Early tests found problems with "porpoising" (repeatedly bouncing back up into the air after touchdown) until the hull shape was redesigned. The heavy plane, with a crew of 11, required a 152-ft (46-m) wingspan to generate enough lift to get airborne.

Ocean traveler

The Boeing 314 Clipper could carry 68 daytime passengers or 38 long-distance travelers in style, with a lounge, fine dining, dressing rooms, and bedrooms. Twelve Clippers were built, making 5,000 transatlantic trips and covering some 12.5 million miles (20.1 million km) before Pan Am retired the fleet in 1946.

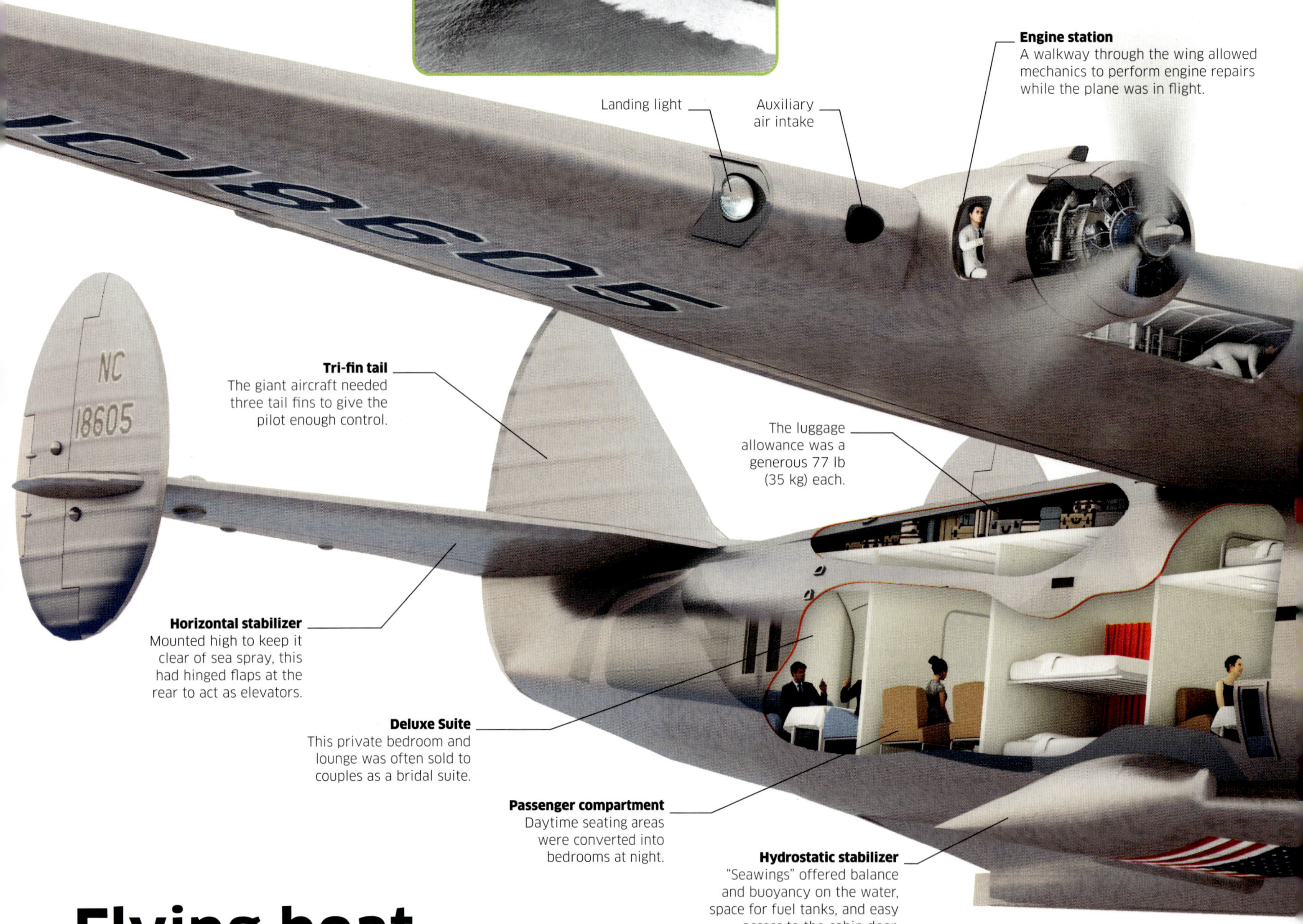

Engine station

A walkway through the wing allowed mechanics to perform engine repairs while the plane was in flight.

Landing light

Auxiliary air intake

Tri-fin tail

The giant aircraft needed three tail fins to give the pilot enough control.

The luggage allowance was a generous 77 lb (35 kg) each.

Horizontal stabilizer

Mounted high to keep it clear of sea spray, this had hinged flaps at the rear to act as elevators.

Deluxe Suite

This private bedroom and lounge was often sold to couples as a bridal suite.

Passenger compartment

Daytime seating areas were converted into bedrooms at night.

Hydrostatic stabilizer

"Seawings" offered balance and buoyancy on the water, space for fuel tanks, and easy access to the cabin door.

Flying boat

Landing on and taking off from water, flying boats can float on their deep, buoyant fuselages, known as hulls. Versatile, capacious, and powerful, these planes set a new standard in luxury air travel.

In the 1930s, growing demand for long-distance flights but limited fuel capacity meant planes had to make multiple short hops, and a lack of airports hampered some routes. Flying boats, which could land on any stretch of calm water, proved a solution. The Boeing Clipper, built for the Pan Am airline, was the largest civilian aircraft in the world when first introduced in 1938, and remains an iconic early plane.

All aboard!

Passengers would arrive by boat or walk along a floating platform to board the aircraft, using the plane's hydrostatic stabilizer as a gangway. Among those who embarked was Franklin D. Roosevelt, who during World War II became the first US President to fly while in office.

209 hours—the **time it took a Clipper** to complete the first around-the-world journey **by a commercial aircraft.**

During **World War II**, Boeing Clippers flew **uranium** from **South Africa** to the US for its **atomic weapons** program.

157

Flight deck
The two pilots had identical setups, with control columns shaped like steering wheels. Behind was a large radio and navigation room, blacked out so that all the instruments could be viewed clearly.

Expanding horizons

With an unrivaled range, Clippers forged regular air routes across the Atlantic and into the Pacific for the first time. Three were sold to a British airline, BOAC, to fly between the UK and Africa. Trips that usually took weeks could now be done in a day, and far-flung destinations were within reach. Dreams of flying between continents captured the public imagination.

Wright Twin Cyclone engine produced 1,600 hp.

Radio direction finder antenna for navigation

Bridge
The pilot used throttles to control the engine power and air speed to cruise at 188 mph (303 kph).

A dark upper panel stopped glare from the sea dazzling the pilot.

Anchor and gear room
Downstairs from the bridge was a store for ropes and equipment to moor the aircraft on water.

Bolstered bow
The front of the aircraft's hull was strengthened for the impact of landing on water.

Crew bunks
The crew slept in shifts in cramped bunk rooms.

Galley kitchen
Staff served regular meals and could also host a cocktail bar.

Fuel pump
3,525 gallons (13,340 liters) of fuel were brought up from the seawings during a flight.

A glass dome
on the Clipper helped the crew to navigate by the
stars at night.

BOEING 314 CLIPPER

Origin: US

Year: 1938

Role: Long-range flying boat

Length: 106 ft (32 m)

158 air and space ○ **WORLD WAR II AIRCRAFT**

300,000 The number of US military aircraft built during the war.

World War II aircraft

From the Battle of Britain and mass bombing raids over Europe to the dropping of two atomic bombs on Japan, the war in the air was a crucial and hotly contested part of World War II (1939–1945).

War accelerated progress in military aircraft design and engineering. Planes rapidly increased in speed, range, and payload and were manufactured in their tens of thousands. New materials, advances in avionics (electronic aids), and widespread use of radar would continue to have a great impact in the decades after the war.

MITSUBISHI A6M ZERO

Carrier-based fighter

Length: 29¾ ft (9.06 m)
Top speed: 331 mph (533 kph)

This lightweight Japanese fighter proved fearsome in traditional dogfights due to its outstanding maneuverability.

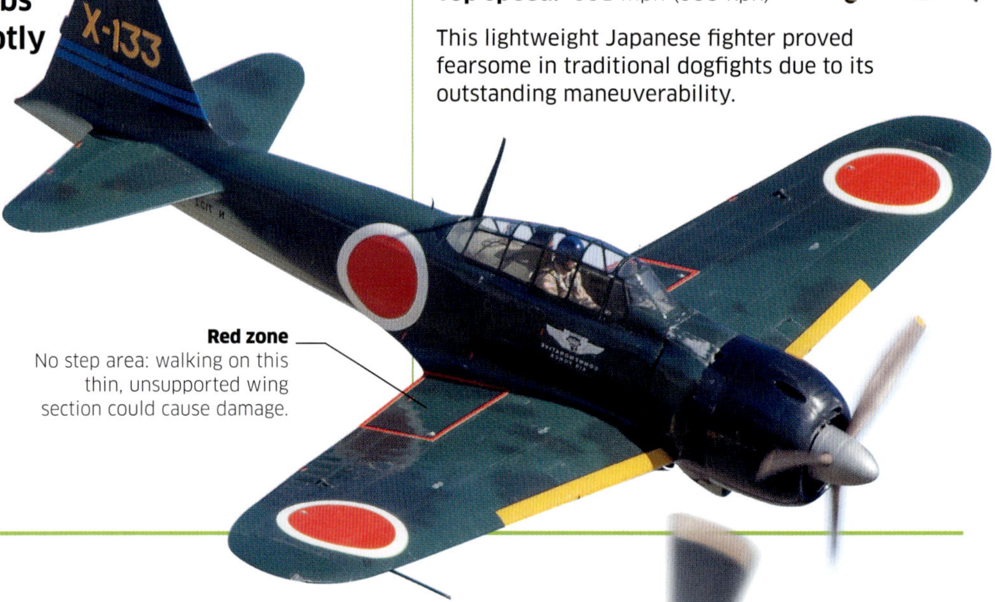

Red zone
No step area: walking on this thin, unsupported wing section could cause damage.

NORTH AMERICAN P-51D MUSTANG

Fighter

Length: 32¼ ft (9.83 m)
Top speed: 437 mph (703 kph)

Flown mostly as a combat fighter or an escort to bomber squadrons, this powerful US fighter possessed an unrivaled range, with disposable "drop tanks" enabling flights up to 1,650 miles (2,660 km). It first entered service in 1942.

Lightweight body
The aluminum fuselage weighed 7,635 lb (3,463 kg).

Cool running
The air scoop channeled cool air over the radiator and produced additional thrust.

Four-bladed variable pitch propeller

MESSERSCHMITT BF 109

Fighter, reconnaissance, and ground attack

Length: 29¼ ft (8.95 m)
Top speed: 386 mph (621 kph)

Hinged rudder on rear of tail

Key supplies
Behind the cockpit were a fuel tank, oxygen cylinders, and a radio pack.

The German Air Force's main fighter plane was easy to construct and maintain, though some variants had a short range and poor sightlines. Over 34,000 of these versatile single-seaters were produced.

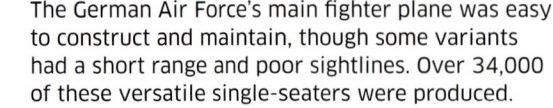

VOUGHT F4U CORSAIR

Carrier-based fighter

Length: 33¾ in (10.26 m)
Top speed: 417 mph (671 kph)

Powered by a large, 18-cylinder radial engine, this fast, high-performance American fighter featured folding wings for storage on ships. It could turn sharply and climb 4,337 ft (1,322 m) per minute.

The **Short S.25** had a **flushing porcelain toilet** for its crew!

The **Ju87's** two sirens **shrieked** as the plane **dived** to frighten soldiers below.

1984 The **year** in which the Dominican Air Force **retired** its **last P51 Mustang**.

159

AVRO LANCASTER
Heavy bomber

Length: 69 ft (21 m)
Top speed: 282 mph (454 kph)

With a weapons bay almost half the length of its body, the RAF's most successful bomber made 156,000 wartime sorties, including the famous Dambusters Raid. Tough and robust, a Lancaster could continue flying with just two of its four engines active.

Driving force
Rolls Royce V12 Merlin engines spun metal propellers 13 ft (4 m) in diameter.

Flight crew
Pilot and flight engineer sat side by side with navigator and radio operator behind. In the nose was the bomb aimer.

Mid gunner could swivel 360 degrees.

Rear gunner
The hydraulically powered rear gun turret held four machine guns.

Control surfaces
Twin tail fins, each with a rudder, provided stability and control at low speeds.

ILYUSHIN IL-2
Ground attack aircraft

Length: 38¼ ft (11.65 m)
Top speed: 257 mph (414 kph)

The Soviet Union built more than 36,000 of these rugged two-seater aircraft. Its armor-plated cockpit and front fuselage earned it the nickname "the flying tank," and it played a key role on the Eastern Front.

Body made of wood and metal

DE HAVILLAND MOSQUITO
Fighter-bomber

Length: 44½ ft (13.56 m)
Top speed: 415 mph (668 kph)

Based on a light wooden frame covered in fabric, Britain's Mosquito was the fastest fighter-bomber of its time. When loaded as a bomber, it flew without guns, using speed and agility to avoid attack.

BOEING B-17 FLYING FORTRESS
Heavy bomber

Length: 74½ ft (22.7 m)
Top speed: 287 mph (462 kph)

Named for its 13 defensive machine guns, the Flying Fortress carried a crew of 10 and had a range of 2,000 miles (3,220 km), with broad wings for high-altitude flight. The US built 12,700 of them.

SHORT S.25 SUNDERLAND
Maritime patrol and bomber

Length: 85½ ft (26 m)
Top speed: 211 mph (340 kph)

This British flying boat was based on the S.23 passenger plane. Its deep hull held two decks, 9 to 11 crew, and fuel for 14-hour patrols. It dropped depth charges to disable submarines.

Powered rear gun turret

JUNKERS JU87B "STUKA"
Dive bomber

Length: 36½ ft (11.1 m)
Top speed: 211 mph (340 kph)

This German bomber terrorized Allied forces early in the war, but later its slow cruising speed— 130 mph (210 kph)—made it a sitting duck for enemy fighters.

The rear gunner also operated the radio.

Jointed wing
An inverted gull-wing design gave the pilot a better view of the ground.

Two 110-lb (50-kg) bombs carried under each wing

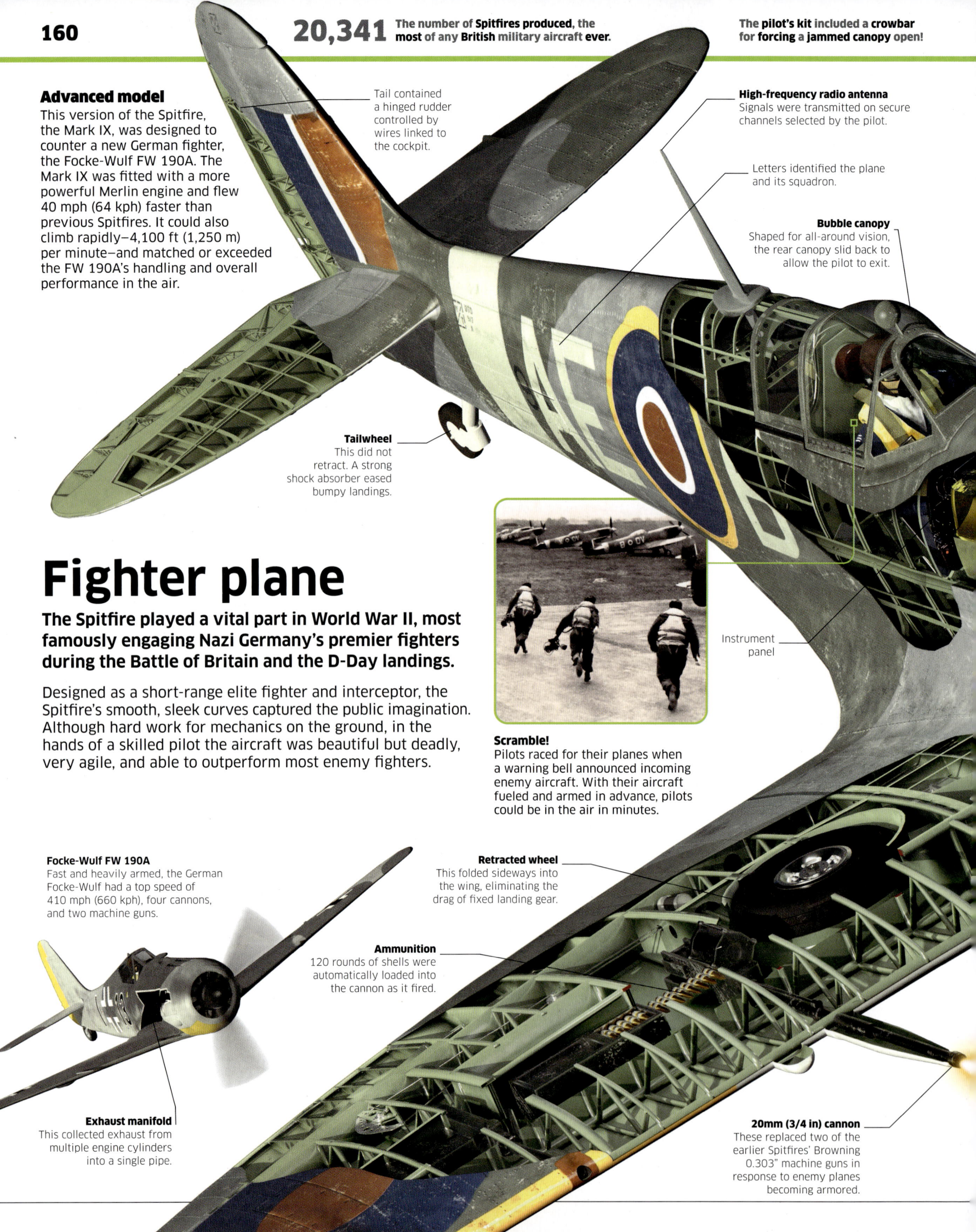

Advanced model

This version of the Spitfire, the Mark IX, was designed to counter a new German fighter, the Focke-Wulf FW 190A. The Mark IX was fitted with a more powerful Merlin engine and flew 40 mph (64 kph) faster than previous Spitfires. It could also climb rapidly—4,100 ft (1,250 m) per minute—and matched or exceeded the FW 190A's handling and overall performance in the air.

Tail contained a hinged rudder controlled by wires linked to the cockpit.

High-frequency radio antenna
Signals were transmitted on secure channels selected by the pilot.

Letters identified the plane and its squadron.

Bubble canopy
Shaped for all-around vision, the rear canopy slid back to allow the pilot to exit.

Tailwheel
This did not retract. A strong shock absorber eased bumpy landings.

Fighter plane

The Spitfire played a vital part in World War II, most famously engaging Nazi Germany's premier fighters during the Battle of Britain and the D-Day landings.

Designed as a short-range elite fighter and interceptor, the Spitfire's smooth, sleek curves captured the public imagination. Although hard work for mechanics on the ground, in the hands of a skilled pilot the aircraft was beautiful but deadly, very agile, and able to outperform most enemy fighters.

Instrument panel

Scramble!
Pilots raced for their planes when a warning bell announced incoming enemy aircraft. With their aircraft fueled and armed in advance, pilots could be in the air in minutes.

Focke-Wulf FW 190A
Fast and heavily armed, the German Focke-Wulf had a top speed of 410 mph (660 kph), four cannons, and two machine guns.

Retracted wheel
This folded sideways into the wing, eliminating the drag of fixed landing gear.

Ammunition
120 rounds of shells were automatically loaded into the cannon as it fired.

Exhaust manifold
This collected exhaust from multiple engine cylinders into a single pipe.

20mm (3/4 in) cannon
These replaced two of the earlier Spitfires' Browning 0.303" machine guns in response to enemy planes becoming armored.

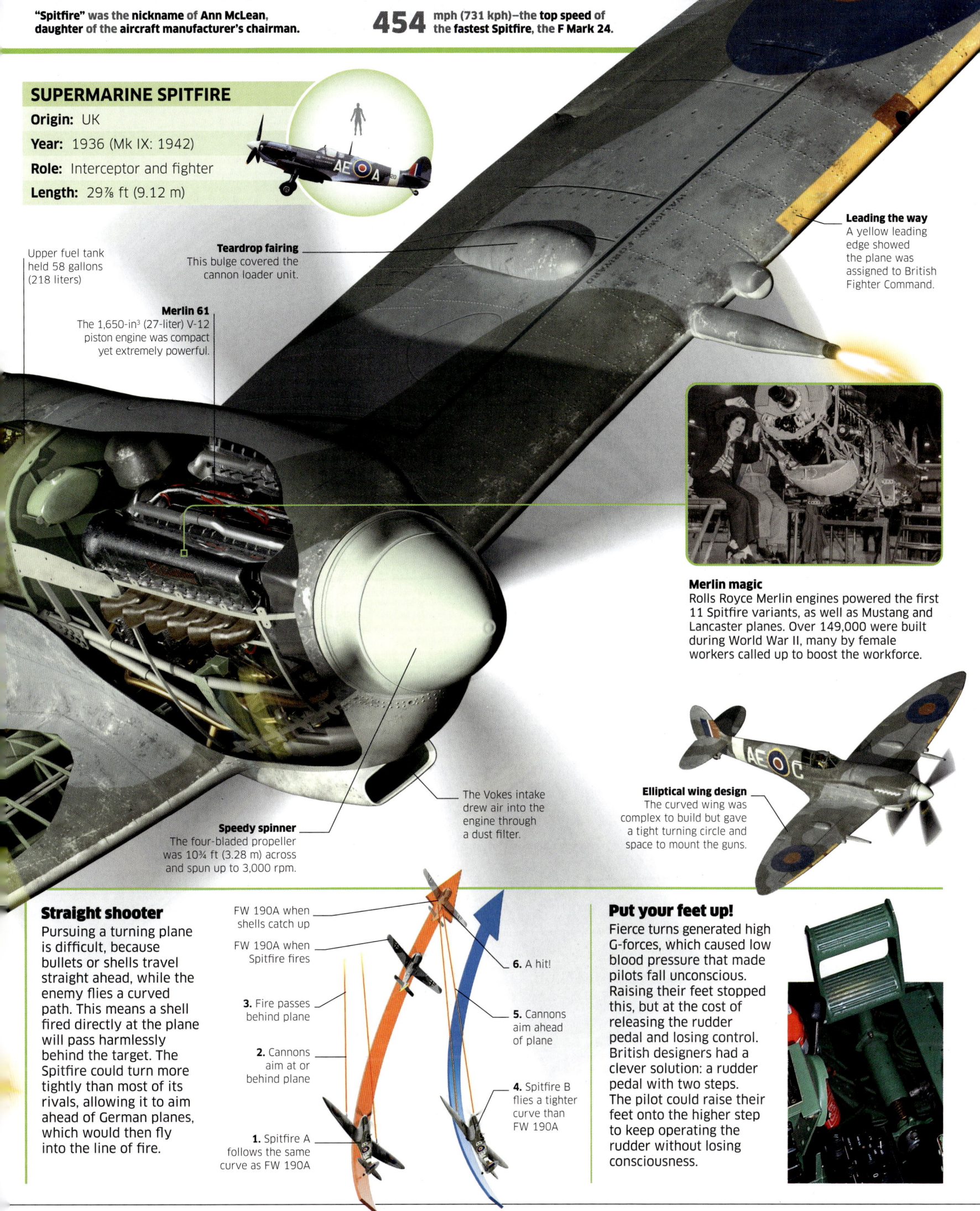

"**Spitfire**" was the nickname of **Ann McLean**, daughter of the aircraft manufacturer's chairman.

454 mph (731 kph)–the **top speed** of the fastest Spitfire, the **F Mark 24.**

SUPERMARINE SPITFIRE

Origin: UK

Year: 1936 (Mk IX: 1942)

Role: Interceptor and fighter

Length: 29⅞ ft (9.12 m)

Upper fuel tank held 58 gallons (218 liters)

Teardrop fairing
This bulge covered the cannon loader unit.

Merlin 61
The 1,650-in³ (27-liter) V-12 piston engine was compact yet extremely powerful.

Leading the way
A yellow leading edge showed the plane was assigned to British Fighter Command.

Merlin magic
Rolls Royce Merlin engines powered the first 11 Spitfire variants, as well as Mustang and Lancaster planes. Over 149,000 were built during World War II, many by female workers called up to boost the workforce.

Speedy spinner
The four-bladed propeller was 10¾ ft (3.28 m) across and spun up to 3,000 rpm.

The Vokes intake drew air into the engine through a dust filter.

Elliptical wing design
The curved wing was complex to build but gave a tight turning circle and space to mount the guns.

Straight shooter
Pursuing a turning plane is difficult, because bullets or shells travel straight ahead, while the enemy flies a curved path. This means a shell fired directly at the plane will pass harmlessly behind the target. The Spitfire could turn more tightly than most of its rivals, allowing it to aim ahead of German planes, which would then fly into the line of fire.

FW 190A when shells catch up

FW 190A when Spitfire fires

3. Fire passes behind plane

2. Cannons aim at or behind plane

1. Spitfire A follows the same curve as FW 190A

6. A hit!

5. Cannons aim ahead of plane

4. Spitfire B flies a tighter curve than FW 190A

Put your feet up!
Fierce turns generated high G-forces, which caused low blood pressure that made pilots fall unconscious. Raising their feet stopped this, but at the cost of releasing the rudder pedal and losing control. British designers had a clever solution: a rudder pedal with two steps. The pilot could raise their feet onto the higher step to keep operating the rudder without losing consciousness.

162 air and space ○ **JET ENGINES**

1.4 tons (1.3 tonnes) of **air** is **sucked into** a **Rolls Royce Trent XWB engine** every second during flight.

Jet engines

All aircraft were powered by piston engines spinning propellers until the invention of jet engines. These new power units revolutionized military and civil aviation.

The arrival of jet engines not only propelled military aircraft to far greater speeds than before, but made them much more reliable. They possessed fewer moving parts than piston engines and proved more fuel-efficient at high speeds. Airliners with jet engines could be built larger to carry more passengers and fly farther distances than before. This slashed the cost of flying and saw airlines boom. In a year, some 4 billion passengers travel by plane, the vast majority in jet aircraft.

ACTION-REACTION

Jet engines work according to Newton's Third Law, which states that for every action, there is an equal and opposite reaction. Around 50 CE, Heron of Alexandria demonstrated this principle with his *aeolipile*.

Water was heated over a fire in a metal boiler, producing steam. This rose through pipes into a hollow metal sphere with two curved spouts on opposite sides. The rising pressure in the sphere caused jets of steam to leave the spouts. These jets resulted in an equal but opposite reaction, causing the sphere to spin around in the opposite direction from the exiting steam.

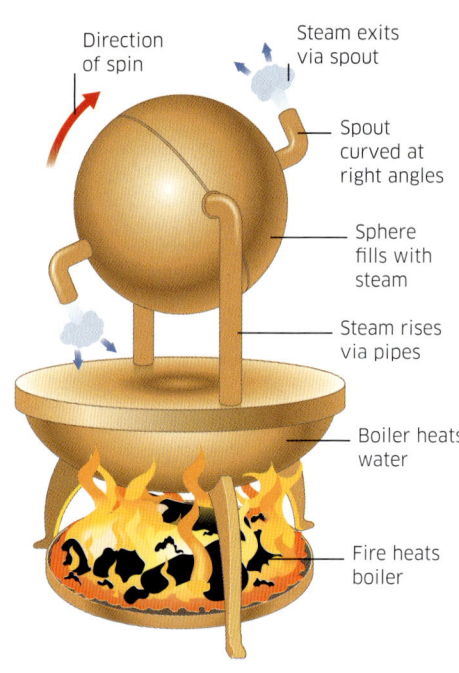

Direction of spin

Steam exits via spout

Spout curved at right angles

Sphere fills with steam

Steam rises via pipes

Boiler heats water

Fire heats boiler

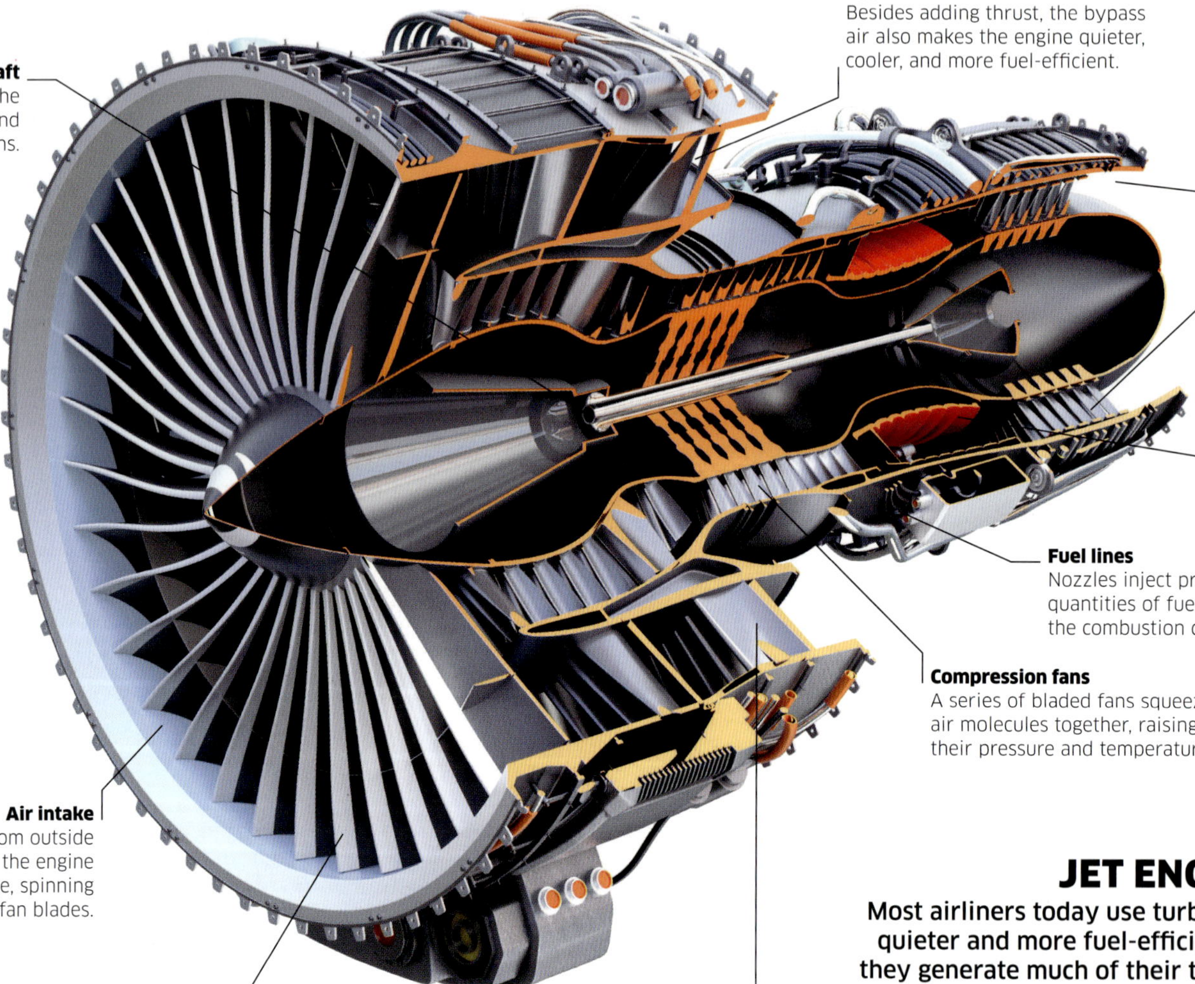

Drive shaft
This turns the compression and duct fans.

Bypass duct
Besides adding thrust, the bypass air also makes the engine quieter, cooler, and more fuel-efficient.

Exhaust nozzle
Gases exit the engine through this opening in the rear, at up to 1,243 mph (2,000 kph).

Low-pressure turbine
The expanding hot gases turn this turbine, which powers the drive shaft that turns the compression fans and duct fans.

Combustion chamber
Air is mixed with aviation fuel and ignited to burn fiercely, producing rapidly inflating gases.

Fuel lines
Nozzles inject precise quantities of fuel into the combustion chamber.

Compression fans
A series of bladed fans squeeze air molecules together, raising their pressure and temperature.

Air intake
Cold air from outside enters the engine via large, spinning fan blades.

Fan blades
These are made from complex metal alloys or composite materials featuring carbon fiber.

Duct fan
The duct fan speeds up cold air traveling through the bypass ducts.

JET ENGINE UNCOVERED

Most airliners today use turbofan jet engines. These are quieter and more fuel-efficient than turbojets, because they generate much of their thrust using bypass air. This means that only a small amount of the air entering the engine goes through the engine core to be compressed, mixed with fuel, and ignited. Instead, most air travels around the outside of the core in bypass ducts, driven and accelerated by the duct fan. This air exits at the rear of the engine alongside the hot exhaust gases from the core, providing extra thrust. See "Different jets" (opposite) to compare types of jet engine.

3,092°F (1,700°C)—the peak temperature in some combustion chambers when a jet engine is running at maximum power.

31,000 The number of parts that make up a single GE9X turbofan engine.

163

JET PIONEERS

In 1922, French engineer Maxime Guillaume patented a simple jet engine featuring fans to compress air, but never built a prototype. It was left to two other engineers—one in Britain, one in Germany—to develop the first working jet engines, which both ran for the first time in 1937. Seven years later, the first military jet was ready for combat. The Messerschmitt Me 262 hit 540 mph (870 kph) in level flight, far faster than any nonjet aircraft of the time.

Frank Whittle (1907–1996)

Whittle (on right) spent the 1930s perfecting his vision of jet-powered aircraft. In April 1937, his WU turbojet became the first jet engine to run successfully on the ground. In 1941, a Whittle W.1 engine powered the first British jet aircraft, the experimental Gloster E28/39. In 1944, the Gloster Meteor became Britain's first jet fighter.

Hans von Ohain (1911–1998)

Von Ohain was just 19 when he first patented a jet engine design. His initial version ran on hydrogen, but was converted to gas in 1937. His engine powered the first jet aircraft, the Heinkel He 178, in 1939, but it was Axel Frank who designed the engine for the first jet fighter. The Messerschmitt Me 262 entered service in 1944.

DIFFERENT JETS

Jet engines vary in type but work on the same basic principles. They draw in and compress air, raising its pressure and temperature. The air then enters a combustion chamber, where it is mixed with fuel and ignited. As it burns fiercely, rapidly expanding gases are produced. These inflate out of the rear of the jet engine, exiting via exhaust nozzles to produce large amounts of forward thrust.

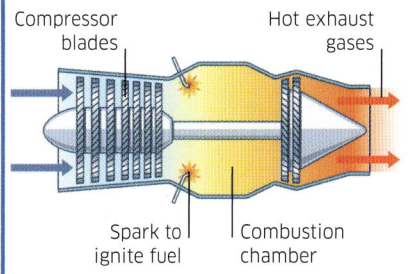

Compressor blades · Hot exhaust gases · Spark to ignite fuel · Combustion chamber

Turbojet

The simplest form of jet engine admits a constant flow of air while combusted gases leave the engine at high speed. Turbojets produce lots of thrust but are loud and consume a lot of fuel.

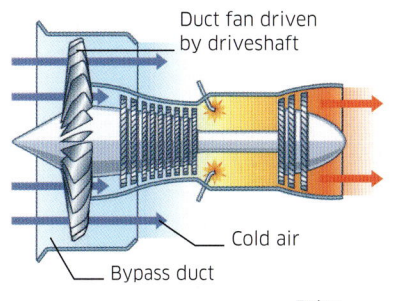

Duct fan driven by driveshaft · Cold air · Bypass duct

Turbofan

Turbofans split incoming airflow. Some passes into the combustion chamber, but cold air also runs through bypass ducts. These ducts add thrust by channeling air around the combustion chamber and out of the engine's rear.

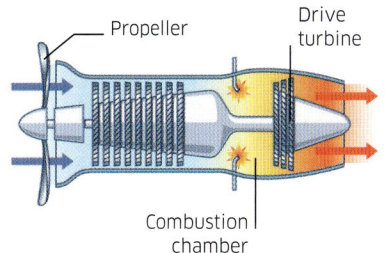

Propeller · Drive turbine · Combustion chamber

Turboprop

In this version, the gases produced by combustion only produce a little thrust. They are mostly used to turn a turbine, which powers a drive shaft that spins a propeller.

SUPER ENGINE

In 2018, tests run on General Electric's GE9X engine for the Boeing 777X airliner revealed it to be the most powerful commercial jet engine ever. The whole unit is wider than a Boeing 737 fuselage and delivers 134,300 lb (597,390 N) of thrust—equivalent to three-and-a-half of the engines used on Concorde (see pages 168–169).

Big unit
The world's most powerful engine, a GE9X, mounted on a Boeing 777X.

THRUST VECTORING

The exhausts of most jet engines are fixed. But vectored thrust engines feature exhaust nozzles that tilt, changing the direction of gases leaving the engine, and thus the direction of thrust acting on the aircraft. Found mostly in military jets, vectored thrust increases a plane's agility in the air.

High roller
The Lockheed Martin F-22 Raptor's exhaust nozzles can tilt 20° up or down to dive, climb, turn, or roll faster.

How it works

Vectored thrust is built into a jet aircraft's flight control systems. When the pilot alters the aircraft's direction using the joystick and rudder pedals, the nozzles move, too, to make sharper climbs, dives, or turns. In this diagram, the nozzle helps push the tail down to climb or thrusts the tail up to dive.

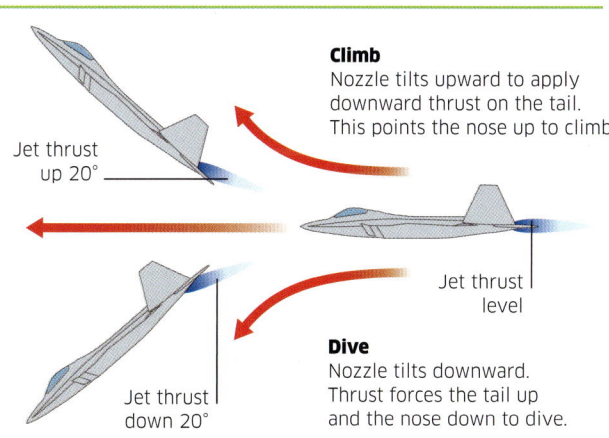

Jet thrust up 20°

Climb
Nozzle tilts upward to apply downward thrust on the tail. This points the nose up to climb.

Jet thrust level

Dive
Nozzle tilts downward. Thrust forces the tail up and the nose down to dive.

Jet thrust down 20°

Jetman

Jet-powered daredevil Yves Rossy joins the eight Dassault/Dornier Alpha Jets of the Patrouille de France aerobatic display team 4,000 ft (1,200 m) above Dubai in 2016.

A former Swiss Air Force pilot, Rossy began to develop wingsuits in the 2000s. His latest carbon-fiber Jetwing's four small jet engines propel him to just over 250 mph (400 kph). For 13 minutes, he soars, swoops, and climbs, steering with his body, before parachuting to earth when the fuel runs low.

CESSNA 172
High-wing light aircraft
Origin: US
Length: 27⅛ ft (8.3 m)

More than 44,000 of this easy-to-handle aircraft have been built since 1955. Thousands of pilots learn to fly in this popular four-seater, with the top-mounted wing giving good visibility below.

Undercarriage
Streamlined fairings reduce drag on the landing gear's wheels.

PIPISTREL VELIS ELECTRO
Pioneering electric light aircraft
Origin: Slovenia
Length: 21¼ ft (6.5 m)

Twin battery packs power this two-seater's electric motor for flights up to 50 minutes long. It produces no emissions while in the air and is much quieter than traditional light aircraft.

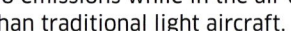

Civil aircraft

All nonmilitary planes are classed as civil aircraft. They range from single-seaters to giant jet airliners that fly hundreds of people between continents.

Commercial aircraft transported over 4 billion passengers on 34 million flights in 2023, as well as cargo ranging from airmail postcards to giant aerospace parts. Other civil aircraft perform a wide range of tasks, from pilot training and medical evacuation to aerial surveying; agricultural tasks; and aerobatics, flying loop-the-loops, barrel rolls, and other daring maneuvers for entertainment at air shows.

ATR 42
Reliable feederliner
Origin: Italy and France
Length: 74½ ft (22.7 m)

Feederliners serve shorter routes and carry fewer passengers than airliners. The ATR 42 can operate off short runways and carry up to 48 people, who enter and exit via a door in the rear of the fuselage.

Prop power
A six-bladed propeller gives a cruising speed of 310 mph (500 kph).

Turboprop engine
Besides spinning a propeller for flight, the engine also drives an electric generator.

BOEING 787
Wide-bodied airliner
Origin: US
Length: 186 ft (57 m)

Lightweight composite materials form half of the 787-8, making it lighter and more fuel-efficient than its predecessors. It can cruise at 561 mph (903 kph) with a range of 8,463 miles (13,620 km).

AIRBUS A380
Double-decker airliner
Origin: Multinational
Length: 238½ ft (72.7 m)

The world's largest airliner weighs up to 634 tons (575 tonnes) fully laden and can carry up to 853 people across two decks. To lift all this off the ground, it has a huge 261½-ft (79.7-m) wingspan.

Flexible wings
The wings flex and bend up to 10 ft (3 m) to reduce drag and improve fuel economy.

Wide body typically holds up to 296 passengers plus luggage

The **wings of an Airbus A380 cover** an area **larger than 3 tennis courts.**

GRUMMAN G-164
Cropduster turned stunt plane

Origin: US
Length: 24¼ ft (7.1 m)

Introduced in 1957, this single-engined biplane was designed to spray crops and drop seeds at low altitude–the cockpit could be sealed to keep out sprayed chemicals. It later found favor with wing walkers and other stunt fliers.

Wing walker
Performers strike daring poses at 90 mph (140 kph) to wow crowds.

AIRBUS BELUGA XL
Outsized cargo carrier

Origin: Multinational
Length: 207 ft (63.1 m)

Based on an Airbus A330, this cargo plane's giant fuselage can hold pairs of airliner wings and is loaded via a flip-up nose. The Beluga XL has a range of up to 2,500 miles (4,000 km).

Piggyback
The jet engine is mounted on top of the plane.

CIRRUS VISION SF50
Innovative mini business jet

Origin: US
Length: 30⅞ ft (9.4 m)

This ultralight jet's carbon-fiber fuselage seats six passengers and a single pilot. If the turbojet engine fails, the safety system can parachute the entire plane down to the ground.

GULFSTREAM G200
Medium-range business jet

Origin: US and Israel
Length: 63¼ ft (19 m)

Also known as the IAI Galaxy, this small jet usually seats 8–10 business passengers on trips of up to 3,900 miles (6,750 km). Its twin jet engines offer a top speed of 560 mph (900 kph).

CASA C-212 AVIOCAR
Multipurpose transporter

Origin: Spain
Length: 53 ft (16.2 m)

Simple and sturdy, this plane's boxy fuselage can carry 5,950 lb (2,700 kg) of cargo, loaded via a rear ramp. It can take off from unpaved airstrips as short as 1,300 ft (400 m).

Turboprop engines
Twin engines supply enough power for short take-off and landing (STOL) capability.

Soft landing
Fixed landing gear has low-pressure tires for use on grass runways.

Air intakes
At supersonic speeds, these slowed incoming air by more than 75 percent to prevent damage to the engines.

Exhaust nozzles
Specially developed variable nozzles reduced weight and engine noise.

Concorde

This supersonic airliner flew at more than twice the speed of sound—around 1,354 mph (2,179 kph)—slashing flight times on some long-distance routes.

Concorde's slim fuselage and streamlined, sweeping, triangular "delta wing" were an elegant way to generate enough lift while minimizing drag in order to fly supersonically. With four powerful engines, the aircraft could reach 2.04 times the speed of sound and cruise at 60,000 ft (18,300 m). Every part of the plane was specially engineered to endure the extreme forces of high-altitude supersonic flight.

Fuel pumped backward

Forward trim fuel tanks

Main tanks

Rear tank

Balancing act
As Concorde reached supersonic speeds, aerodynamic forces shifted its center of lift backward, causing the tail to rise and the nose to tilt down. To correct this, fuel was pumped to the rear when gaining speed, and forward when slowing for descent and landing.

CONCORDE	
Origin:	UK and France
Year:	1969
Role:	Supersonic airliner
Length:	203¾ ft (62.1 m)

Galley 1
One of seven kitchen areas on the plane, this had two electric ovens.

Streamliner
The aluminum-alloy body was as long as a Boeing 747 but only one third of the width.

Zoom with a view

Of the 20 Concordes built, 14 went into service with British Airways and Air France, operating between 1976 and 2003. Despite high fares, Concorde carried more than 2.5 million passengers in this period, including UK businessman Fred Finn, who made a record 718 flights, always sitting in seat 9A. When cruising at almost twice the altitude reached by other airliners, Concorde's passengers could view the curve of Earth below.

Movable nose
The nose tilted down 5° for take-off and 12.5° for landing to give the pilot a clear view of the runway.

Flight deck
The pilot, co-pilot, and flight engineer used innovative fly-by-wire controls.

Visor
A heat-resistant glass shield protected the cockpit at high speeds, but was lowered to improve visibility when landing.

Reflective paint
At top speed, friction heated the nose up to 260°F (127°C). Reflective white paint reduced this by around 18°F (10°C).

Concorde flew from London to New York in a **record-breaking** 2 hours, 52 mins, 59 seconds.

106 tons (96 tonnes)—the **amount** of **fuel Concorde** carried in its **13 tanks:** around **1.1 tons (1 tonne)** per **passenger!**

In the **15 seconds** it took to **pour a drink** on **board, Concorde** could travel **5½ miles (9 km)** at full **speed!**

169

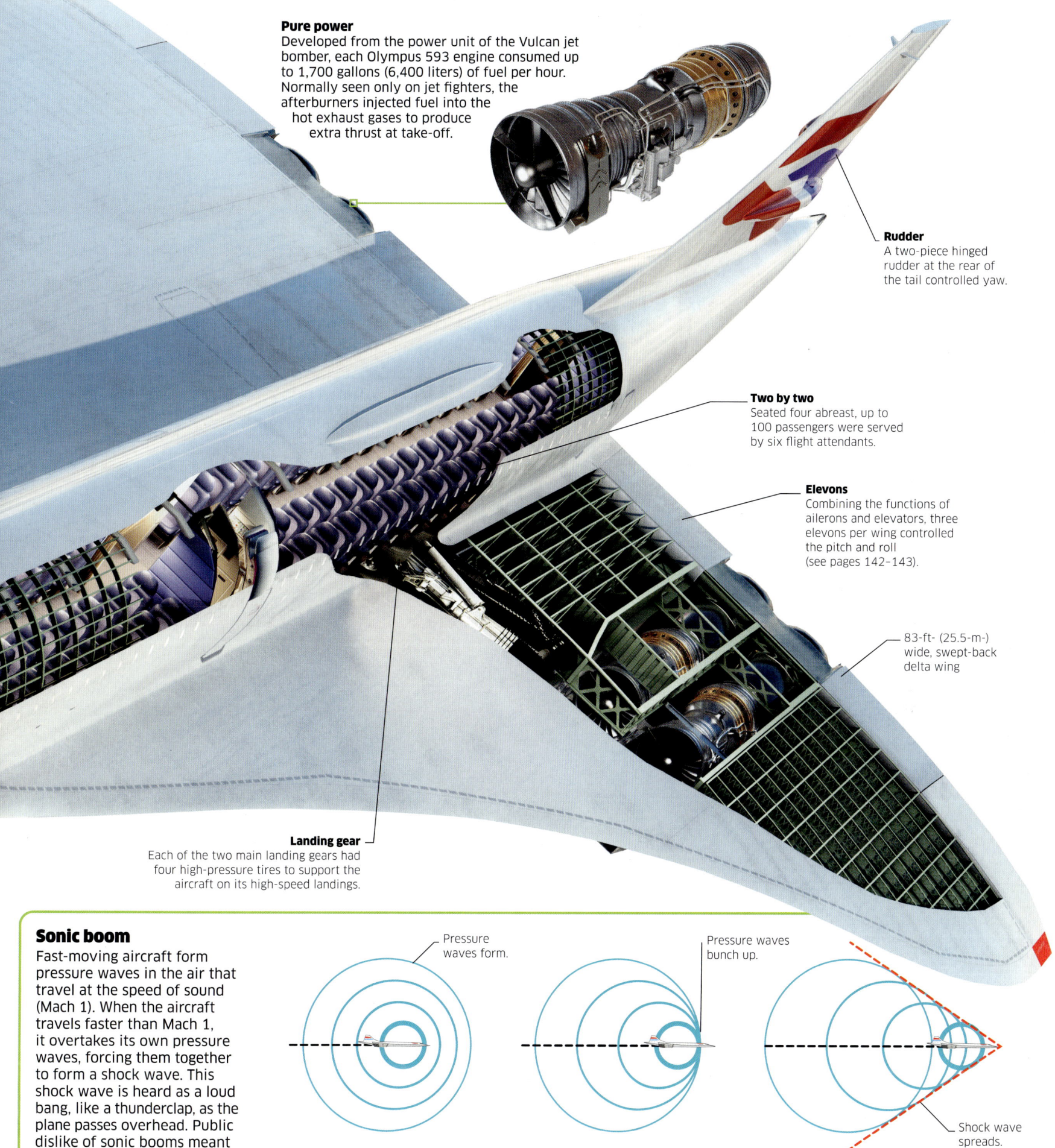

Pure power
Developed from the power unit of the Vulcan jet bomber, each Olympus 593 engine consumed up to 1,700 gallons (6,400 liters) of fuel per hour. Normally seen only on jet fighters, the afterburners injected fuel into the hot exhaust gases to produce extra thrust at take-off.

Rudder
A two-piece hinged rudder at the rear of the tail controlled yaw.

Two by two
Seated four abreast, up to 100 passengers were served by six flight attendants.

Elevons
Combining the functions of ailerons and elevators, three elevons per wing controlled the pitch and roll (see pages 142–143).

83-ft- (25.5-m-) wide, swept-back delta wing

Landing gear
Each of the two main landing gears had four high-pressure tires to support the aircraft on its high-speed landings.

Sonic boom

Fast-moving aircraft form pressure waves in the air that travel at the speed of sound (Mach 1). When the aircraft travels faster than Mach 1, it overtakes its own pressure waves, forcing them together to form a shock wave. This shock wave is heard as a loud bang, like a thunderclap, as the plane passes overhead. Public dislike of sonic booms meant Concorde flew more slowly over land and only reached full speed above the ocean.

Pressure waves form.

Pressure waves bunch up.

Shock wave spreads.

Below Mach 1
Aircraft creates pressure waves, which spread like ripples on a pond.

Mach 1
Aircraft travels at the same speed as its pressure waves, causing them to bunch up.

Above Mach 1
Aircraft goes supersonic. Pressure waves combine to form a shock wave, like the bow wave of a boat.

Wheels down

Beachgoers greet the extremely loud and incredibly close arrival of a Boeing 747 airliner coming in to land on St. Maarten.

Large, heavy jet airliners need plenty of runway—often 4,900 ft (1,500 m) or more—to land, slow down, and taxi away. On this small Caribbean island, the airport begins right by the shore, so pilots skim low over Maho Beach to get their plane's wheels down on the tarmac quickly. Below, people mass on the beach to enjoy a close encounter with these awesome aircraft.

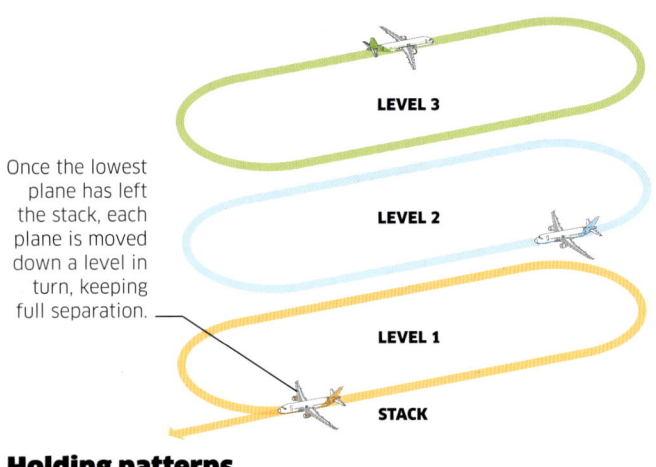

LEVEL 3

LEVEL 2

LEVEL 1

STACK

Once the lowest plane has left the stack, each plane is moved down a level in turn, keeping full separation.

Airport

Airports are major transportation hubs, handling thousands of flights carrying many tons of cargo and vast numbers of people.

As well as the terminal buildings and transportation links that most arriving and departing passengers see, airports have large, dedicated cargo handling, security, catering, refueling, cleaning, and repair facilities. The hundreds of aircraft that use an airport each day must be marshaled and directed both in the air and on the ground in a carefully planned series of movements that ensure swift, efficient, and safe operation.

Holding patterns

When many planes arrive at once, air traffic controllers direct aircraft to a safe area of sky called a stack. There, aircraft fly in oval holding patterns, each separated by 1,000 ft (305 m) of altitude, and await their turn to descend and land.

High-speed taxiway
allows aircraft to exit the runway faster

Compass bearing in tens of degrees helps pilots land in the right place: 27 is 270°, or due west)

Hit the brakes
After touching down, the airliner brakes hard to reduce speed.

Reversible runway
Touchdown markings at both ends allow the runway to be used in either direction, depending on winds.

Taxiway centerlines mark safe routes around the airport

Tanker truck
Aircraft take on fuel from special trucks or from underground pumps.

Newly landed aircraft taxis off the runway toward the apron and terminals

Rotating radar
The antenna sweeps the local area to track moving aircraft.

Local controllers instruct pilots through take-off and landing.

Ground controllers direct aircraft around the airport.

Air traffic control
From the tower, air traffic controllers coordinate all take-offs, landings, and aircraft movements on the ground to avoid crashes. They also monitor local weather, track aircraft on radar screens, and are in constant radio contact with pilots.

Visual control room
The top of the control tower offers 360° views of the airport.

Denver International Airport, Colorado, has the **world's largest baggage system**, with **28 miles (44 km)** of track and belts.

4,200 The number of flights that pass through **Gibraltar Airport every year**, despite a **four-lane highway** crossing the runway!

173

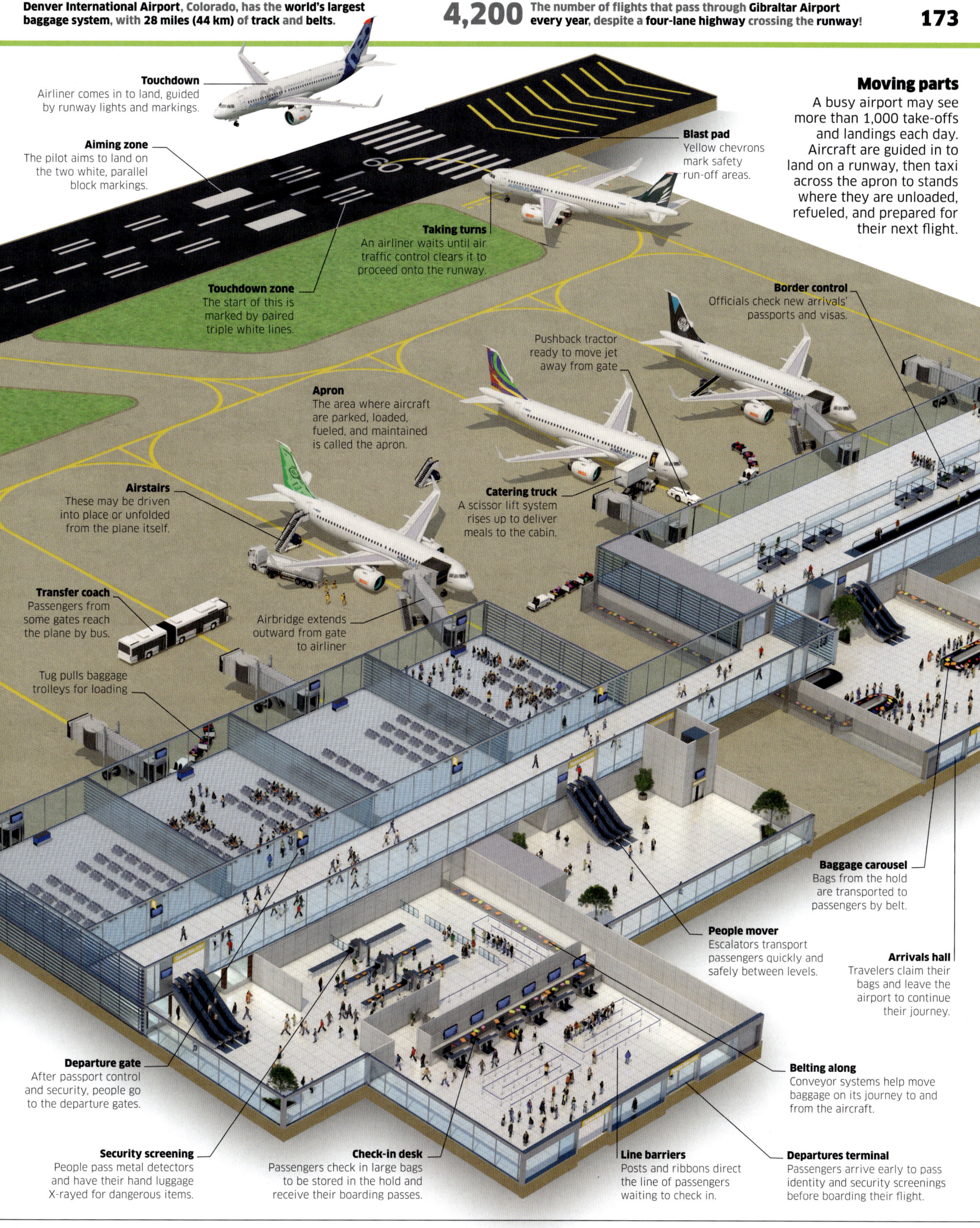

Touchdown
Airliner comes in to land, guided by runway lights and markings.

Aiming zone
The pilot aims to land on the two white, parallel block markings.

Touchdown zone
The start of this is marked by paired triple white lines.

Blast pad
Yellow chevrons mark safety run-off areas.

Taking turns
An airliner waits until air traffic control clears it to proceed onto the runway.

Moving parts
A busy airport may see more than 1,000 take-offs and landings each day. Aircraft are guided in to land on a runway, then taxi across the apron to stands where they are unloaded, refueled, and prepared for their next flight.

Border control
Officials check new arrivals' passports and visas.

Pushback tractor
ready to move jet away from gate

Apron
The area where aircraft are parked, loaded, fueled, and maintained is called the apron.

Catering truck
A scissor lift system rises up to deliver meals to the cabin.

Airstairs
These may be driven into place or unfolded from the plane itself.

Transfer coach
Passengers from some gates reach the plane by bus.

Airbridge extends
outward from gate to airliner

Tug pulls baggage
trolleys for loading

Baggage carousel
Bags from the hold are transported to passengers by belt.

People mover
Escalators transport passengers quickly and safely between levels.

Arrivals hall
Travelers claim their bags and leave the airport to continue their journey.

Departure gate
After passport control and security, people go to the departure gates.

Belting along
Conveyor systems help move baggage on its journey to and from the aircraft.

Security screening
People pass metal detectors and have their hand luggage X-rayed for dangerous items.

Check-in desk
Passengers check in large bags to be stored in the hold and receive their boarding passes.

Line barriers
Posts and ribbons direct the line of passengers waiting to check in.

Departures terminal
Passengers arrive early to pass identity and security screenings before boarding their flight.

Rotorcraft

The long, thin rotor blades that spin atop helicopters and other rotorcraft are aerofoils (see page 143). Turning at high speed, these blades act much like plane wings, generating lift as air travels around them.

Most rotorcraft are helicopters. Their powered rotors generate lift without the whole craft moving forward at high speed. This means that, unlike most fixed-wing aircraft, helicopters can take off and land vertically, maneuver in tight spaces, and hover motionless in midair with ease. Recently, remote-controlled drones with multiple rotors have emerged as an alternative vehicle for search-and-rescue missions, city taxi services, and urgent deliveries.

EARLY PROTOTYPES

Ukrainian–American inventor Igor Sikorsky first attempted to build helicopters in 1909 and 1910 without success. He returned to the challenge in the 1930s, creating the first practical helicopter with a single tail rotor, the VS-300, followed by the first mass-produced machine, the R-4, which entered service in 1943.

Test flight
Igor Sikorsky pilots his own invention, the VS-300 helicopter, on a flight in 1940.

Helicopter rotor assembly
Control rods, connected by two discs called a swashplate, alter the pitch of the rotor blades to generate more or less lift. Ball bearings between the discs allow the upper disc to spin with the rotors while the lower one remains fixed.

Rotor blade

Mast nut secures rotor to mast

Rotor mast

Blade grip

Control rod

Swashplate (upper disc)

Ball bearings

Swashplate (lower disc)

Driveshaft

HOW A HELICOPTER WORKS

Most helicopters are made up of a tail boom and body containing a cockpit; passenger or cargo space; one or more engines; and, above the body, a set of main rotor blades. These typically spin 225–500 times per minute, acting like aircraft wings to generate lift.

At the controls

The pilot can alter the pitch of all the rotors at the same time to climb or descend, or tilt the swashplate left, right, forward or back, so that the helicopter moves in the direction of the tilt. The rotor blades are hinged where they join the rotor mast. This enables a moving swashplate to alter the rotors' pitch or angle of attack (see page 142). Greater pitch produces more lift.

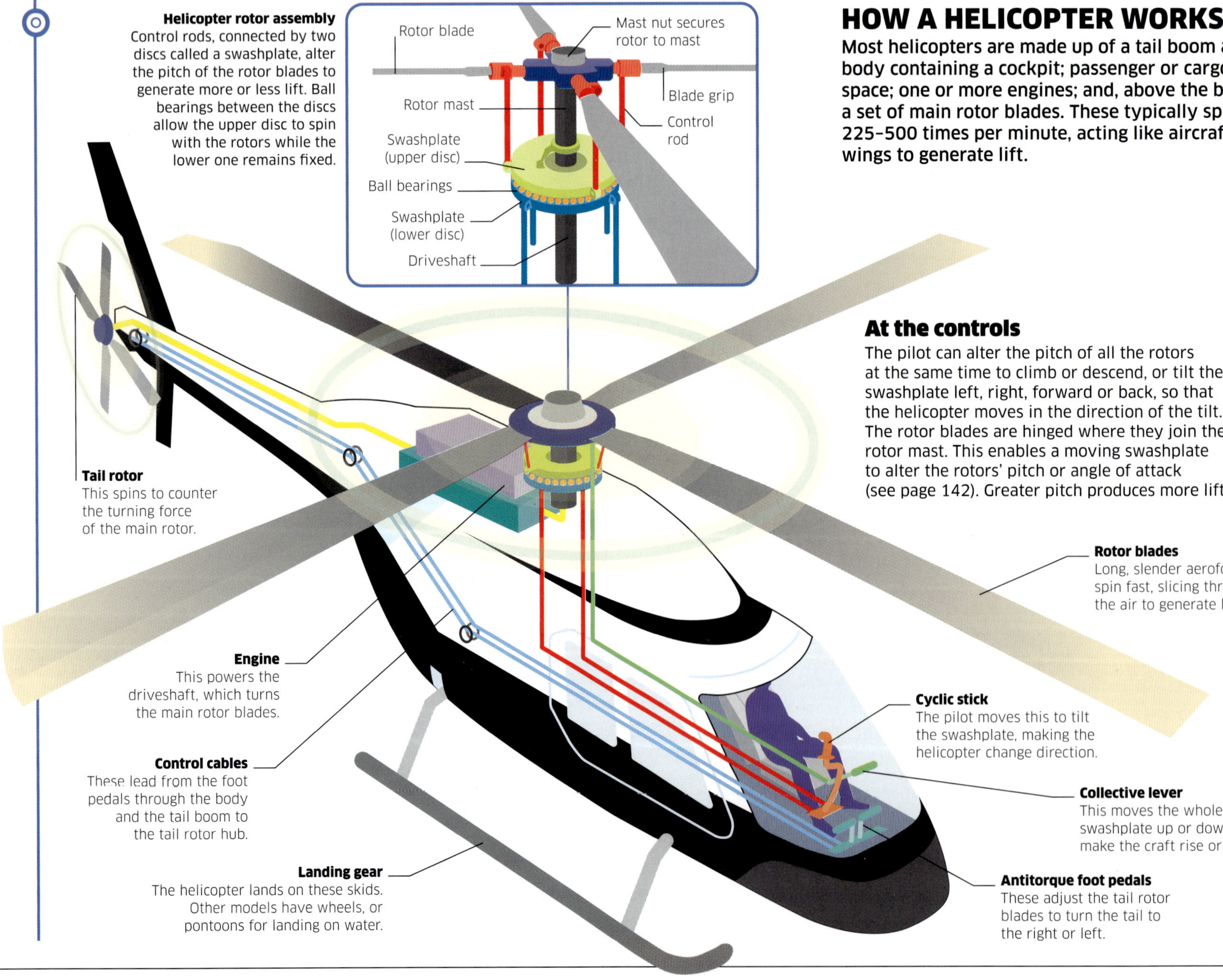

Tail rotor
This spins to counter the turning force of the main rotor.

Rotor blades
Long, slender aerofoils spin fast, slicing through the air to generate lift.

Engine
This powers the driveshaft, which turns the main rotor blades.

Control cables
These lead from the foot pedals through the body and the tail boom to the tail rotor hub.

Cyclic stick
The pilot moves this to tilt the swashplate, making the helicopter change direction.

Collective lever
This moves the whole swashplate up or down to make the craft rise or fall.

Landing gear
The helicopter lands on these skids. Other models have wheels, or pontoons for landing on water.

Antitorque foot pedals
These adjust the tail rotor blades to turn the tail to the right or left.

115 tons (105 tonnes)–the **maximum take-off weight** of the **world's heaviest helicopter, the Mil Mi-12.**

29,030 ft (8,848 m)–the **highest landing** made by a **helicopter,** an **Airbus AS350 B3,** on the **summit of Mount Everest.**

175

CONTROLLING SPIN

When a helicopter's main rotor spins, it generates an equal and opposite turning force on the helicopter's body. This could spin the craft dizzyingly fast in the opposite direction, making it impossible to fly. To counter this, most helicopters require a second rotor.

Tail rotor

Smaller than the main rotor, this is mounted on the tail boom, at right angles to the main rotor. It generates sideways thrust, which balances out the turning force from the main rotors. By changing the blade angle, the pilot can adjust the sideways thrust to point the tail left or right–a similar effect to an airplane's rudder.

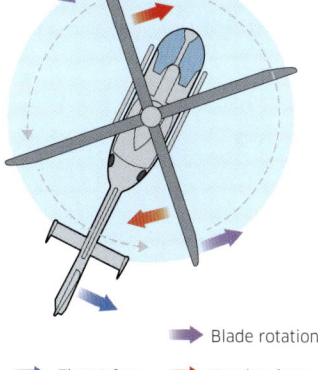

▶ Blade rotation

▶ Thrust from tail rotor ▶ Turning force on body

Contra-rotation

Some helicopters have two main rotors, which spin on the same axle but in opposite directions. Known as coaxial rotors, they cancel out the turning forces.

Fenestron

This spinning ducted fan is set inside the tail bodywork and produces less noise than a regular tail rotor.

Fenestron

AUTOGYRO

These simple aircraft have a backward-facing powered propeller that generates forward thrust, but the rotor is not powered. As the craft moves forward, the flow of air causes the rotors to spin, generating lift.

Autogyros are usually small, single- or tandem-seat craft with a pusher propeller mounted behind the cockpit.

MANY ROTORS

Some rotorcraft have multiple sets of main rotors, each spinning on its own axle. Pairs of rotors spin in opposite directions to each other in order to balance the turning forces, so no tail rotor is needed. Tandem rotorcraft such as the Chinook helicopter (see page 177) have two rotors, quadcopters have four, while the electric VoloDrone (see page 185) has 18!

Quadcopter

A quadcopter (see page 184) controller alters the speed of different configurations of the drone's rotors to steer. To move forward, the back two rotors spin faster to produce more lift, tilting the quadcopter down at the front. This points all four rotor sets at an angle so that they produce both lift and forward thrust. To turn on the spot, one of the pairs of diagonally opposite rotors must spin faster.

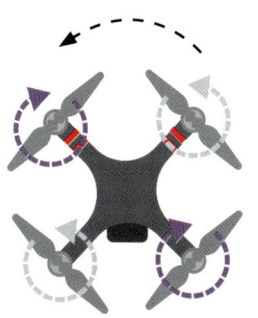
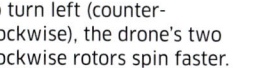
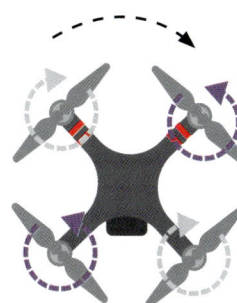

To turn left (counter-clockwise), the drone's two clockwise rotors spin faster.

To turn right (clockwise), the two counterclockwise rotors spin faster.

MANEUVERING

Helicopter pilots maneuver using three key controls. The throttle controls the engine power, and with it the rotor spin speed. The collective lever raises or lowers the whole swashplate, which alters the pitch of all the rotor blades by the same amount. The cyclic stick tilts the swashplate in one direction so that the rotor blades generate unequal lift, which sends the helicopter forward, backward, or sideways.

▶ Lift
▶ Weight

Take-off

The pilot increases the pitch and speed of the rotor blades, using the collective lever and throttle. This produces enough lift to overcome the craft's weight and get airborne.

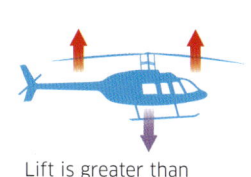

Lift is greater than weight, so craft rises

Blades all have increased pitch, generating maximum lift

Swashplate raised by collective lever

Hovering

When all four forces of flight are balanced, the helicopter stays stationary in midair. Lift equals weight and no forward or sideways thrust or drag is produced.

Blades all have equal pitch

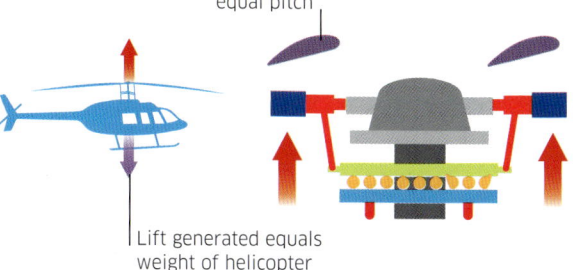

Lift generated equals weight of helicopter

Forward flight

The pilot moves the cyclic stick forward to tilt the swashplates up at the back. This increases blade pitch at the rear, angling the helicopter so that all of its blades produce forward thrust as well as lift.

Lift increased at rear, tilting the helicopter forward

Rear blades' pitch is greater

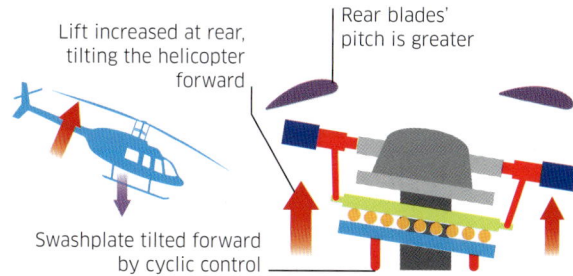

Swashplate tilted forward by cyclic control

176 air and space ○ **WORKING HELICOPTERS**

84.9 mph (136.7 kph)—the **average speed** of the **fastest** around-the-world helicopter journey, in an AW109.

AÉROSPATIALE SA-313 ALOUETTE II
First jet-powered helicopter
Origin: France
Top speed: 115 mph (185 kph)

This versatile French five-seater first flew in 1955. Powered by a turboshaft jet engine pioneered in planes, it was widely used for mountain rescue, photography, observation, evacuation, and more.

SIKORSKY R-4
First practical helicopter
Origin: US
Top speed: 75 mph (121 kph)

The first working helicopter entered service with UK and US forces from 1943 for observation and sea rescue. A two-seater with wooden rotor blades, the R-4 was slow and difficult to fly.

Working helicopters

With their ability to take off and land vertically, hover in midair, and maneuver in tight spaces, helicopters can perform tasks that would be impossible for other aircraft.

Thousands of helicopters are employed every day transporting military or civilian passengers, performing medical evacuations, or flying search and rescue missions. Others are employed by police, land surveyors, filmmakers, and firefighters—or for transporting bulky cargoes such as 330-ft- (100-m-) long wind turbine blades.

AS365 DAUPHIN N2
Air ambulance
Origin: France
Top speed: 190 mph (306 kph)

The Dauphin's cabin is equipped for emergency medical treatment on board, while a range of 514 miles (827 km) helps it fly multiple patients to hospital at life-saving speed without refueling.

ERICKSON-SIKORSKY S-64 AIR CRANE
Heavy lifter and firefighter
Origin: US
Top speed: 125 mph (202 kph)

Two powerful 4,050-hp engines drive six blades, giving this craft supreme lifting ability. Air Cranes can carry a 2,642-gallon (10,000-liter) water tank, refillable in just 45 seconds, to fight raging wildfires.

MIL MI-26
Super-heavy lifter and transporter
Origin: Russia
Top speed: 183 mph (295 kph)

The world's biggest operational helicopter is Russian, has a crew of five, and carries up to 90 troops or 22 tons (20 tonnes) of cargo. Its eight rotor blades span a mighty 105 ft (32 m).

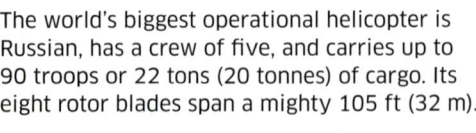

BELL UH-1 IROQUOIS
Military utility helicopter
Origin: US
Top speed: 127 mph (204 kph)

Its easy handling and simple, robust design made the "Huey" a mainstay of many military forces. More than 16,000 were built from 1956 to 1987.

Net weight
Food and medical supplies can be carried in a sling below the helicopter.

Easy access
Sliding cabin doors provided entry and exit for up to 11 troops.

Cockpit

BOEING CH-47 CHINOOK
Heavy lifter and troop transporter
Origin: US
Top speed: 196 mph (315 kph)

Over 1,250 of these tandem-rotor helicopters have been built. Its fuselage holds up to 55 troops or 11 tons (10 tonnes) of cargo. The overlapping rotors spin in opposite directions to keep the craft stable.

SCHWEIZER 300C
Light helicopter
Origin: US
Top speed: 109 mph (176 kph)

This affordable, long-serving light helicopter weighs just 1,100 lb (499 kg). It is used for traffic patrols, aerial surveys, training, sightseeing, crop-spraying, pipeline inspections, and more.

AGUSTAWESTLAND AW109
Multirole medium-lift helicopter
Origin: UK and Italy
Top speed: 193 mph (311 kph)

From TV filming to ferrying VIPs and medical patients, this popular Anglo-Italian, twin-engine craft seats eight people and has been in production since 1971. Its wheels retract to reduce drag.

KAMAN K-MAX K1200
Super-strong synchrocopter
Origin: US
Top speed: 120 mph (193 kph)

This unusual cargo-lifter has two intermeshing rotors that turn in opposite directions to shift more than double its own weight. It can even be fitted with a wrecking ball for demolition jobs!

SIKORSKY X2
Experimental high-speed helicopter
Origin: US
Top speed: 287 mph (463 kph)

This advanced prototype featured fly-by-wire controls, a pusher propeller, and co-axial rotor blades spinning in opposite directions to balance the craft. The X2 was retired in 2011 after setting speed records.

Rigid rotor blades prevent clash between upper and lower blades.

Vertical stabilizer and rudder

1,275 The number of AH-64 Apaches operating around the world.

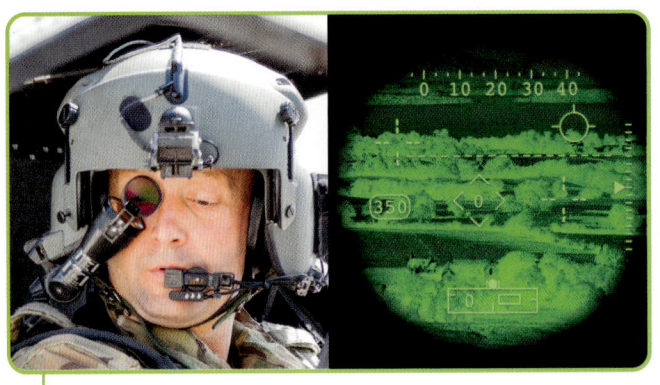

Attack helicopter

Helicopter gunships are like flying battle tanks, designed to overwhelm ground forces and strike key targets using heavy weapons and cutting-edge technology.

Faster than a tank and tougher than a fighter jet, the attack helicopter fills a crucial role in the armed forces. A Boeing AH-64 Apache can strike fast and disappear, or linger as air support, using brute firepower to dominate the battlefield with missiles, rockets, and a formidable cannon.

High-tech monocle
Both Apache pilots wear a Helmet Mounted Display over one eye. This shows key data to save them from looking down at control panels. Instead, pilots rapidly switch focus between eyes to change view. The gunner can aim and fire the chain gun simply by looking at the target.

Back-seat driver
The pilot sits above and behind the gunner for a broader view. Both can fly the helicopter and fire weapons.

Blast screen
A transparent shield separates the two pilots, protecting one even if the other is hit.

Co-pilot gunner (CPG)
The CPG sits at the front to aim weapons and navigate.

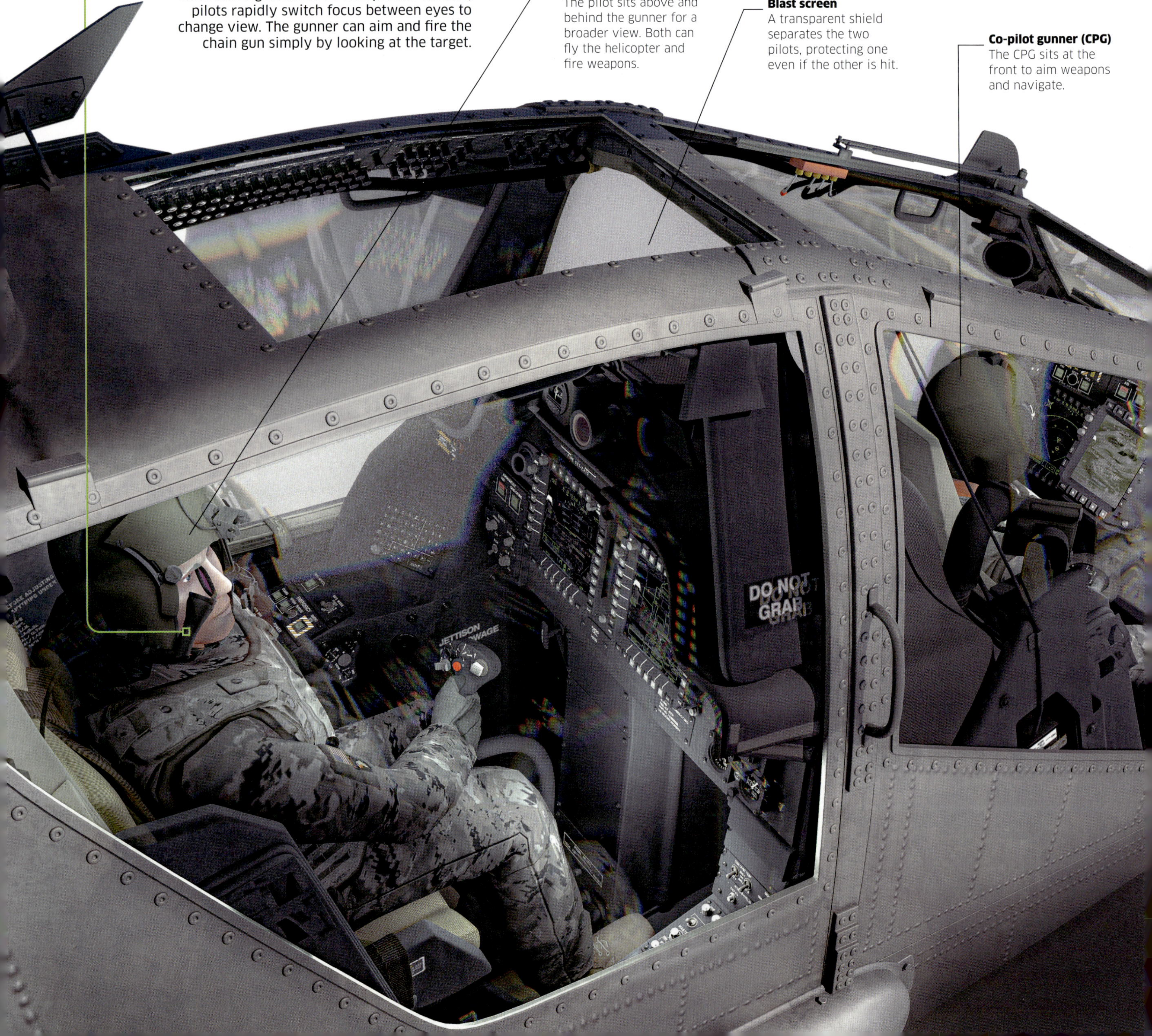

The **crew, weapons, and fuel** of an AH-64 can **weigh more** than the aircraft itself!

179

Apache intelligence

The AH-64 Apache is one of the world's most advanced attack helicopters. Its Longbow fire-control radar can identify 256 separate threats in 30 seconds, prioritize them, and offer a shortlist of 16 targets to the gunner. Information is also shared with nearby allies and drones to give a full and clear picture of the battlefield.

Precision striker
The latest Longbow Hellfire missiles can destroy a tank or bunker up to 6.8 miles (11 km) away in one hit. They are self-guided, using high-frequency radar to find their target through cloud or smoke, so the Apache can take cover as soon as it fires.

Power unit
Two 1,700-hp engines give a cruising speed of 173 mph (265 kph).

Longbow radar
This sensor-packed dome can track enemy forces up to 5 miles (8 km) away.

Gunner's night sight

Laser rangefinder

Hydra launcher
holds 19 rockets.

Chain gun
The 30mm cannon fires up to 10 rounds per second.

Night sight
The Pilot's Night Vision Sensor (PNVS) sends infrared images to the pilot's headset.

Enhanced gun sights
A telescope, laser sight, and night sight help the gunner aim weapons in all conditions.

Display and control
The panels are designed for rapid reading. Side buttons switch views.

Avionics pod
These house much of the AH-64's advanced computing power.

Countermeasures

The Apache is armored against bullets and has a self-sealing fuel tank, but it needs other tools to counter enemy missiles. Chaff launchers on the tail release a cloud of glass fibers coated in aluminum (above), which tricks enemy radar systems into believing the chaff cloud is the aircraft. Likewise, bright flares blast away from the body like fireworks to distract heat-seeking missiles.

Nimble beast

Approaching a target, sensors help the pilot skim the ground so that the AH-64 is only visible moments before it attacks. A rigid rotor system allows it to turn sharply or even roll upside down and back again. Twin turbojet engines lift the heavy Apache at up to 2,790 ft (850 m) per minute, as high as 20,000 ft (6,100 m) above sea level. The system cleverly mixes hot engine exhaust with cool air from outside to avoid heat-seeking missiles.

BOEING AH-64D APACHE

Origin: US

Year: 1997

Top speed: 173+ mph (279+ km/h)

Length: 58⅛ ft (17.73 m)

180   air and space ○ MILITARY AIRCRAFT

9G The force—nine times the force of gravity—that fighter pilots are trained to withstand when turning.

Military aircraft

A modern military requires aircraft capable of a range of functions—from aerial combat and high-altitude reconnaissance to troop support and heavy lifting.

Air superiority requires fast, maneuverable fighters to win the battle in the air. These clear the skies for ground attack and troop support aircraft, which may be slower but more effective against certain targets. In contrast, long-range bombers and observation planes are designed to avoid detection, while transporters of varying sizes airlift tanks, troops, weapons, or supplies into or out of conflict zones.

The B2 Spirit is the world's **most expensive** aircraft. To develop, build, and maintain them costs **$2.1 billion** per aircraft.

B-2 SPIRIT
Stealth bomber
Year: 1989
Top speed: 630 mph (1,010 kph)

An unusual flying wing design helps the US Air Force's 172-ft- (52.4-m-) wide B-2 travel deep into enemy territory undetected. The crew of two can fly 6,900 miles (11,000 km) without refueling.

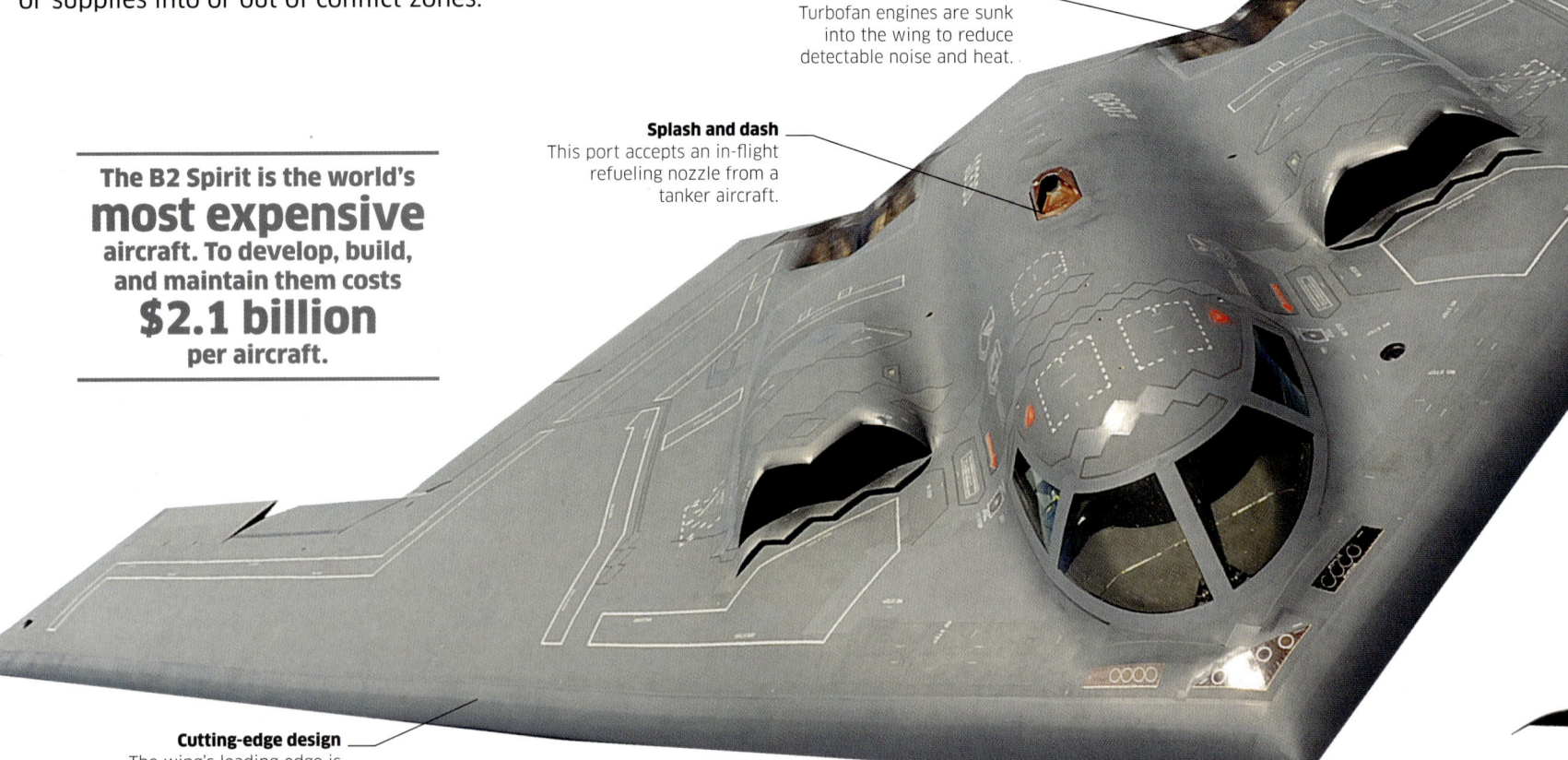

Quiet exhaust
Turbofan engines are sunk into the wing to reduce detectable noise and heat.

Splash and dash
This port accepts an in-flight refueling nozzle from a tanker aircraft.

Cutting-edge design
The wing's leading edge is reinforced with light-but-strong diagonal ribs.

RAFALE
Multirole combat fighter
Year: 1991
Top speed: 1,188 mph (1,912 kph)

This versatile French jet fighter can also operate as a bomber, ground attack, or reconnaissance plane. Its rear-mounted twin engines and triangle-shaped delta wing give it a climb rate of up to 1,000 ft (305 m) per second.

A-10 THUNDERBOLT
Ground attack aircraft
Year: 1972
Top speed: 439 mph (706 kph)

This robust US plane can take heavy punishment: over 1,100 lb (500 kg) of titanium armor protects the cockpit, while parts on its left and right sides are interchangeable, allowing easy repairs. Long wings generate plenty of lift for short take-offs.

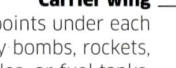

Carrier wing
Four hard points under each wing can carry bombs, rockets, missiles, or fuel tanks.

114 minutes—the time an SR-71 Blackbird took to fly from New York to London.

Each wing of a C-5 Galaxy is longer than a basketball court.

5,200 The number of aircraft in the US Air Force—the largest in the world.

181

CV-22B OSPREY
Half-plane, half-helicopter
Year: 1989
Top speed: 351 mph (565 kph)

The CV-22B's tilting engines turn propellers into rotors so that the US craft can take off and land vertically from tight spaces such as forest clearings and ships.

Full speed ahead!
Engines rotate from vertical to horizontal for faster forward flight.

Upwardly mobile
Tilted upward, the propeller blades act as helicopter rotors, generating lift.

Material advantage
Surface covered in radar-absorbent materials to reduce the chance of detection.

Take-off
Engines in vertical position.

E-3 SENTRY
Early warning system
Year: 1972
Top speed: 531 mph (854 kph)

This US craft's distinctive rotodome houses a spinning, highly sensitive radar system, which detects incoming missiles, planes, and shipping to give advance warning of possible attacks.

Rotodome packed with reconnaissance instruments

SR-71 BLACKBIRD
Reconnaissance plane
Year: 1964
Top speed: 2,193 mph (3,530 kph)

The world's fastest jet aircraft was unarmed but flew fast and high to evade detection as it spied on the surface below. The US jet cruised at three times the speed of sound and altitudes up to 83,661 ft (25,500 m).

Hot stuff
The titanium skin reached 600°F (316°C) at Mach 3.0.

EUROFIGHTER TYPHOON C-16
Multirole jet fighter
Year: 1994
Top speed: 1,320 mph (2,125 kph)

Low weight and a small delta wing make this advanced jet an agile and formidable force in air-to-air combat. More than 600 are in service in the UK, Germany, and seven other nations.

C-5 GALAXY
Heavyweight transport
Year: 1968
Top speed: 532 mph (856 kph)

Taking off and landing on 28 wheels, this cargo plane can carry 2 Abrams tanks, 6 Apache helicopters, or 15 Humvees in its vast hold. A second deck seats up to 75 people. it can fly 2,300 miles (4,260 km) without refueling.

Open up!
Hydraulic pistons raise the nose to allow access to the cargo hold.

Big fans
Each of the four turbofan engines generates 50,580 lb (225 kN) of thrust.

Sliding floors
Cargo is loaded and unloaded on ramps and rails running the full width of the fuselage.

U.S. AIR FORCE
6025

Multirole machine

The F-35B's powerful liftfan helps it take off and land vertically on small ships and in forest clearings, making it supremely nimble. In the cockpit, a single pilot operates several touchscreens and an augmented reality headset, while a host of sophisticated electronic aids assess threats and lock weapons onto targets.

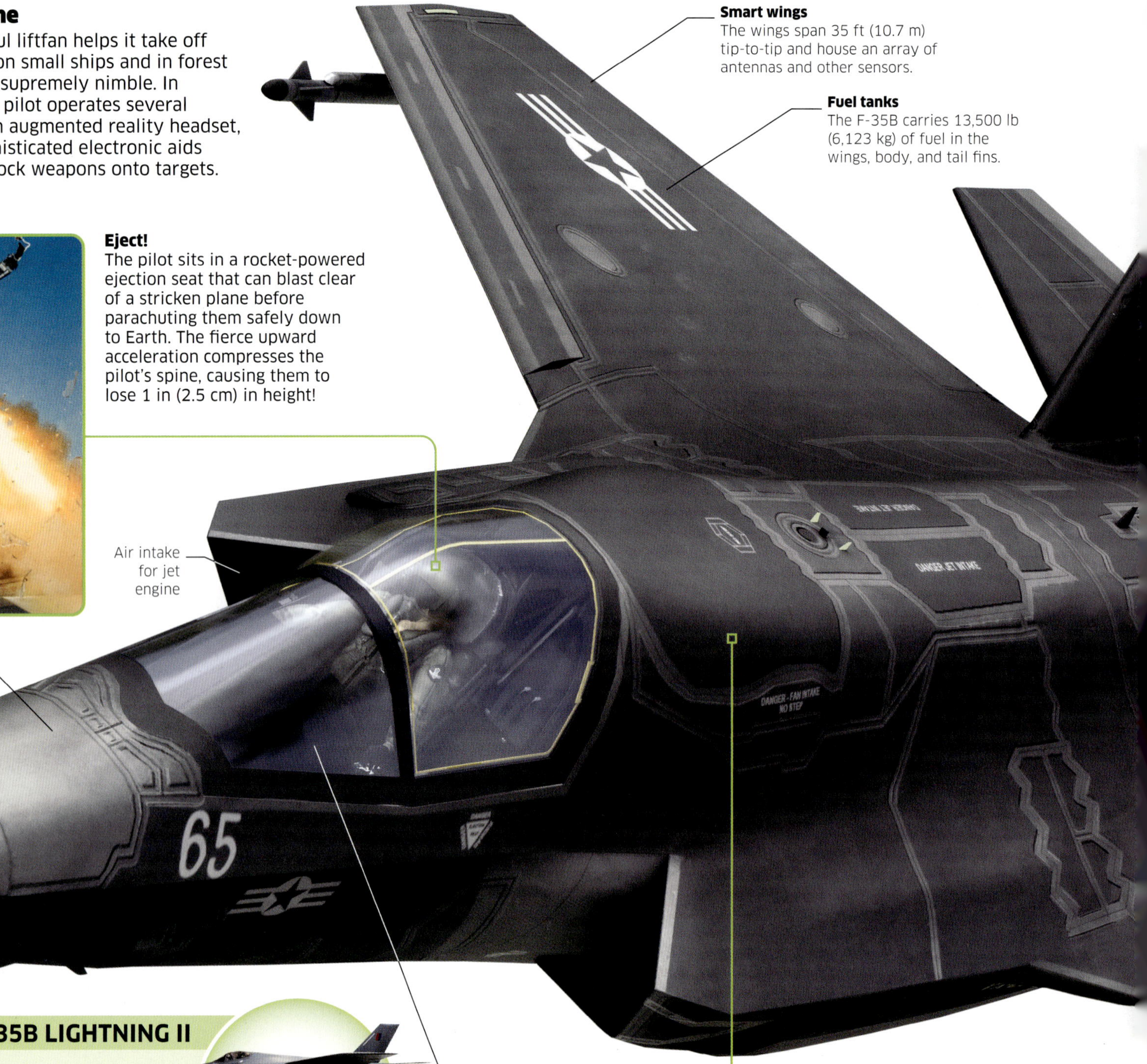

Smart wings
The wings span 35 ft (10.7 m) tip-to-tip and house an array of antennas and other sensors.

Fuel tanks
The F-35B carries 13,500 lb (6,123 kg) of fuel in the wings, body, and tail fins.

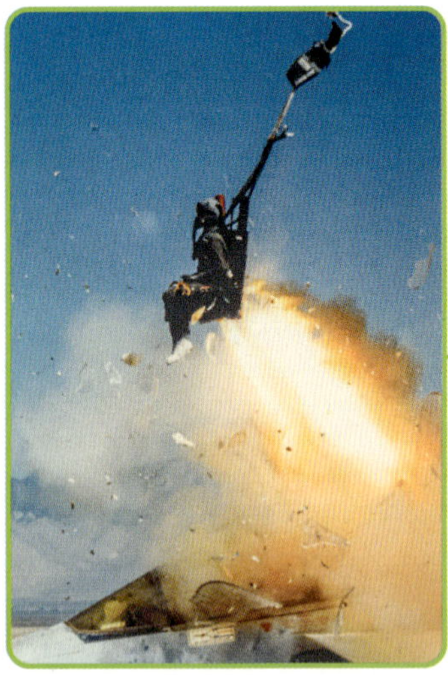

Eject!
The pilot sits in a rocket-powered ejection seat that can blast clear of a stricken plane before parachuting them safely down to Earth. The fierce upward acceleration compresses the pilot's spine, causing them to lose 1 in (2.5 cm) in height!

Air intake for jet engine

Sensitive nose
Agile radar uses 1,600 antennas to track threats.

65

LOCKHEED F-35B LIGHTNING II

Origin: US

Year: 2015

Speed: 1,228 mph (1,976 kph)

Length: 51 ft (15.7 m)

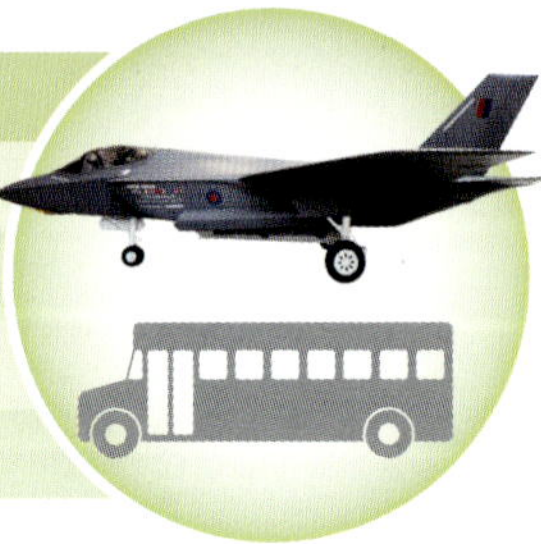

Bubble canopy
The pilot lifts the canopy to enter the cockpit. If they eject, the rear section is blown off with explosives.

Liftfan hatch
During short take-offs and vertical landings, this hatch opens to draw air into the liftfan, which supplies almost half the required vertical thrust. It closes only once the transition to forward flight is complete.

Fighter jet

The world's most advanced jet fighters combine supersonic speed with the latest avionics (flight computers) and stealth technology to conduct an array of different military missions.

From reconnaissance and electronic warfare to all-out ground attack or air combat, the F-35 is ready for anything. Three variants make it even more versatile—a conventional jet (F-35A); the F-35B shown here, for vertical take-offs; and the F-35C with folding wings, flown from aircraft carriers. More than 950 F-35s are currently in service with 17 different nations.

Liftfan

65

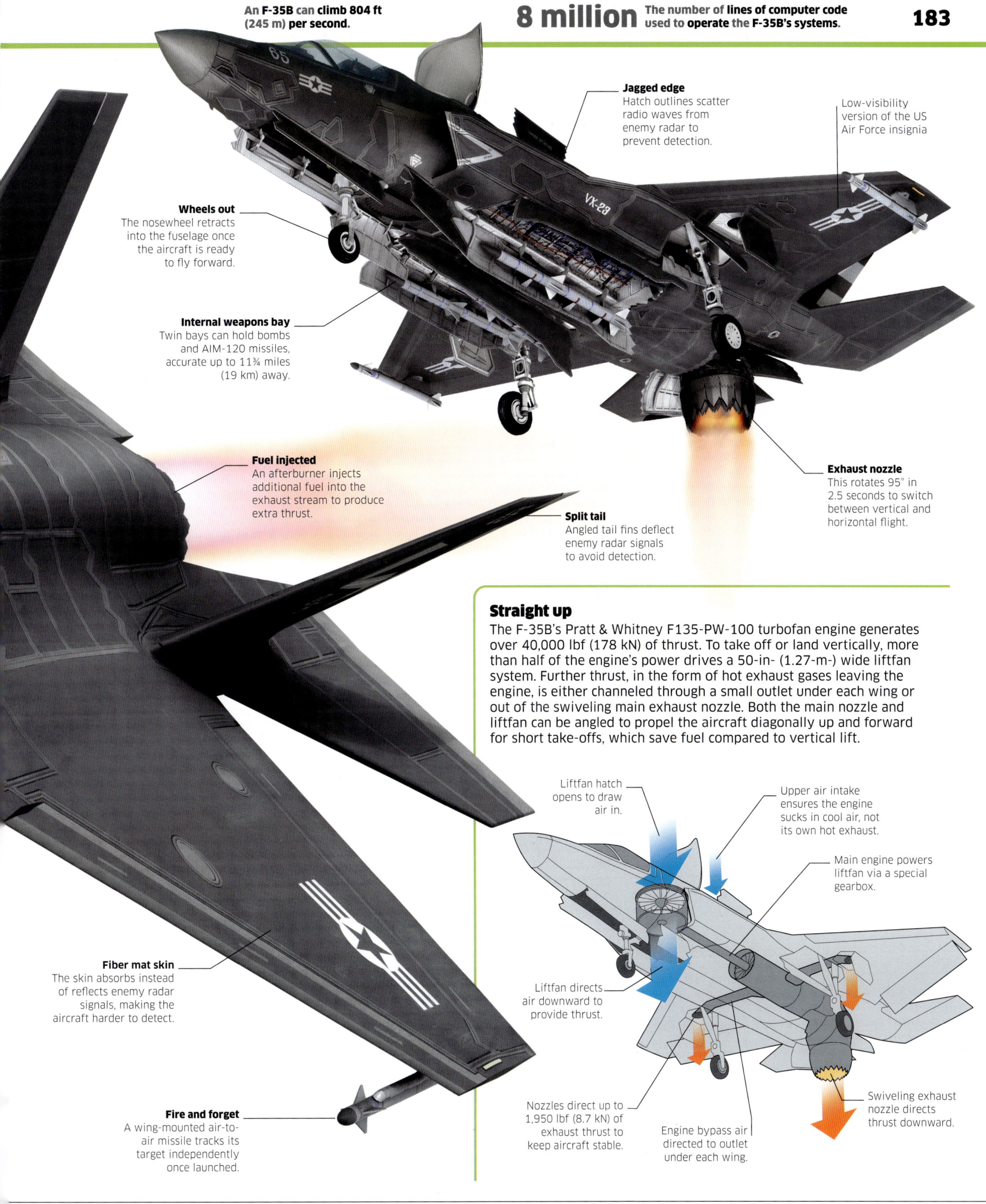

Jagged edge
Hatch outlines scatter radio waves from enemy radar to prevent detection.

Low-visibility version of the US Air Force insignia

Wheels out
The nosewheel retracts into the fuselage once the aircraft is ready to fly forward.

Internal weapons bay
Twin bays can hold bombs and AIM-120 missiles, accurate up to 11¾ miles (19 km) away.

Fuel injected
An afterburner injects additional fuel into the exhaust stream to produce extra thrust.

Split tail
Angled tail fins deflect enemy radar signals to avoid detection.

Exhaust nozzle
This rotates 95° in 2.5 seconds to switch between vertical and horizontal flight.

Fiber mat skin
The skin absorbs instead of reflects enemy radar signals, making the aircraft harder to detect.

Fire and forget
A wing-mounted air-to-air missile tracks its target independently once launched.

Straight up

The F-35B's Pratt & Whitney F135-PW-100 turbofan engine generates over 40,000 lbf (178 kN) of thrust. To take off or land vertically, more than half of the engine's power drives a 50-in- (1.27-m-) wide liftfan system. Further thrust, in the form of hot exhaust gases leaving the engine, is either channeled through a small outlet under each wing or out of the swiveling main exhaust nozzle. Both the main nozzle and liftfan can be angled to propel the aircraft diagonally up and forward for short take-offs, which save fuel compared to vertical lift.

Liftfan hatch opens to draw air in.

Upper air intake ensures the engine sucks in cool air, not its own hot exhaust.

Main engine powers liftfan via a special gearbox.

Liftfan directs air downward to provide thrust.

Nozzles direct up to 1,950 lbf (8.7 kN) of exhaust thrust to keep aircraft stable.

Engine bypass air directed to outlet under each wing.

Swiveling exhaust nozzle directs thrust downward.

184 air and space • **DRONES**

190 miles (300 km)—the **maximum range** of a Zipline drone on a single **charge**.

1.16 oz (33 g)—the **weight** of a **Black Hornet** drone.

ZIPLINE PLATFORM 1
Emergency delivery
Top speed: 63 mph (101 kph)
Length: 6 ft (1.8 m)

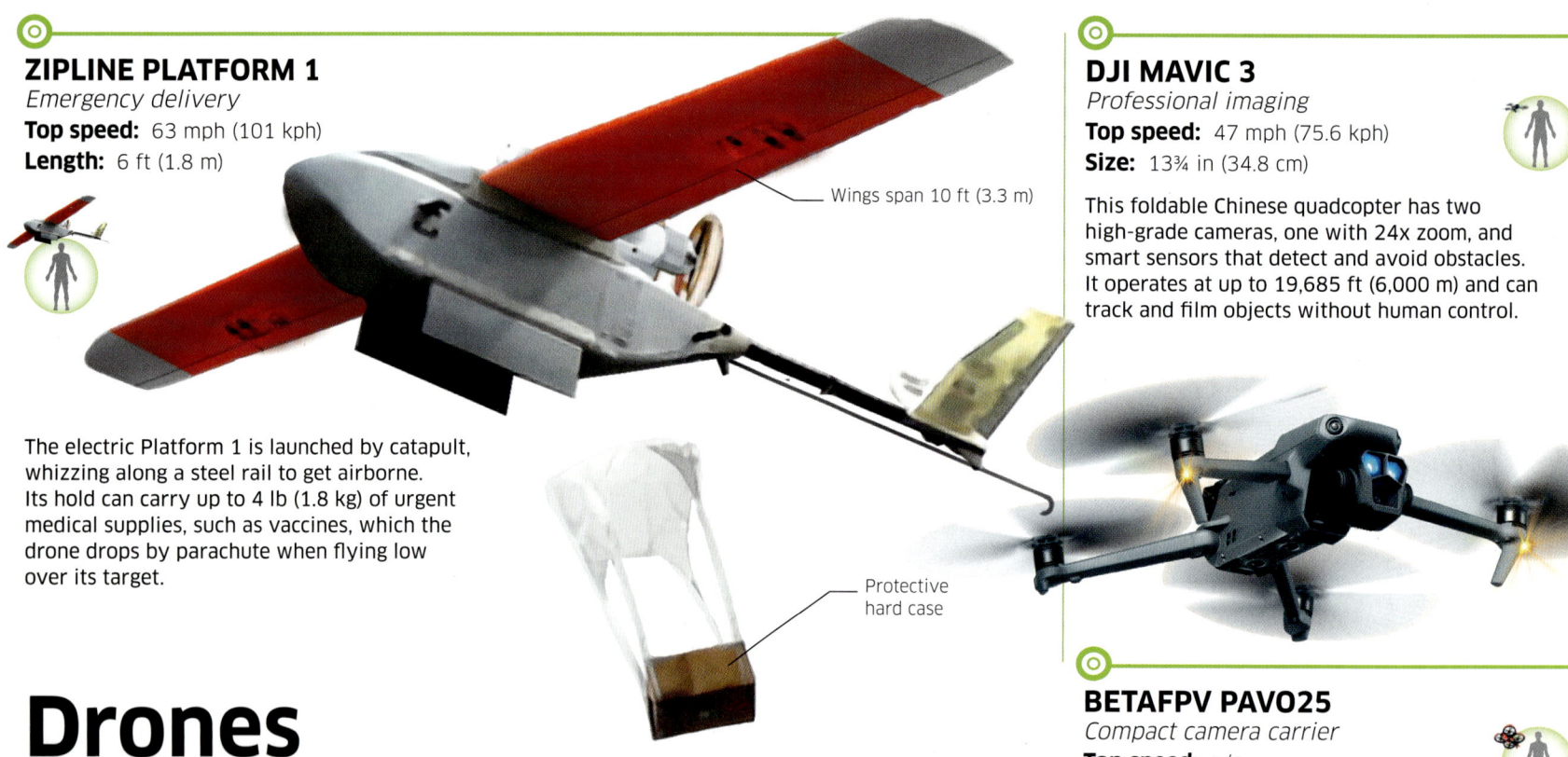

Wings span 10 ft (3.3 m)

The electric Platform 1 is launched by catapult, whizzing along a steel rail to get airborne. Its hold can carry up to 4 lb (1.8 kg) of urgent medical supplies, such as vaccines, which the drone drops by parachute when flying low over its target.

Protective hard case

Drones

Drones and uncrewed aerial vehicles (UAVs) dispense with an onboard human pilot. Some are flown for fun, while others are deployed to perform specific tasks.

First developed for military use in hostile situations, drones are now manufactured in the tens of thousands. They can be built and operated far more cheaply than larger, piloted planes and helicopters. Most drones are remote-controlled from the ground by human operators, who transmit commands via radio waves. Others fly autonomously, navigating with GPS, obstacle avoidance sensors, and an onboard computer.

DJI MAVIC 3
Professional imaging
Top speed: 47 mph (75.6 kph)
Size: 13¾ in (34.8 cm)

This foldable Chinese quadcopter has two high-grade cameras, one with 24x zoom, and smart sensors that detect and avoid obstacles. It operates at up to 19,685 ft (6,000 m) and can track and film objects without human control.

BETAFPV PAVO25
Compact camera carrier
Top speed: n/a
Size: 6¼ in (15.4 cm)

This small, "pusher"-style Chinese quadcopter's four rotors direct thrust downward. Suitable for indoor or outdoor flights of seven minutes between battery charges, it has a mount for HD cameras and is often used for filming in confined spaces.

Cradle for battery

Camera mount

Enclosed rotors
Molded plastic blade guard prevents damage or injury in tight flying areas.

TELEDYNE FLIR BLACK HORNET
Local reconnaissance
Top speed: 13.4 mph (21.6 kph)
Size: 6½ in (16.8 cm)

Developed in Norway, this military drone is just 6½ in (16.8 cm) long yet can fly for 25 minutes, sending night-vision video back to soldiers or police on the ground.

RQ4 GLOBAL HAWK
Long-range surveillance
Top speed: 357 mph (575 kph)
Size: 47½ ft (14.5 m)

This large US military drone has a wingspan of 131 ft (39.8 m) and can fly for 34 hours at a time. On board is a high-tech suite of radar, infrared, and optical sensors that can map up to 38,600 sq miles (100,000 sq km) per day.

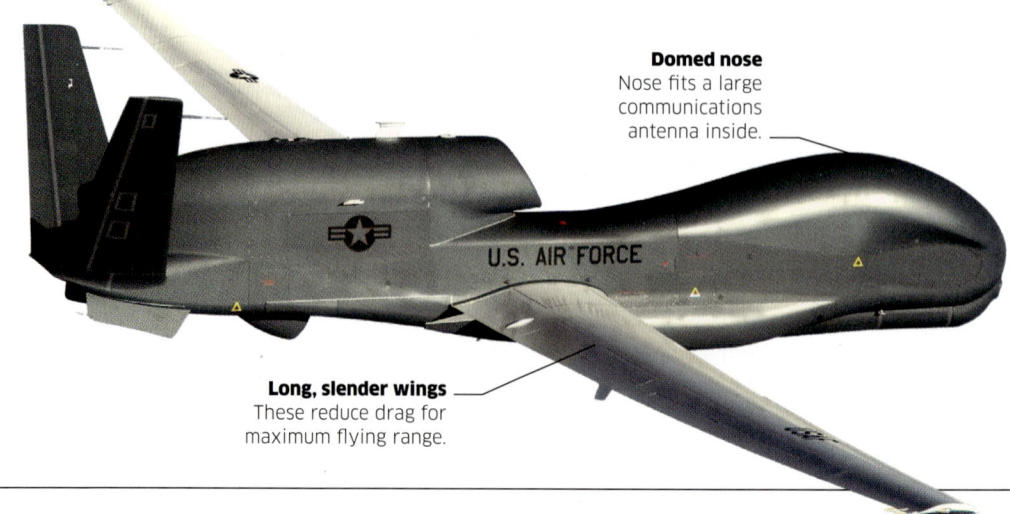

Domed nose
Nose fits a large communications antenna inside.

U.S. AIR FORCE

Long, slender wings
These reduce drag for maximum flying range.

2,000 rpm—the maximum speed of the MQ-9 Reaper's propellers.

A **Zephyr UAV** flew continuously for almost **26 days** in 2018.

2,600 The number of RMAX helicopters that operate **around the world**.

185

GENERAL ATOMICS MQ-9 REAPER
Precision striker

Top speed: 300 mph (482 kph)
Size: 36 ft (11 m)

This remotely operated American UAV stages precision strikes using weapons carried on seven hard points on its body and wings. It has a maximum range of 1,150 miles (1,850 km).

BOEING MQ-25 STINGRAY
Midair refueling

Top speed: 385 mph (620 kph)
Size: 50 ft (15.5 m)

This large US Navy drone carries up to 15,000 lb (6,800 kg) of aviation fuel, which it supplies in midair to carrier-based naval aircraft such as the F/A-18 and F-35C to extend their range. Its wings fold in half for storage on ships.

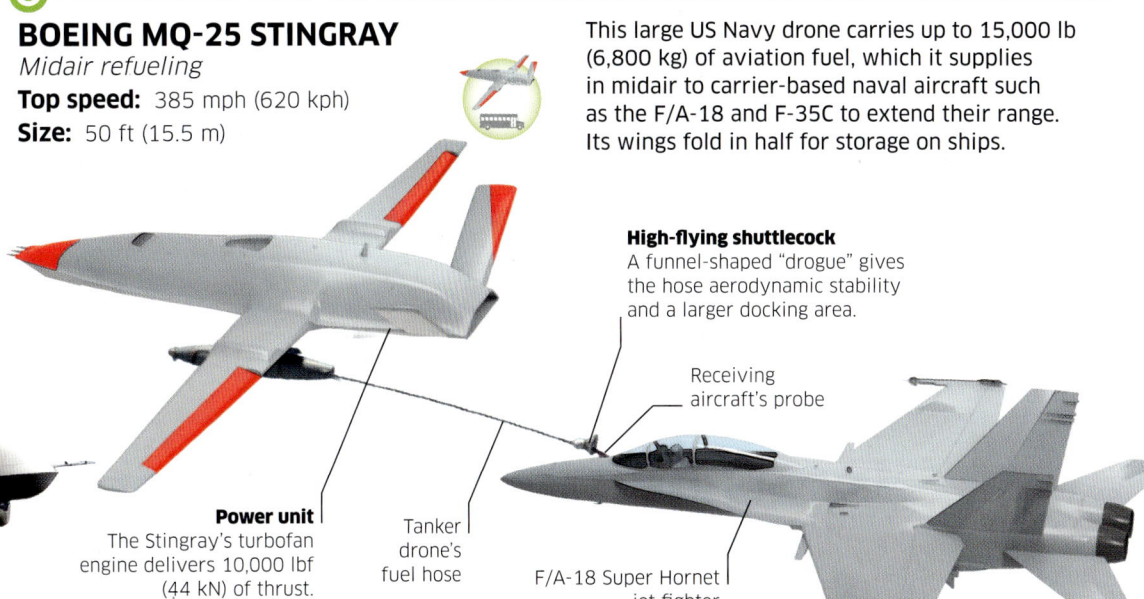

High-flying shuttlecock
A funnel-shaped "drogue" gives the hose aerodynamic stability and a larger docking area.

Receiving aircraft's probe

Power unit
The Stingray's turbofan engine delivers 10,000 lbf (44 kN) of thrust.

Tanker drone's fuel hose

F/A-18 Super Hornet jet fighter

REGA DRONE
Mountain search and rescue

Top speed: 37 mph (60 kph)
Size: 7¼ ft (2.2 m)

This Swiss drone is equipped with cameras, thermal imagers, and other sensors for finding people in rugged terrain. The Rega can scour a 6.2-sq-mile (16-sq-km) area in two hours.

Tail rotor
Stabilizes the drone by counteracting torque created by the main rotor blades.

LITTLE RIPPER DRONE
Life-saver

Top speed: 40 mph (64 kph)
Size: 5¼ ft (1.6 m)

These Australian drones fly for 25–30 minutes a time above the water, where their cameras spot sharks or find people in trouble. They can drop flotation aids or medicines to those in need.

Inflatable rescue pod

YAMAHA RMAX
Crop sprayer

Top speed: 12 mph (20 kph) (spraying)
Size: 12 ft (3.6 m)

This remote-controlled Japanese helicopter offers cheaper and more precise spraying than full-size, piloted aircraft. Its tank holds enough gas for a 60-minute flight.

Chemical tank
Two tanks each hold 2.38 gallons (8 liters).

VOLODRONE
Multirotor utility drone

Top speed: 68 mph (110 kph)
Size: 30 ft (9.2 m)

This German drone is designed to ferry payloads up to 440 lb (200 kg). Its aluminum and composite frame holds 18 electric rotors, all powered by lithium-ion battery packs.

Battery power
All nine battery packs can be swapped in under five minutes.

Hang tight!
Loads are slung beneath the craft.

186 air and space ○ SPACEFLIGHT

1958 The year **Vanguard 1, the oldest satellite still in orbit,** was launched.

Spaceflight

Outer space is a unique environment, unlike anything encountered on Earth. In order to perform in these rare conditions, spacecraft are designed to be very different from other forms of transportation.

To reach space, a spacecraft must first overcome the enormous pull of Earth's gravity. For the craft to reach a particular destination or travel on a planned trajectory (flight path), it needs to be designed and operated by people who are specialists in physics and understand how the laws of motion and gravity affect objects outside the atmosphere and far from Earth's surface. In addition, craft have to function in the hostile conditions of space, presenting another challenge for designers and engineers.

⊚ GETTING INTO SPACE

Once in space, where there is little or no air resistance, a spacecraft can keep moving without constant additional propulsion. However, to reach space a craft must both overcome the downward pull of gravity and pass through the many layers of Earth's atmosphere. Each layer has different properties, from the high drag of the dense troposphere, to the subzero cold of the mesosphere and the blistering heat of the thermosphere. Spacecraft require powerful upward thrust and resilient materials to reach and survive this harsh environment.

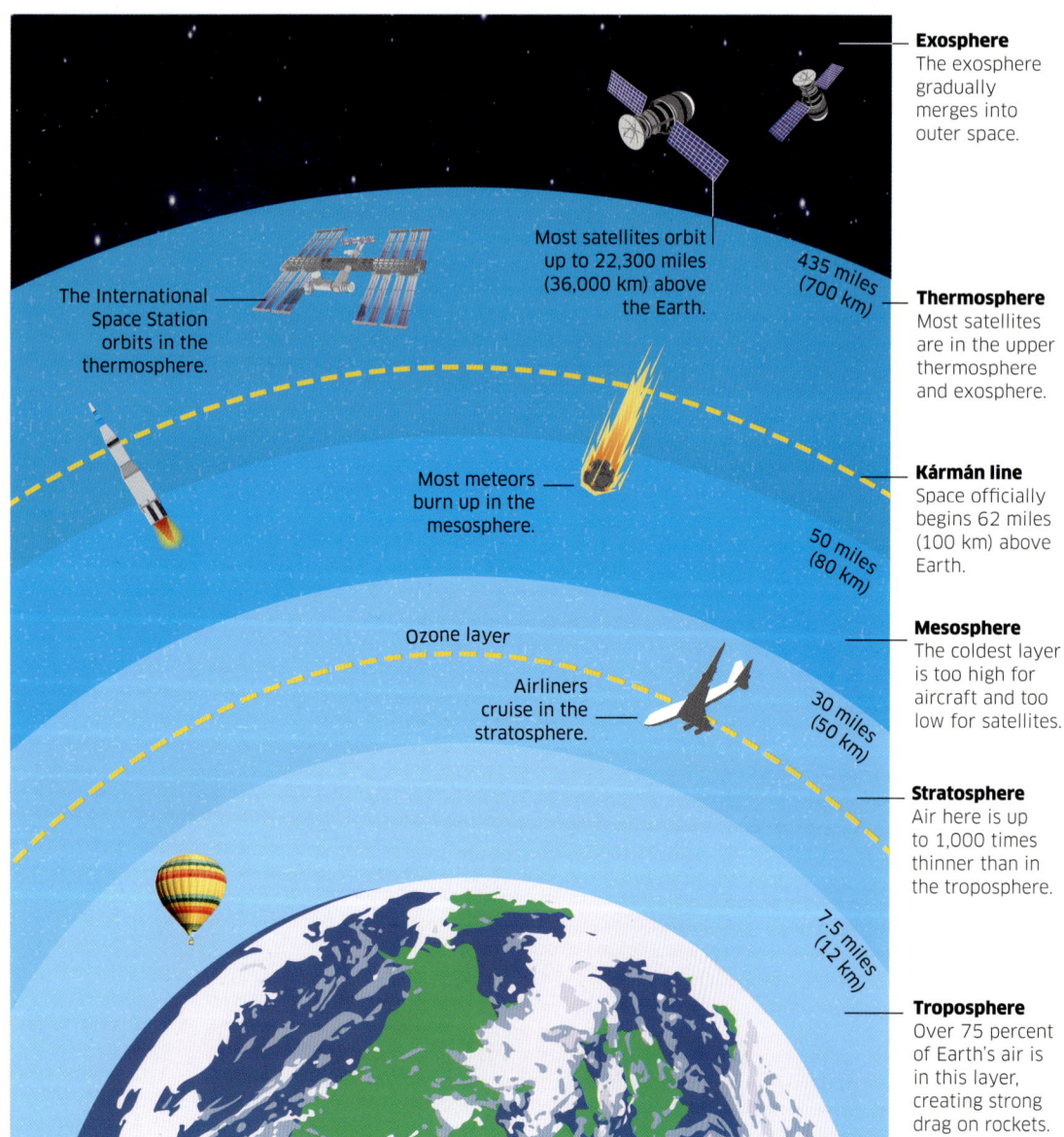

The International Space Station orbits in the thermosphere.

Most satellites orbit up to 22,300 miles (36,000 km) above the Earth.

Most meteors burn up in the mesosphere.

Ozone layer

Airliners cruise in the stratosphere.

435 miles (700 km)

50 miles (80 km)

30 miles (50 km)

7.5 miles (12 km)

Exosphere
The exosphere gradually merges into outer space.

Thermosphere
Most satellites are in the upper thermosphere and exosphere.

Kármán line
Space officially begins 62 miles (100 km) above Earth.

Mesosphere
The coldest layer is too high for aircraft and too low for satellites.

Stratosphere
Air here is up to 1,000 times thinner than in the troposphere.

Troposphere
Over 75 percent of Earth's air is in this layer, creating strong drag on rockets.

⊚ ORBITS

An orbit is a path in space that one object takes around another due to the force of gravity. To reach a stable orbit around Earth, a spacecraft must travel so fast at right angles to the force of gravity that its outward motion balances the inward pull. The ISS travels at around 17,900 mph (8 km/second) to maintain an orbit 250 miles (400 km) above the surface.

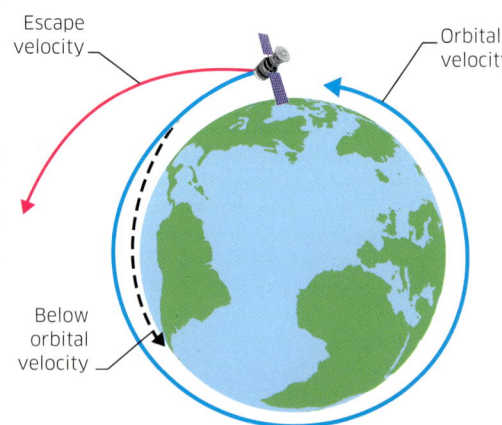

Escape velocity

Orbital velocity

Below orbital velocity

Entering and leaving orbit
At each altitude, there is a set speed an object must travel to remain in orbit—this is called orbital velocity. Orbits "decay," or slow, as there is still a little drag in inner space. If an orbiting craft slows down too much, it will fall out of orbit and plunge toward Earth. If a craft speeds up enough, it will reach escape velocity—at this speed, a craft is able to overcome gravity and fly toward outer space.

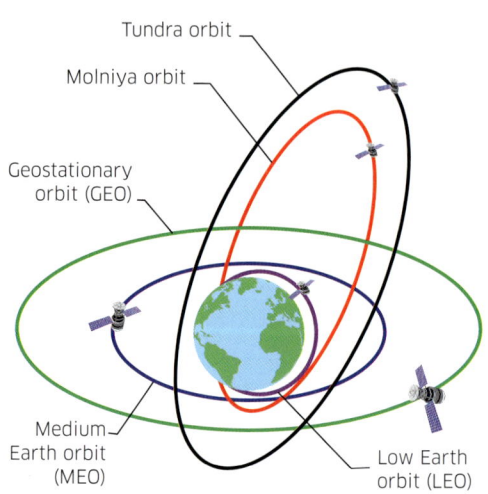

Tundra orbit

Molniya orbit

Geostationary orbit (GEO)

Medium Earth orbit (MEO)

Low Earth orbit (LEO)

Types of orbit
There are many types of orbit, each with different uses. Most artificial satellites and crewed craft fly in a low Earth orbit (LEO), circling the world every 90–128 minutes. This orbit is easiest to reach and good for observing the Earth's surface. A high geostationary orbit (GEO) above the equator allows communications satellites to stay over a single location, as their orbit matches the Earth's rotation below. Tundra and Molniya paths focus on polar regions, while medium Earth orbits (MEOs) are often used for navigation systems.

25,000 mph (40,233 kph)—the **speed** a spacecraft **needs to reach** **to escape Earth's gravity** and travel farther into space.

5,000°F (2,760°C)—the **temperature** spacecraft may face when **trying to reenter Earth's atmosphere.**

187

PROPULSION SYSTEMS

On Earth, vehicles can use traction, screw, or jet propulsion to push themselves forward using land, water, or air. In empty space, where there is not even air, spacecraft must rely on different methods.

Reaction engines

In order to generate forward thrust in a vacuum, spacecraft rely on reaction engines. These create thrust through the physical principle that any force acting on an object (an action) is balanced by an equal reaction force in the opposite direction. Reaction engines expel mass in one direction, which pushes the spacecraft in the other.

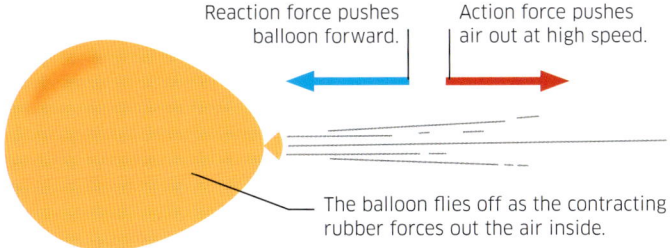

Reaction force pushes balloon forward.

Action force pushes air out at high speed.

The balloon flies off as the contracting rubber forces out the air inside.

Chemical rockets

Launch vehicles today create forward thrust through a controlled burning of chemical fuel. This produces expanding hot gases that drive the vehicle forward. To work in airless space, rockets must mix fuel with a chemical called an oxidizer, which supplies oxygen so that the fuel can combust (burn).

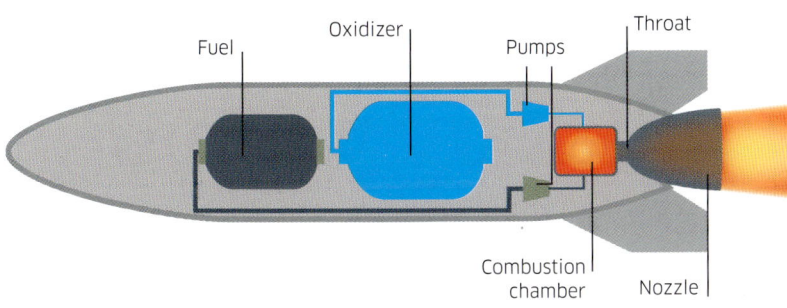

Fuel

Oxidizer

Pumps

Throat

Combustion chamber

Nozzle

Ion engines

Uncrewed space probes may carry a reaction engine called an ion thruster. It uses solar electricity to expel electrically charged particles (ions) at high speeds, driving the spacecraft in the opposite direction. Ion thrusters generate weak thrust but are fuel-efficient, boosting spacecraft to high speeds over many months.

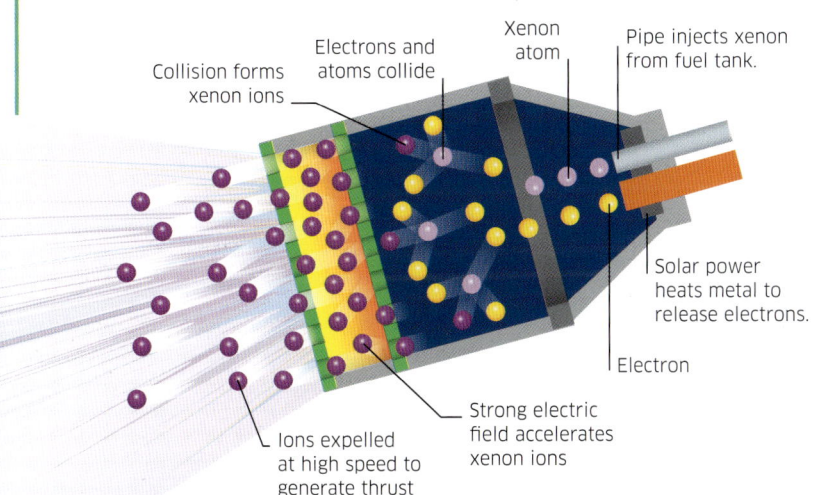

Electrons and atoms collide

Xenon atom

Pipe injects xenon from fuel tank.

Collision forms xenon ions

Solar power heats metal to release electrons.

Electron

Ions expelled at high speed to generate thrust

Strong electric field accelerates xenon ions

ROCKET STAGES

Rockets require large amounts of heavy propellant to reach orbit. For this reason, launchers are designed in multiple stages, each with its own fuel tanks and engines. As each stage runs out of fuel, it separates and falls away to save weight, leaving the upper stages to continue into space.

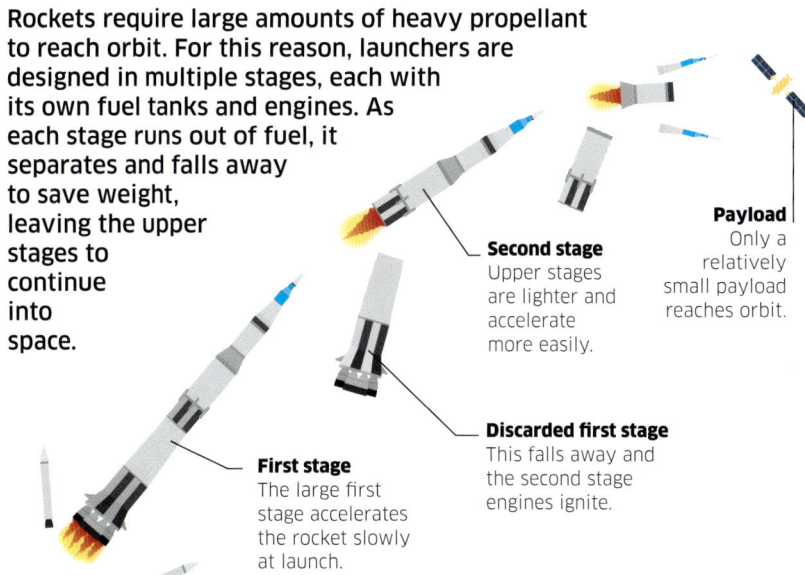

Payload
Only a relatively small payload reaches orbit.

Second stage
Upper stages are lighter and accelerate more easily.

Discarded first stage
This falls away and the second stage engines ignite.

First stage
The large first stage accelerates the rocket slowly at launch.

DESIGNING FOR SPACE TRAVEL

Space is a very hostile environment with many unique challenges for spacecraft designers. These include exposure to radiation (intense X-rays and damaging high-speed particles from the Sun) and the extremes of hot and cold. Spacecraft electronics are designed to survive radiation damage, while insulation and coolant pipes help even out extreme temperatures.

Aerodynamics
Spacecraft can be designed in any shape because they mostly operate in airless space. For launch, they are folded under a streamlined cover. Here, the Apollo 11 command module (top half) carries the first Moon lander.

Reentry challenges
Spacecraft that reenter the atmosphere experience extreme heat as they hit gas particles at high speeds. Protective shields are designed to resist heat, or burn off to carry the heat energy away from the craft, as on this Apollo capsule.

Crewed and robot spacecraft
Crewed spacecraft are mostly larger and heavier than uncrewed ones because they carry all the equipment needed to support and protect the astronauts on board. Compact robot vehicles are smaller and lighter, so they are easier to send on long-distance missions, such as the Galileo probe that studied Jupiter and its moons (above).

Launch vehicles

Designed to overcome the pull of gravity and deliver objects and people into outer space, rocket-boosted launch vehicles are the most powerful forms of transportation ever built.

A launch vehicle is any form of transportation capable of reaching outer space–more than 62 miles (100 km) above Earth's surface. These craft need to generate huge amounts of thrust to escape the pull of Earth's gravity and must also be able to work in the vacuum of space. The only practical launch vehicles capable of meeting these challenges so far have all been chemical-fueled rockets powered by the violent combustion of liquid or solid fuel.

V-2
Origin: Germany
Years: 1942–1945
Launch mass: 13.8 tons
(12.5 tonnes)

A weapon of war designed by scientists working for the German military during World War II, the V-2 missile became the first vehicle to enter space. In the late 1940s, captured V-2s were taken to the US and used to carry animals and scientific instruments on suborbital test flights.

Tail fins with steering rudders

R-7 SEMYORKA
Origin: Soviet Union
Years: 1957–1961
Launch mass: 310 tons
(280 tonnes)

The R-7 was converted from a missile originally designed for the Soviet Union's nuclear program and had an additional upper rocket added to allow it to reach orbit. On October 4, 1957, it launched Sputnik 1–the first ever artificial satellite– into orbit.

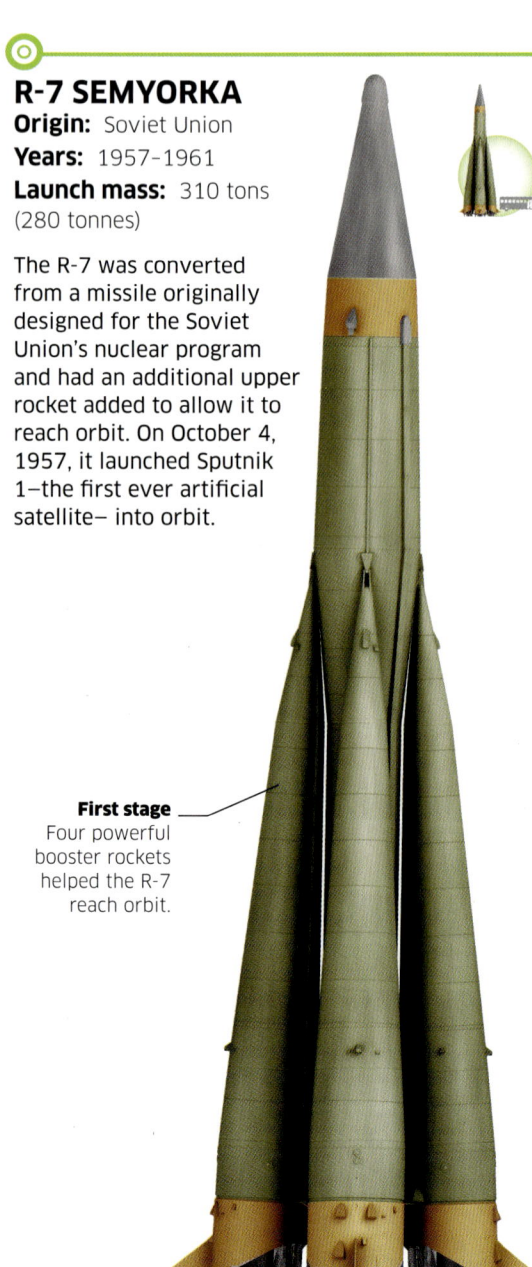

First stage
Four powerful booster rockets helped the R-7 reach orbit.

MERCURY-REDSTONE
Origin: US
Years: 1960–1961
Launch mass: 33 tons
(30 tonnes)

The first rocket to put an American in space was developed by a team of German rocket scientists working in the US. A modified Redstone was used to launch the first US satellite, Explorer 1, in 1958.

SOYUZ-FG
Origin: Soviet Union/Russia
Years: 2001–2019
Launch mass: 336 tons
(305 tonnes)

The workhorse of the Soviet–and later Russian–space program, the Soyuz rocket family has made more than 1,900 launches since its debut in the 1960s. The latest Soyuz rockets still use a core first stage aided by four large boosters.

Lower stages burn
refined kerosene fuel with liquid oxygen.

SATURN V
Origin: US
Years: 1967–1973
Launch mass: Up to 3,257 tons (2,955 tonnes)

To put astronauts on the Moon, NASA developed a giant rocket. The Saturn V consisted of three stages–the first two powered by five engines. The final, upper stage had a single engine, which carried Apollo spacecraft from Earth orbit toward the Moon.

Upper stages used
lighter hydrogen fuel to reduce mass.

Heavy fuel
The first stage rocket engines used energy-rich and dense kerosene as fuel.

500,000 The **number of cars** that it would take to generate the same **power** as the **SLS** at **liftoff**.

3 days—the **length of the journey** from **Earth** to the **Moon**.

189

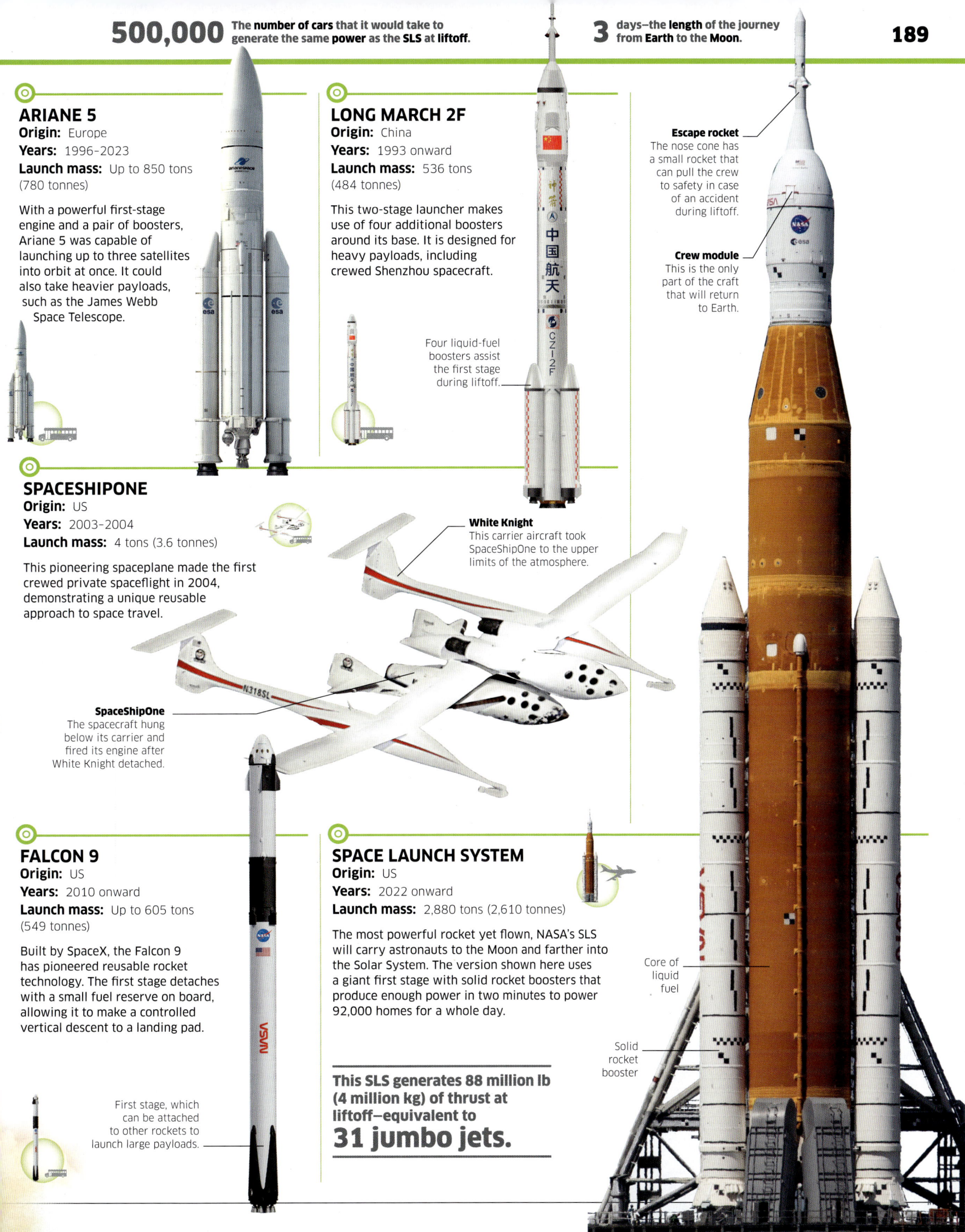

ARIANE 5
Origin: Europe
Years: 1996–2023
Launch mass: Up to 850 tons (780 tonnes)

With a powerful first-stage engine and a pair of boosters, Ariane 5 was capable of launching up to three satellites into orbit at once. It could also take heavier payloads, such as the James Webb Space Telescope.

LONG MARCH 2F
Origin: China
Years: 1993 onward
Launch mass: 536 tons (484 tonnes)

This two-stage launcher makes use of four additional boosters around its base. It is designed for heavy payloads, including crewed Shenzhou spacecraft.

Four liquid-fuel boosters assist the first stage during liftoff.

Escape rocket
The nose cone has a small rocket that can pull the crew to safety in case of an accident during liftoff.

Crew module
This is the only part of the craft that will return to Earth.

SPACESHIPONE
Origin: US
Years: 2003–2004
Launch mass: 4 tons (3.6 tonnes)

This pioneering spaceplane made the first crewed private spaceflight in 2004, demonstrating a unique reusable approach to space travel.

White Knight
This carrier aircraft took SpaceShipOne to the upper limits of the atmosphere.

SpaceShipOne
The spacecraft hung below its carrier and fired its engine after White Knight detached.

FALCON 9
Origin: US
Years: 2010 onward
Launch mass: Up to 605 tons (549 tonnes)

Built by SpaceX, the Falcon 9 has pioneered reusable rocket technology. The first stage detaches with a small fuel reserve on board, allowing it to make a controlled vertical descent to a landing pad.

First stage, which can be attached to other rockets to launch large payloads.

SPACE LAUNCH SYSTEM
Origin: US
Years: 2022 onward
Launch mass: 2,880 tons (2,610 tonnes)

The most powerful rocket yet flown, NASA's SLS will carry astronauts to the Moon and farther into the Solar System. The version shown here uses a giant first stage with solid rocket boosters that produce enough power in two minutes to power 92,000 homes for a whole day.

Core of liquid fuel

Solid rocket booster

This SLS generates 88 million lb (4 million kg) of thrust at liftoff—equivalent to

31 jumbo jets.

Rolling rocket

Spacefaring vehicles are built to be huge but must be transported from their construction site to the launchpad—traveling by planes, boats, and large land-crawling vehicles.

In this 2020 photograph, part of NASA's Space Launch System rocket is being transferred to the Stennis Space Center, Mississippi, for a test run. Workers prepare to load the rocket core stage onto a specially adapted barge called Pegasus, which is 310 ft (94 m) long—longer than an American football field. The barge was originally built in 1999 and has carried equipment for many NASA missions.

192 air and space ○ **CREWED SPACECRAFT**

24 The number of people who **traveled to the Moon** from 1968 to 1972. All were **US astronauts** and **12 walked on the Moon**.

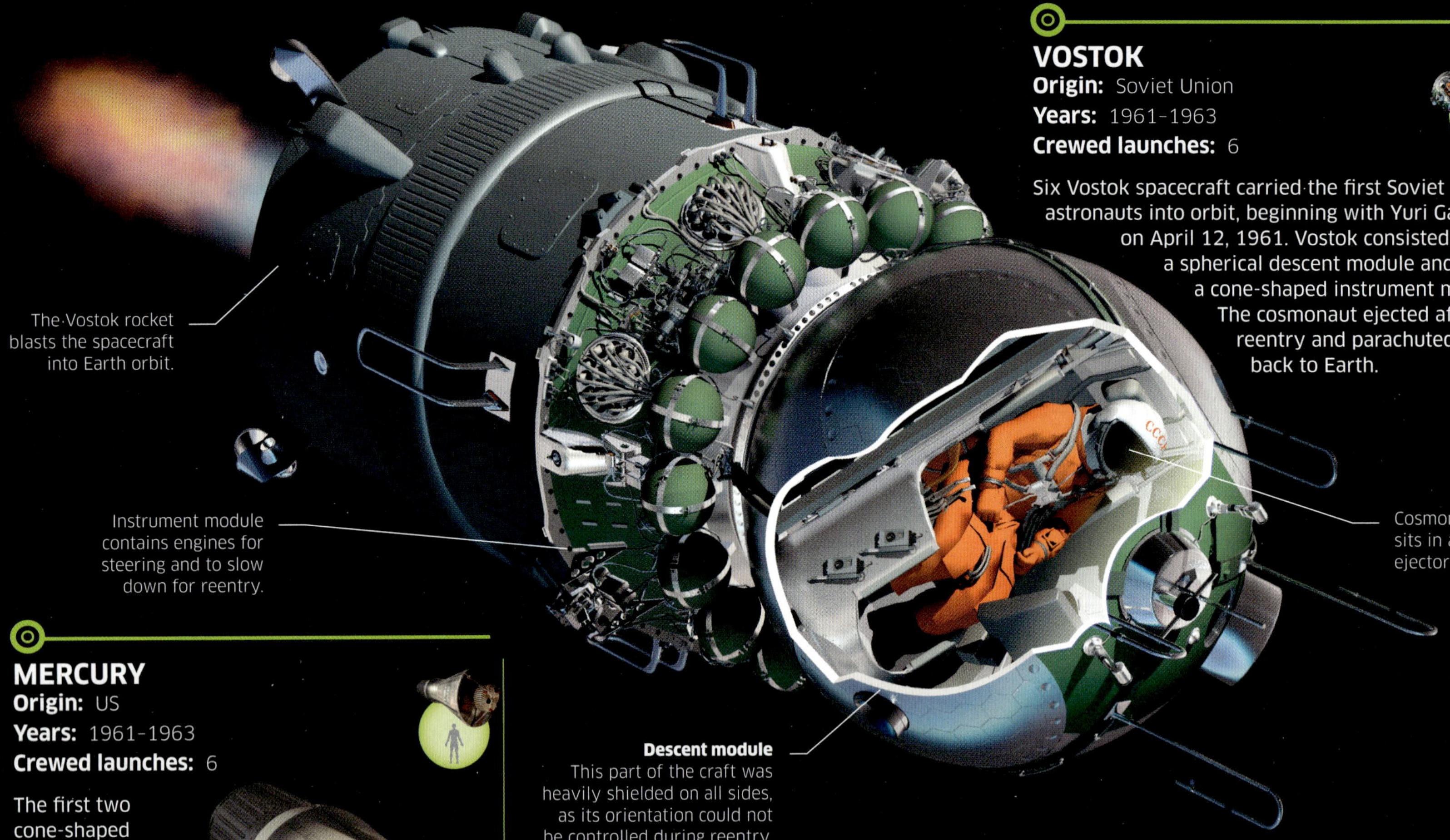

The Vostok rocket blasts the spacecraft into Earth orbit.

Instrument module contains engines for steering and to slow down for reentry.

VOSTOK

Origin: Soviet Union
Years: 1961–1963
Crewed launches: 6

Six Vostok spacecraft carried the first Soviet astronauts into orbit, beginning with Yuri Gagarin on April 12, 1961. Vostok consisted of a spherical descent module and a cone-shaped instrument module. The cosmonaut ejected after reentry and parachuted back to Earth.

Cosmonaut sits in an ejector seat.

Descent module
This part of the craft was heavily shielded on all sides, as its orientation could not be controlled during reentry.

MERCURY

Origin: US
Years: 1961–1963
Crewed launches: 6

The first two cone-shaped Mercury capsules made suborbital flights in 1961. Four later missions used the Atlas rocket to reach orbit. Once the capsule had been slowed in reentry, parachutes eased it to an ocean splashdown.

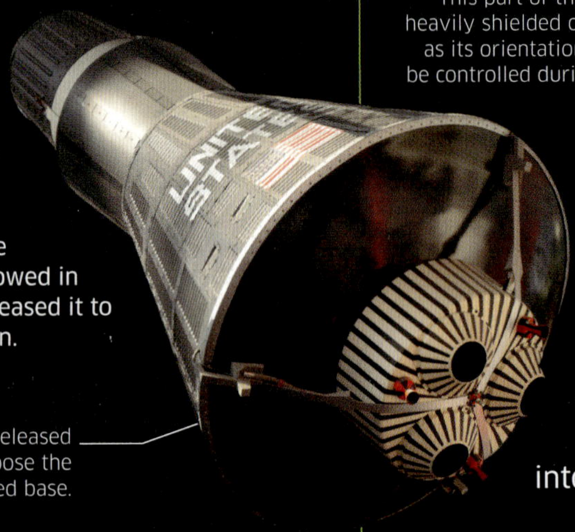

Rocket pack was released before reentry to expose the broad, heat-shielded base.

Crewed spacecraft

Some of the most complex vehicles ever designed, crewed spacecraft are made up of multiple connected modules, each of which does a different job, such as propulsion or storage.

A spacecraft must not only withstand the stresses of a rocket launch, but also carry supplies and life-support systems. In addition, it needs propulsion devices for steering, shields to withstand the heat of reentry into Earth's atmosphere, and parachutes or braking retrorockets for landing.

VOSKHOD

Origin: Soviet Union
Years: 1964–1965
Crewed launches: 2

Voskhod 1 crammed three cosmonauts aboard in 1964, while 1965's two-man Voskhod 2 (shown here) had an airlock, permitting cosmonaut Alexei Leonov to make the first spacewalk.

GEMINI

Origin: US
Years: 1964–1966
Crewed launches: 10

Gemini was capable of long flights and could rendezvous and dock with other vehicles. It carried engines in an adapter module mounted behind the cone-shaped reentry module.

Adapter module

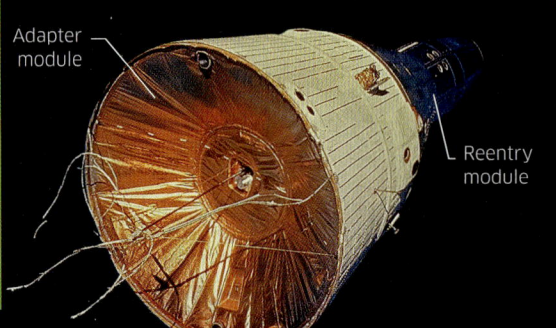

Reentry module

SOYUZ

Origin: Soviet Union/Russia
Years: 1967 onward
Crewed launches: Around 150

Often used to take astronauts to the International Space Station (ISS), Soyuz is made up of a spherical orbital module, a bell-shaped reentry module, and a cylindrical service module.

Solar panel "wings" provide additional power.

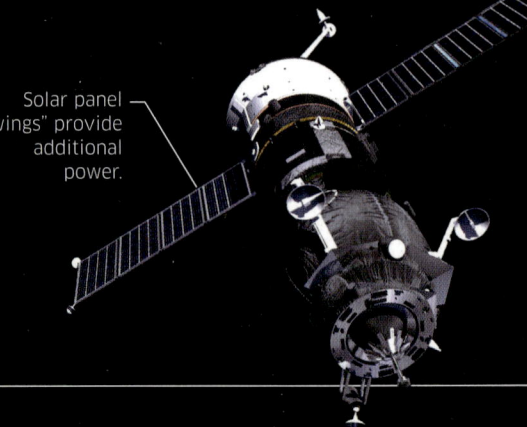

1 month–the length of time **life-support** systems on the **Soyuz spacecraft** can keep the crew alive.

300 The total number of hours **Apollo astronauts** spent on the **Moon**.

193

APOLLO
Origin: US
Years: 1968–1975
Crewed launches: 15

Six Apollo craft made successful landings on the Moon between 1969 and 1972. At the Moon, two astronauts boarded the lunar module for the surface mission, while a third remained in orbit with the command and service modules.

Lunar lander
After flying to the Moon's surface and back, this module was discarded.

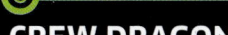

The service module housed the engines used for entering and leaving lunar orbit.

The cone-shaped command module was the only part of the craft that returned to Earth, with the astronauts.

Large observation windows in capsule for up to six passengers.

Booster rocket, designed for reuse following a controlled upright landing.

BLUE ORIGIN

SHENZHOU
Origin: China
Years: 2003 onward
Crewed launches: 11+

Shenzhou has a similar overall design to Soyuz but is larger, with a cylindrical orbital module that carries its own thrusters and solar panels for independent orbital operations.

CREW DRAGON
Origin: US
Years: 2020 onward
Crewed launches: 11+

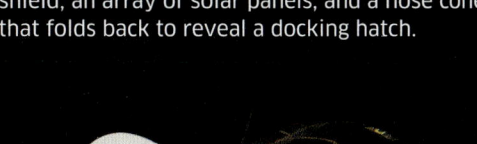

SpaceX's craft can support a crew of four and is partially reusable, with a replaceable heat shield, an array of solar panels, and a nose cone that folds back to reveal a docking hatch.

NEW SHEPARD
Origin: US
Years: 2021 onward
Crewed launches: 11+

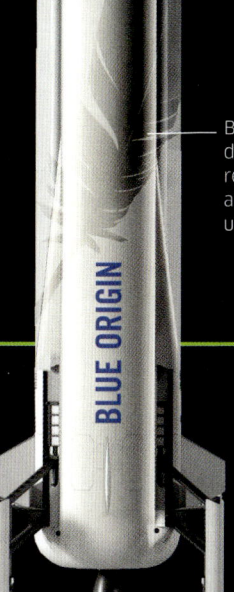

Designed for space tourism, this reusable launch vehicle uses a first-stage "booster" rocket that reaches heights of more than 62 miles (100 km). It releases a crew capsule that makes a brief trip into space before falling back for a controlled landing.

ORION
Origin: US
Years: Crewed launches from 2024
Crewed launches: None yet

NASA's Orion spacecraft is designed for missions to the Moon and beyond. It can carry six people for up to 21 days (or longer if docked to another spacecraft).

European Service Module provides power and propulsion as well as cargo storage space.

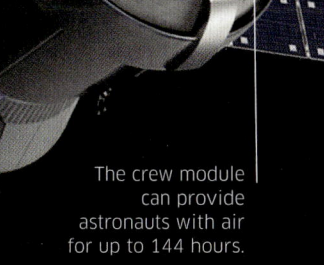

The crew module can provide astronauts with air for up to 144 hours.

BOEING STARLINER
Origin: US
Years: Crewed launches from 2024
Crewed launches: None yet

Starliner is capable of carrying a full crew of seven astronauts, or a mix of crew and cargo, to the ISS. It uses parachutes, rockets, and airbags to cushion the impact of a touchdown on solid ground.

Observation window

542 million miles (872 million km)—the **distance traveled** by the 135 space shuttle missions between 1981 and 2011.

Mission profile
The complete shuttle system consisted of an orbiter, an external tank (ET) containing liquid fuel for the orbiter's main engines, and a pair of solid rocket boosters (SRBs) for extra thrust at liftoff. The shuttle launched vertically, with fuel from the ET pumped to the orbiter's main engines. Following reentry, the vehicle flew as an unpowered glider, using its delta-wing design to provide control as it descended to a runway landing.

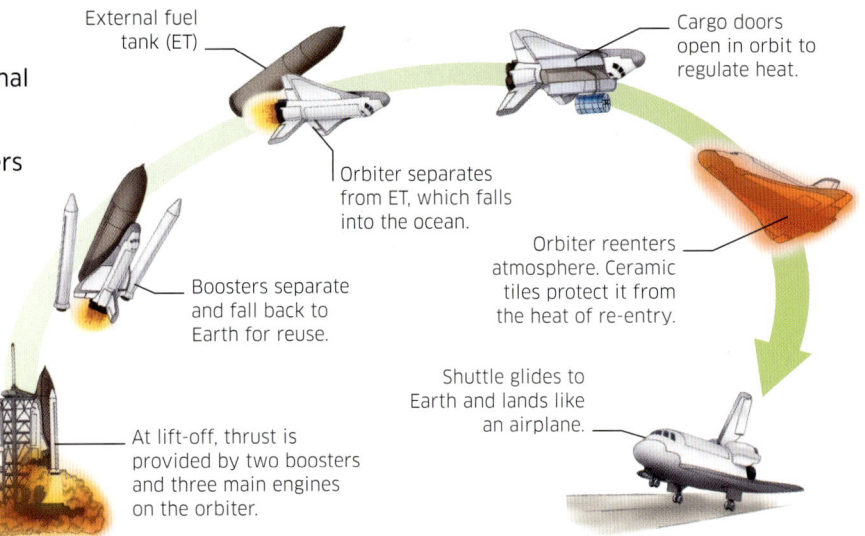

External fuel tank (ET)

Cargo doors open in orbit to regulate heat.

Orbiter separates from ET, which falls into the ocean.

Boosters separate and fall back to Earth for reuse.

Orbiter reenters atmosphere. Ceramic tiles protect it from the heat of re-entry.

At lift-off, thrust is provided by two boosters and three main engines on the orbiter.

Shuttle glides to Earth and lands like an airplane.

Hubble Space Telescope
This was the shuttle's most famous cargo, launched from *Discovery* in 1990 and retrieved for servicing on five later shuttle missions.

SPACE SHUTTLE ORBITER
Origin: US

Year: 1981

Length: 122 ft (37.2 m)

Speed: 17,500 mph (28,200 kph)

Windows
Three layers of glass protected the interior from heat and space impacts.

Seating
The shuttle typically launched with four crew members on the flight deck and three on the middeck.

Reaction Control System (RCS) engines
These small thrusters controlled the shuttle's orientation in orbit.

Crew access hatch
This included an emergency escape system, installed after seven astronauts lost their lives when *Challenger* blew apart in 1986.

Space shuttle

The space shuttle was a revolutionary system devised by US space agency NASA to launch and repair satellites, conduct experiments in orbit, and study Earth and the universe from space.

Developed in the 1970s, the shuttle was an ambitious attempt to make space travel cheaper, easier, and more routine than costly single-use rocket launches. Most of its components were designed to fly into space many times, enabling regular launches.

On the flight deck
The original space shuttle flight deck had around 2,100 individual controls, but upgrades in the early 2000s introduced a simpler layout, as shown here in *Atlantis*. Each front seat had a rotational hand controller "joystick" and rudder pedals to control the shuttle.

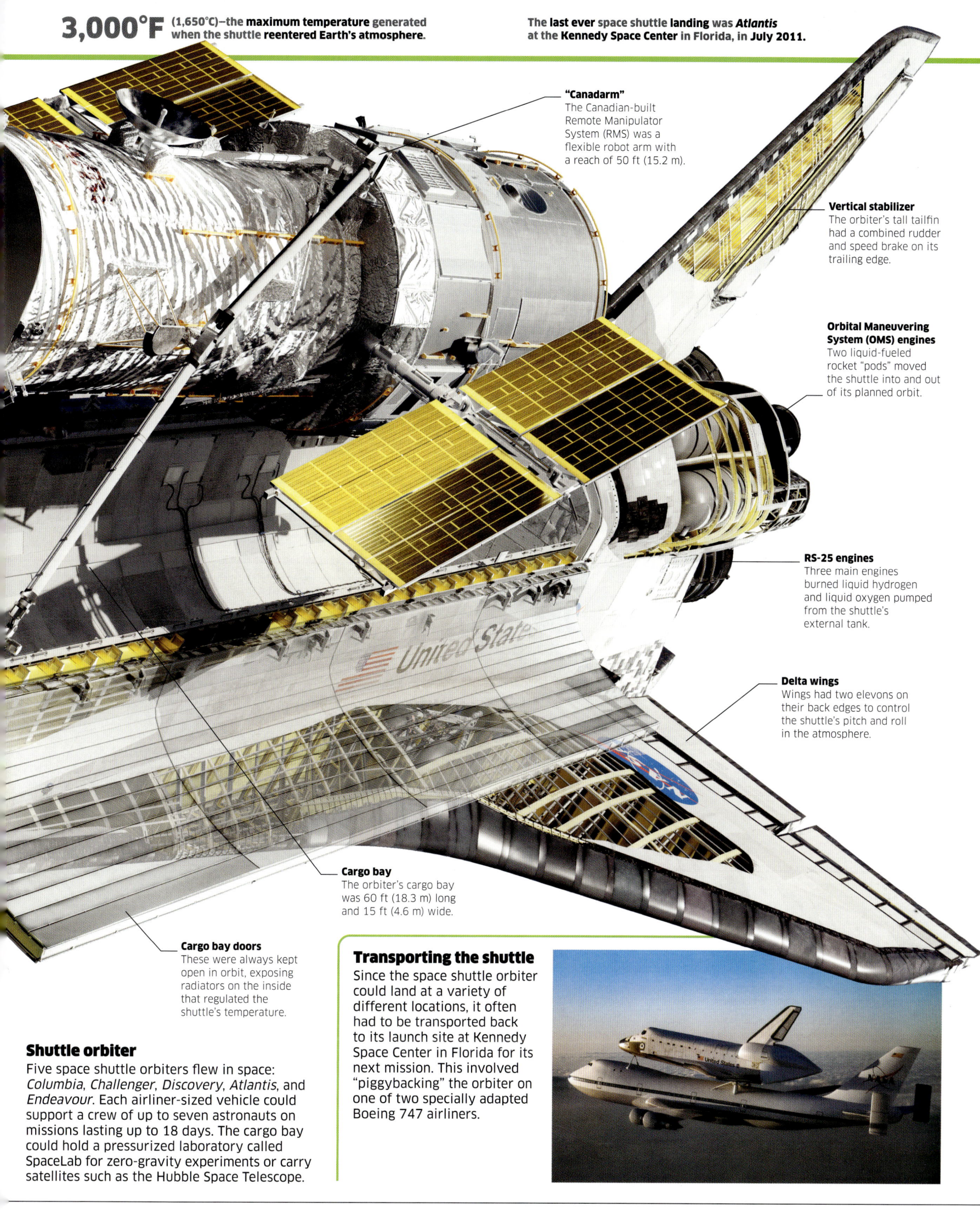

3,000°F (1,650°C)—the **maximum temperature** generated when the shuttle **reentered Earth's atmosphere.**

The **last ever** space shuttle landing was *Atlantis* at the Kennedy Space Center in Florida, in July 2011.

"Canadarm"
The Canadian-built Remote Manipulator System (RMS) was a flexible robot arm with a reach of 50 ft (15.2 m).

Vertical stabilizer
The orbiter's tall tailfin had a combined rudder and speed brake on its trailing edge.

Orbital Maneuvering System (OMS) engines
Two liquid-fueled rocket "pods" moved the shuttle into and out of its planned orbit.

RS-25 engines
Three main engines burned liquid hydrogen and liquid oxygen pumped from the shuttle's external tank.

Delta wings
Wings had two elevons on their back edges to control the shuttle's pitch and roll in the atmosphere.

Cargo bay
The orbiter's cargo bay was 60 ft (18.3 m) long and 15 ft (4.6 m) wide.

Cargo bay doors
These were always kept open in orbit, exposing radiators on the inside that regulated the shuttle's temperature.

Shuttle orbiter

Five space shuttle orbiters flew in space: *Columbia, Challenger, Discovery, Atlantis,* and *Endeavour.* Each airliner-sized vehicle could support a crew of up to seven astronauts on missions lasting up to 18 days. The cargo bay could hold a pressurized laboratory called SpaceLab for zero-gravity experiments or carry satellites such as the Hubble Space Telescope.

Transporting the shuttle

Since the space shuttle orbiter could land at a variety of different locations, it often had to be transported back to its launch site at Kennedy Space Center in Florida for its next mission. This involved "piggybacking" the orbiter on one of two specially adapted Boeing 747 airliners.

196 air and space ○ UNCREWED SPACECRAFT

453,048 The **number of images** taken by **Cassini** over its 20-year mission.

Uncrewed spacecraft

Sending humans into space far beyond Earth's orbit presents huge challenges. It is much easier for robot space probes to transport complex scientific instruments across these vast distances.

Crewed spacecraft must carry large amounts of equipment and supplies simply to keep astronauts alive. Automated probes face no such constraints and can be sent on one-way trips to hostile environments. Their relatively low mass can help them make high-speed journeys to destinations where human exploration is impossible. Advances in electronics, computer control, power systems, and the miniaturization of instruments allow them to act as increasingly versatile robot explorers.

Thermal control shutters
These open and close to control the internal temperature of the craft.

Sensors
Sun and Moon sensors detect the probe's orientation in space.

LUNA 3
Launched: 1959
Destination: The Moon
Mass: 614 lb (278.5 kg)

This Soviet space probe sent back the first pictures of the Moon's far side. Photos were taken on film before being scanned and sent back to Earth as radio signals.

Antenna
The many antennas allow communication with Earth.

PIONEER 5
Launched: 1960
Destination: Interplanetary space
Mass: 95 lb (43.2 kg)

Studying space between Earth and Venus, NASA's Pioneer 5 probe spun once every 12 seconds to remain stable in space. It was powered by batteries recharged by solar-electric cells.

SURVEYOR 1
Launched: 1966 
Destination: The Moon
Mass: 644 lb (294.3 kg)

NASA's Lunar Surveyors paved the way for the Apollo Moon missions. With the help of radar systems, plus rockets and thrusters to slow their descent, they could land softly on the surface.

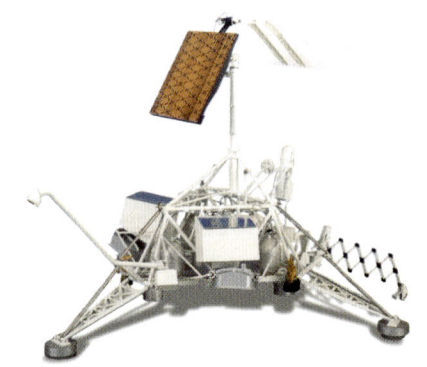

LUNOKHOD 1
Launched: 1970 
Destination: The Moon
Mass: 1,667 lb (756 kg)

The first wheeled vehicle on the Moon, Lunokhod 1 was a large robot rover with eight wheels. Batteries were topped up using solar cells on a tilting lid that closed at night.

1998 The year **Voyager 1** traveled past Pioneer 10—making it the **most distant human-made object in space.**

4 ft (1.2 m)—the **length of the rotors** on Ingenuity, which can spin up to 2,400 revolutions per minute.

197

VENERA 9
Launched: 1975
Destination: Venus
Mass: 3,440 lb (1,560 kg)

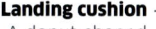

The first probe to return pictures from the hostile surface of Venus, Soviet craft Venera 9 used a heavily armored pressure sphere to protect its cameras and instruments. It survived for 53 minutes before contact was lost.

Landing cushion
A donut-shaped ring with flexible metal suspension cushioned the probe's landing.

CASSINI
Launched: 1997
Destination: Saturn
Mass: 12,593 lb (5,712 kg)

The bus-sized Cassini orbited Saturn for 13 years from 2004, making flybys of the planet's moons and rings. On arrival, it released Huygens, a lander, onto Saturn's moon Titan.

Magnetometer
This measured the magnetic field around Saturn.

VOYAGER 2
Launched: 1977
Destination: Outer solar system
Mass: 1,592 lb (721.9 kg)

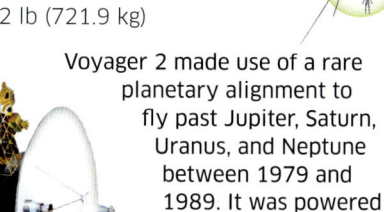

Voyager 2 made use of a rare planetary alignment to fly past Jupiter, Saturn, Uranus, and Neptune between 1979 and 1989. It was powered by a generator that used heat from a radioactive source.

Radio dish
The 12-ft- (3.7-m-) wide dish enabled long-distance communication with Earth.

STARDUST
Launched: 1999
Destination: Comet Wild 2
Mass: 849 lb (385 kg)

Most probes make a one-way trip, but a few return to Earth. Stardust flew past Comet Wild 2 in 2004, collecting comet dust before dropping a sample capsule back through Earth's atmosphere.

Sample capsule

DAWN
Launched: 2007
Destination: Vesta, Ceres
Mass: 2,685 lb (1,217.7 kg)

Thanks to its ion engine (see pages 186–187), Dawn was able to travel far into the asteroid belt, becoming the first probe to survey two large extraterrestrial bodies: Vesta and Ceres.

Solar panels
These generated electricity to power the engine.

PARKER SOLAR PROBE
Launched: 2018
Destination: The Sun
Mass: 1,510 lb (685 kg)

The Parker Solar Probe orbits close to the Sun's outer atmosphere. Instruments analyze material crossing their path, while a lightweight shield keeps them in permanent shade.

Shield
The carbon-composite shield resists temperatures of up to 2,500°F (1,370°C).

PERSEVERANCE
Launched: 2020
Destination: Mars
Mass: 2,260 lb (1,025 kg)

Perseverance is the most advanced of all the wheeled rovers to explore Mars. An array of cameras help it steer, while a robot arm can analyze the soil and drill out rock samples.

Suspension
The suspension system keeps the craft's body level on rocky terrain.

INGENUITY
Launched: 2020
Destination: Mars
Mass: 4 lb (1.8 kg)

Perseverance's Ingenuity helicopter is the first vehicle to fly on another world. Its rotors spin in opposite directions at 2,900 rpm, lifting it off the ground for up to 90 seconds at a time.

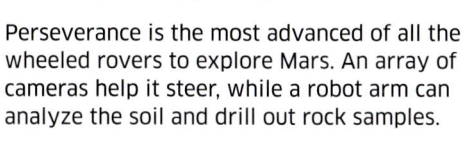

15 The **number of times** a SpaceX
Dragon capsule can be **reused**.

9 **Merlin engines** in the **Falcon rocket** generate
1.71 million lb (776,000 kg) of **thrust at launch.**

Rocket launch

With the astronauts strapped into their seats in the Crew Dragon capsule, the massive Falcon 9 rocket begins to propel itself off the launchpad and into Earth's orbit.

To produce enough thrust to escape Earth's atmosphere and gravitational pull, the Falcon 9 has two sections. The first, lower part is for initial take-off and is designed to return to Earth for future missions. The second stage propels the capsule into Earth's orbit.

Crew section
Passengers sit in custom-molded seats for lift-off.

SuperDraco thrusters
Eight large engines are designed for emergency separation and landing.

Draco thrusters
These 16 small engines maneuver the capsule in space.

Expendable trunk
The trunk carries solar panels and cargo such as satellites.

Second stage
This burns for up to seven minutes, taking the rocket to 124 miles (200 km) above the Earth.

Interstage collar connects first and second stages.

"Chopstick" arms hold rocket in place until launch

Integration tower for access to the spacecraft

Inside the capsule

The Crew Dragon can carry up to four astronauts on trips to the International Space Station or independent flights to orbit the Earth. Flight is mostly automatic, but there are touchscreen controls to indicate the spacecraft's status and allow the crew to take over if necessary. Each spacesuit is custom-made, with a port on one thigh to connect to life support systems such as oxygen. The spacesuit's gloves are designed to be touchscreen compatible, as there are no buttons. In the ceiling is a special space toilet with a suction hose, because in orbit gravity will not pull the waste away.

Separation
The Crew Dragon usually separates from the upper-stage rocket at a height of around 124 miles (200 km), but in emergencies, eight SuperDraco rocket motors can blast the capsule free. These may in time be able to help the Crew Dragon touch down on land instead of splashing into the sea.

SPACE X FALCON 9 BLOCK 5

Year: 2018

Height: 229 ft (70 m)

Stages: 2

Engines: 9 (first stage) + 1 (second)

13,200 lb (6,000 kg)–the **Crew Dragon's** maximum payload (total cargo).

2,900°F (1,600°C)–the **peak temperature** of the capsule's heat shield **during reentry.**

199

Propellant stores
Liquid oxygen and RP-1 kerosene are kept in tanks. The rocket is only fueled once on the launchpad.

Water towers
Water is sprayed at the launchpad to absorb vibrations, protecting the concrete base.

How to catch a space station

Once in orbit, Crew Dragon can maneuver using 16 Draco thrusters. In order to reach the International Space Station (ISS) at its average altitude of 258 miles (415 km), the spacecraft fires these engines several times in a series of "phasing burns." It reaches the space station about 16 hours after launch and is designed to remain docked to the ISS for six months or more.

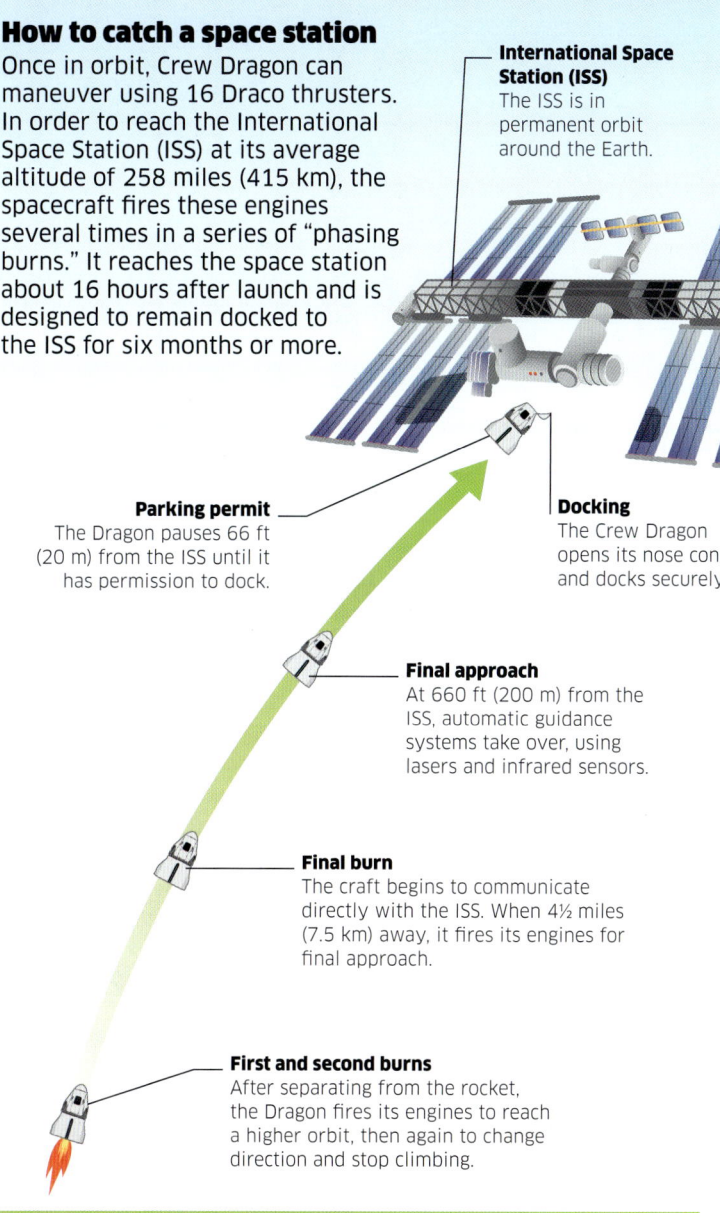

International Space Station (ISS)
The ISS is in permanent orbit around the Earth.

Parking permit
The Dragon pauses 66 ft (20 m) from the ISS until it has permission to dock.

Docking
The Crew Dragon opens its nose cone and docks securely.

Final approach
At 660 ft (200 m) from the ISS, automatic guidance systems take over, using lasers and infrared sensors.

Final burn
The craft begins to communicate directly with the ISS. When 4½ miles (7.5 km) away, it fires its engines for final approach.

First and second burns
After separating from the rocket, the Dragon fires its engines to reach a higher orbit, then again to change direction and stop climbing.

Blast-off!

At the spaceport, a "strongback" vehicle has delivered the rocket to the pad and raised it from horizontal to vertical. Clamps hold the rocket securely on the pad, while astronauts board via the integration tower. Engineers perform final checks and fill the fuel tanks. A final countdown from 10 begins. The engines fire for three seconds, then the clamps release: lift-off!

Back to Earth with a splash

At the end of the mission, the spacecraft undocks from the ISS and discards its trunk, which burns up in the upper atmosphere. It then turns so that it faces "backward" along its orbit and fires its engines for 15 minutes to reduce speed. This sends it spiraling earthward for a fiery reentry. Two small parachutes open to stabilize flight once the heat from reentry has subsided, before four larger parachutes slow the capsule's descent for a gentle splashdown.

First stage
Nine Merlin engines power the rocket 50 miles (80 km) high in 158 seconds. The first stage then separates and descends to Earth for reuse.

Flame catcher
Tunnels under the launchpad channel hot exhaust gases away to prevent them from damaging the rocket.

Glossary

ACCELERATION
The rate at which something speeds up and goes faster.

AERODYNAMICS
The study of how objects travel through air. It is often used to design vehicles and objects that create lower air resistance so they move smoothly through the air, saving energy.

AEROFOIL
A streamlined object that is more curved on one side than the other in order to affect the flow of air across its surfaces. Usually refers to the cross-section of a wing, tail plane, or rotor blade, but some modern racer yachts also have them.

AILERON
Hinged surfaces, usually on an aircraft's wing, that can be raised or lowered to help an aircraft roll or turn.

AIR RESISTANCE
A force that slows down moving objects as they travel through the air. Also called drag.

AIRSHIP
A large balloon or "envelope" filled with gas and powered by an engine. Passengers may be carried in a separate compartment within or below the balloon.

ALL-WHEEL DRIVE
See Four-wheel drive.

AMIDSHIPS
Toward or at the middle of a ship.

AMPHIBIOUS
A vehicle that can travel both on land and in water.

ARTICULATED
A vehicle, such as a bus or truck, made in two or more sections that are joined by a central pivot point so that the vehicle can turn more easily.

ASPHALT
A sticky, black, tarlike substance used to surface roads and pavements.

AUTONOMOUS
A machine that can carry out a job or function on its own.

AVIATION
The production and operation of aircraft.

AVIONICS
An aircraft's electronic and navigation equipment.

AXLE
A sturdy metal rod on which one or more wheels turn. The axle may be fixed to the wheel so that the wheel turns when the axle rotates, or the wheel may spin freely on the axle.

BATTERY
A store of chemicals in a case that, when connected to a circuit, supplies electricity.

BLIMP
An airship with no rigid frame that can be deflated for compact storage or surface transportation.

BODY
The outside parts of a car, including its roof, hood, trunk lid, bumpers, doors, and windows.

BOGIE
In trains, a wheeled chassis or framework that a locomotive or railway carriage sits atop and that runs along the rails. In planes, a mechanism that attaches any more than two wheels to a landing gear strut.

BOILER
A closed container in which water is heated. The part of a steam engine that produces steam.

BOOM
A long fixed or hydraulic arm that extends out from a vehicle.

BOW
The front section of a ship or boat.

BRIDGE
The command center of a ship or submarine. On ships, it is often high up, with unobstructed views, and is also the navigation center.

BUFFER
A shock-absorbing pad that cushions the impact of rail vehicles as they come together.

BUMPER
A metal, rubber, or plastic bar fitted along the front and, sometimes, the back of a truck to limit damage if it bumps into something.

CAB
The part of a train or truck where the driver sits and controls the vehicle.

CAPSULE
The part of a spacecraft in which people or cargo travel and which usually separates from the main rocket.

CARGO
The load carried by an aircraft, ship, truck, or other vehicle.

CHASSIS
The frame of a vehicle upon which the engine and main body are mounted.

CHERRY PICKER
A hydraulic crane, such as one mounted on a truck, that has an elbow joint or telescopic arm supporting a basketlike attachment that is large enough to carry a worker.

CLASS
A group of vehicles built to a common design.

CLUTCH
A device operated by a foot pedal (or hand lever on most motorcycles) that allows the driver to change gears.

CONTROL SURFACE
The moving parts of an aircraft that change the airflow around its trailing edges—e.g., ailerons, elevators, rudders—causing it to roll, pitch, or yaw.

CONVOY
A group of ships or vehicles traveling together in formation.

COSMONAUT
A Russian (or Soviet) astronaut.

COUPE
A two-door car with a fixed roof that slopes down to the rear.

COUPLING
The parts, or mechanism, that allow railway locomotives to be joined together.

COWL OR COWLING
A cover around an aircraft's engine.

CRANKSHAFT
An axle inside an engine that converts the up-and-down movement of the pistons into the rotation of wheels or a propeller.

CUBIC CENTIMETERS (CC)
A measurement of the capacity, or space, inside all the cylinders of an engine. A larger cc usually means a more powerful engine.

CYLINDER
A strong metal tube inside an engine in which steam expands or a fuel ignites to push pistons up and down to generate power.

DECK
One of the horizontal sections, or floors, running along a bus or ship. Often used to refer to the top deck of a ship, which is open to the elements.

DEFIBRILLATOR
A machine that starts the heart beating normally again after a heart attack by giving it an electric shock.

DELTA WING
A triangular-shaped airplane wing.

DEPTH CHARGE
An antisubmarine weapon dropped from ships or planes and set to explode at a certain depth.

DIESEL
A type of fuel made from oil, used in many vehicle engines.

DIRIGIBLE
Another word for an airship.

DISC BRAKES
A type of brake that uses pads to press against a turning disc, creating friction to slow the vehicle down.

DOWNFORCE
The force that pushes down on an aerodynamically shaped vehicle, such as a Formula One car, when air moves rapidly over it. Downforce is the opposite of lift (the force that makes an airplane take off as its wings move quickly through the air).

DRAG
The resistance force formed when an object pushes through a fluid, such as air or water.

DRONE
See UAV.

DROP TANK
An external aircraft tank, usually containing fuel, that can be detached and dropped midflight once empty.

EFFICIENCY
A measure of the amount of energy put into a machine that is transformed into useful work. An energy-efficient engine generates more motion from the fuel it uses.

ELECTROMAGNET
Magnets powered by electricity that can be turned on or off.

ELEVATOR
A control surface on an aircraft that causes the plane to raise or lower its nose and climb or dive.

EMISSIONS
The production or release of something, such as gas. Internal combustion engines produce greenhouse gas emissions, which contribute to global warming.

ENERGY
What enables work to be done. Energy exists in many different forms and cannot be created or destroyed, only transferred from one form to another.

ENGINE
A machine that burns fuel to produce motion.

EXHAUST
A tube that channels waste gases away from a vehicle's engine and out into the open air.

FAIRING
Aerodynamic bodywork designed to improve top speed or user comfort.

FIREBOX
The section at the rear of a steam locomotive boiler where the fuel is burned to heat the water in the boiler.

FLANGE
A lip, rim, or ridge that projects from the edge of an object and helps it connect to other objects. On railways, it helps train wheels stay on the track.

FLAPS
Movable parts of the rear edge of a wing that are used to increase lift at slower air speeds.

FLY-BY-WIRE
An electronic flight control system used in aircraft instead of mechanical or machine-operated controls.

FOREMAST
The mast nearest the bow (front) of a ship.

FOUR-STROKE ENGINE
The most common type of car engine. Each piston in the engine works in four stages, or strokes: Intake (taking in a mixture of air and gas), compression (squeezing the mixture), ignition (a spark ignites the mixture, which burns rapidly and pushes the piston down), and exhaust (the spent mixture leaves the cylinder).

FOUR-WHEEL DRIVE (4WD)
Where power from the engine is used to turn both the front and back wheels of a vehicle.

FREIGHT
Goods transported in bulk by truck, train, ship, or aircraft.

FRESNEL LENS
A lens that is flat on one side and has concentric ridges on the other. It can produce a powerful beam of light in a lighthouse, headlight, or aircraft-carrier landing guide.

FRICTION
The force that slows movement between two objects that rub together. Brakes create lots of friction to slow down a vehicle.

FRONT-WHEEL DRIVE
A system in which power from a vehicle's engine is sent only to the front two wheels.

FUEL CELL
A device in which a chemical reaction, usually involving hydrogen and oxygen, produces electricity and water.

FUSELAGE
The main body of an aircraft, designed to hold crew, passengers, and cargo.

GAUGE
The measurement between the rails of a track. The world's most prevalent gauge is 4ft 8½ in (1,435 mm) and is known as standard gauge. However, many larger and smaller gauges are used on the world's rail systems.

GEAR
A wheel with "teeth" around the outside that intermeshes with another toothed wheel. Gears change how speed of a driving mechanism, such as the engine of a motor car, converts into the speed of the driven parts, such as its wheels.

GENERATOR
A machine that creates electricity for use by electrical devices.

GLIDEPATH
The line of descent for an aircraft coming in to land, reducing forward speed and altitude for a safe landing.

GPS
Short for global positioning system, a navigation system using signals from a group of space satellites to work out a vehicle's position on Earth's surface.

GRADIENT
A slope, or the degree to which the ground slopes.

GROUND CLEARANCE
The distance between the underside of a car and the ground.

HANDLING
How easy a vehicle is to control—for example, when going around bends or driving in difficult conditions.

HATCHBACK
A small car with a rear hood and window covering the trunk area.

HORSEPOWER (HP)
A commonly used measure of the power of a vehicle's engine.

HOVERCRAFT
A vehicle that moves across land, water, and other surfaces on a cushion of air produced by fans and held in by a heavy-duty rubber "skirt."

HULL
The main body of a boat or a ship.

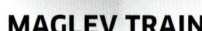

HYBRID
A vehicle that has both an internal combustion engine and a second source of power, such as an electric motor.

HYDRAULICS
A power system that uses liquid to transfer force from one place to another to, for example, operate a vehicle's brakes.

INTERNAL COMBUSTION ENGINE
A type of engine in which fuel and air are mixed and burned (combusted) inside cylinders to produce power.

JET ENGINE
An engine that creates forward motion by forcing hot air out of a nozzle at the back.

KEEL
A long, generally straight assembly of timber, metal, plastic, or carbon fiber that forms the lowest part of the hull of a ship along most of its length. It might be made up of several pieces of wood joined together.

KNOT
The unit for measuring the speed of boats through the water. One knot is equivalent to 1.15 mph (1.852 kph).

LEVER
A simple machine that magnifies or reduces forces, making it easier to move a load. An oar is an example of a lever.

LIFT
The force created by air moving over a wing or rotor blade to keep an aircraft rising through the air.

LOCK
On a canal or river, a place where walls have been built with gates at each end so that boats can move to a higher or lower section of the canal or river as the water level is gradually raised or lowered.

LOCOMOTIVE
A wheeled vehicle used for pulling trains. Electric locomotives usually rely on electricity provided by an external source, while steam and diesel locomotives generate their own power.

MAGLEV TRAIN
Short for magnetic levitation, a train that works by being raised above special tracks and moved forward by the power of electromagnets. Maglevs produce virtually no friction and are very quiet in operation at high speed.

MASS
A measure of the amount of matter in an object.

MODULE
Part of a larger spacecraft that can detach and operate independently—for example, the Apollo 11 Moon lander.

MONORAIL
A train that runs on a single rail.

MOTOR
A mechanical device that uses electricity to produce motion.

OFFROAD
To travel in a vehicle away from roads and over tracks, trails, or open ground.

ORBIT
The path of one object around another due to the force of gravity, such as that of a spacecraft around a planet.

OUTRIGGERS
Bars that extend out from the side of vehicles like cranes or canoes to provide support and help the vehicle balance.

PANTOGRAPH
A spring-loaded frame that passes electricity to vehicles, such as trains and trolley buses, from overhead power lines.

PAYLOAD
The load carried by an aircraft or space launch vehicle.

PERISCOPE
A tube containing mirrors that lets a submarine captain see what is above the water. Modern submarines use electronic equipment instead.

PISTON
A tightly fitting plunger that moves up and down inside a cylinder due to the pressure of steam or as fuel burns within an internal combustion engine.

PLOW
A piece of agricultural equipment with sharp blades for cutting or turning over earth.

POLLUTION
The effect when harmful waste products reach the air, water, or land and damage the environment or the health of living things.

POWERPLANT
The equipment (including engine and transmission) that provides a vehicle with power for propulsion.

PRESSURE
The force of air or water pushing on things.

PRESSURIZED
An object containing air or another gas at a pressure higher inside the container than outside it—such as a spacesuit, an airliner's fuselage, or an oxygen cylinder.

PROBE
An uncrewed spacecraft traveling to a planet, moon, comet, or other body to collect information.

PRODUCTION VEHICLE
A vehicle that is mass-produced for sale to the public. This differs from experimental prototypes made for internal tests.

PROPELLER
A set of blades spun by a plane or boat engine to create thrust.

PROTOTYPE
An experimental version of a design, which is not yet ready to be mass-produced and sold.

RADAR
A system used to locate an object or vehicle by bouncing radio waves of them and measuring how long it takes for these to return.

RADIO WAVES
A type of invisible wave that can be transmitted through the air, often used for communication.

RAIL TRACKS
The permanent rails that provide a runway for a train's wheels.

RAILCAR
A self-propelled passenger vehicle, usually with the engine located under the floor.

REAR-WHEEL DRIVE
A system in which power from a vehicle's engine is sent only to the rear wheels.

REVOLUTION
One complete turn of a rotating object.

ROADHOLDING
A vehicle's roadholding affects how easy it is to control safely in difficult driving conditions or when going around bends.

ROCKET ENGINE
An engine that burns fuel along with oxygen or oxidizer (oxygen-producing chemicals) to produce a stream of gases that propels the rocket. The rocket engine carries its own supply of oxygen or oxidizer.

ROTOR BLADES
Long, thin aerofoils that are spun by a helicopter or other rotorcraft to produce lift.

RPM
Revolutions ("revs") per minute. The number of times a wheel rotates in one minute.

RUDDER
A vertical plate or board that can be moved to steer avessel or help turn an aircraft.

RUNABOUT
A small car or boat used mainly for short journeys.

SADDLE
The seat on a bicycle or motorcycle where the cyclist or rider sits.

SATELLITE
An object orbiting a larger body in space. Artificial satellites orbit around Earth or another planet or moon to collect or relay data. Natural satellites are usually called moons.

SEDAN
Any type of car with a fixed metal roof and a separate, enclosed trunk.

SHELL
A weapon consisting of a metal container filled with explosives that can be fired from a large gun over long distances.

SONAR
A system for detecting and locating objects, particularly underwater, using sound waves.

SOVIET UNION (USSR)
Officially the "Union of Soviet Socialist Republics," a Russian-controlled group of countries between 1922 and 1991.

SPOILER
A small wing mounted on the back of a car to produce downforce at high speeds, which helps the car grip the road better.

SPORT-UTILITY VEHICLE (SUV)
A rugged vehicle designed for offroad use but often used on normal roads in urban areas.

STATION WAGON
A car with an extended body and a large floor-to-roof trunk space.

STERN
The rear part of a boat or vessel.

STREAMLINED
A streamlined object has smooth curves so that air or water flows easily, reducing drag.

SUBMERSIBLE
A small, mobile undersea vehicle often used for underwater science or commercial operations such as exploration or salvage work. It can either be crewed or uncrewed.

SUPERSONIC
Faster than the speed of sound, which is about 768 mph (1,236 kph) at sea level but slower at altitude.

SUSPENSION
A system that cushions the car's structure (and occupants) from the motion of the wheels as they traverse uneven roads and surfaces.

TAXI
To move an aircraft at low speed, under its own power, on the ground or water. Also a car that will take people where they want to go in return for money.

THERMAL IMAGER
A piece of equipment that uses the heat emitted by people and things to produce an image.

THRUST
The force that pushes a powered aircraft through the air or boat through the water, usually generated by an engine.

TORPEDO
A self-propelled underwater weapon that is launched from a ship or submarine.

TORQUE
The twisting force created by a turning component (such as a propeller), powered by an engine. Vehicles can use different gears to change how much torque they produce.

TRACK
A loop of rubber or metal links that are turned around by sets of wheels or rollers.
For rail tracks, see Rail track.

TRANSMISSION
The system that brings the power of the engine to the wheels or propeller of a vehicle, usually via a gearbox. It can convert the relatively high speed of the engine into a range of useful forward and reverse speeds

TREAD
The raised patterns on a tire's surface that increase friction between the wheels of a vehicle and the ground.

TURBINE
A set of paddles turned by steam, water, or air.

TURBOCHARGER
A device that uses waste gases to boost an engine's power.

TURBOFAN
A type of gas-turbine engine (jet engine) in which some of the power drives a fan that pushes out cold air with the exhaust, thereby increasing thrust. Turbofans are used in most airliners because they are more economical and less noisy than turbojet engines.

TURBOJET
A type of gas-turbine engine (jet engine) in which all propulsion is generated from rapidly expanding, high-speed hot exhaust gases.

TWO-STROKE ENGINE
A type of engine that fires the spark plug with every revolution of the crankshaft instead of every other revolution, as in a four-stroke engine.

UAV
Short for uncrewed aerial vehicle and also known as drones, these are flying machines which either control themselves or are remote controlled by a human operator.

ULTRASOUND
Sound waves that cannot be detected by human ears.

VARIABLE PITCH PROPELLER
A propeller in which the angle of the blades in relation to airflow can be adjusted while it is spinning.

VENTILATION
An installation in a building or vehicle that provides a supply of fresh air.

VTOL
Short for Vertical Take-off and Landing. Aircraft that use VTOL can use their thrust to lift straight up into the air like a helicopter and do not require a long runway as a result.

WATERLINE
The level normally reached by the water on the side of a ship.

WHEELBASE
The exact distance between the axles of the front and rear wheels.

Index

Acknowledgments

The publisher would like to thank the following people for their assistance in the preparation of this book: Rona Skene for additional text, Simon Mumford for maps and airplane knowledge, Danielle Cluer Gee for proofreading; Elizabeth Wise for the index, and Paige Towler and her team at the Smithsonian. Special thanks to Tim Meldrum at Emirates Team NZ for help with **Te Rehutai;** Tobias Bender at Marigraph and AG Ems for use of **Ostfriesland** on the ferry feature; Robert Burke for his kind permission to adapt his computer generated model of a Mk IX Spitfire.

The publisher would like to thank the following for their kind permission to reproduce their photographs:

(Key: a-above; b-below/bottom; c-centre; f-far; l-left; r-right; t-top)

2 Dorling Kindersley: Tina Chambers / IFREMER, Paris (clb). **3 Shutterstock.com:** Mechanik (tl). **4 Alamy Stock Photo:** RGB Ventures / SuperStock / NASA (l). **6 Dreamstime.com:** David Fowler (cl). **8 Alamy Stock Photo:** Album (cra); Chuck Eckert (tc); kostas koufogiorgos (ca); Rich Gold (crb). **Dreamstime.com:** Kseniya Abramova / Tristana (cb). **Getty Images / iStock:** Dorling Kindersley (tr). **Shutterstock.com:** Dmitro (bc). **TurboSquid:** 3d_molier International (cla). **9 Alamy Stock Photo:** David Askham (ftl); Phillip Bond (tc); Hugh Williamson (cra). **Bridgeman Images:** Photo © CCI (fcl). **Dorling Kindersley:** Dave King / Motorcycle Heritage Museum, Westerville, Ohio (fcla); Gary Ombler, Courtesy of Deutsches Fahrradmuseum, German (clb). **Dreamstime.com:** Chinaview (bl); Vampy1 (tr); Dvmsimages (cl); Veniamin Kraskov (cr). **Getty Images / iStock:** Rysownik (c). **Getty Images:** Science & Society Picture Library (br). **Shutterstock.com:** AlexLMX (tl); Haggardous50000 (cb). **10 Depositphotos Inc:** travelarium (bl). **Shutterstock.com:** Evgeny Subbotsky (clb); Dale Warren (cl). **11 123RF.com:** photo5963 (c). **Alamy Stock Photo:** philipus (cra). **Dorling Kindersley:** Dave King / Science Museum, London (tc); Dave King / The Science Museum (cla). **Getty Images / iStock:** SOPHIE-CARON (crb). **Shutterstock.com:** Captain Wang (br). **12 Alamy Stock Photo:** Peak Images (ca). **Dreamstime.com:** Fernando Corts (bc/chariot); Edwardj (bc). **Getty Images:** Science & Society Picture Library (tr). **Getty Images / iStock:** VUSLimited (bl). **13 Bridgeman Images:** Chatsworth Settlement Trustees (ca, tc). **Dorling Kindersley:** Gary Ombler / Milestones Museum (cb). **Dreamstime.com:** David Fowler (br); Sborisov (bc); Hanna Tsiarleyeva (tr). **Getty Images:** Science & Society Picture Library (crb/x2). **14 Alamy Stock Photo:** Painters (br). **Bridgeman Images:** Look and Learn (tl). **15 Alamy Stock Photo:** Heritage Image Partnership Ltd (bl). **16-17 TurboSquid:** CG ARTStudio (c). **16 Getty Images / iStock:** getty_dumy67 (bl). **19 Alamy Stock Photo:** Associated Press (br). **20 Dorling Kindersley:** Gary Ombler / R. Florio (clb). **Dreamstime.com:** Dvmsimages (bl); David Park (br). **Shutterstock.com:** Gertan (cl); StevenK (bc). **21 Alamy Stock Photo:** NPC Collectiom (cr); ZarkePix (c). **Dorling Kindersley:** Matthew Ward (fcla, clb). **Dreamstime.com:** Ajdibilio (tr); Vladimir Masilko (tc); Cristina Ionescu (cla); Milos Ruzicka (cra); Artzzz (bl); Eva Diana Lopez (crb). **Shutterstock.com:** canadianPhotographer56 (br); Cristina Ionescu (cl). **22-23 CGTrader:** Dragos Burian (c/chassis). **TurboSquid:** HKV Studios (c). **23 Alamy Stock Photo:** CTK (crb). **24 Alamy Stock Photo:** Malcolm Haines (bc). **Dorling Kindersley:** Matthew Ward (tr, cb); James Mann / Colin Laybourn / P&A Wood (tl). **Dreamstime.com:** Artzz (cra); Dmitry Orlov (cla, br); Michel Bussieres (c). **Getty Images:** ullstein bild (ca). **Shutterstock.com:** GUIDO BISSATTINI (bl). **25 Dorling Kindersley:** Matthew Ward (tl). **Dreamstime.com:** Robert Diepenbrock (cl); Dmitry Orlov (cla); Mmc12 (b). **Shutterstock.com:** Danila2332 (cb); Simlinger (ca); James Hime (cra); Sid0601 (crb). **26-27 TurboSquid:** 3D Horse (c/engine); anasyakoub (c). **26 Dreamstime.com:** Hollandog (tl). **27 Alamy Stock Photo:** Brian Jannsen (t). **28-29 Alamy Stock Photo:** Sipa US. **30-31 TurboSquid:** OpticalDreamSoft (c). **30 Alamy Stock Photo:** The Canadian Press (cra). **31 Getty Images:** MLADEN ANTONOV (br). **32 Alamy Stock Photo:** G.P.Essex (ftr); Sueddeutsche Zeitung Photo (tr); James King-Holmes (cr). **Dreamstime.com:** Framestock Footages (bc). **Shutterstock.com:** Frame Stock Footage (br). **33 Alamy Stock Photo:** Hans Blossey / imageBROKER.com GmbH & Co. KG (tc); John Henderson (cl); Matthew Richardson (bl, bc). **Science Photo Library:** TRL LTD. (tr); LOUISE MURRAY (tl). **34-35 TurboSquid:** fisherman3d. **34 Alamy Stock Photo:** Rolf Schulten / imageBROKER.com GmbH & Co. KG (bl). **36-37 Getty Images:** fStop Images - Caspar Benson. **38-39 TurboSquid:** AT_studio Pro Models_3D (c). **38 Alamy Stock Photo:** Gwendoline Defente / Associated Press (bl). **39 Dreamstime.com:** Hamik (br). **41 Alamy Stock Photo:** Martin Bond (cr). **Getty Images:** Spencer Platt / Staff (crb); DIRK WAEM / Stringer (cra). **42-43 TurboSquid:** 3d_molier International (c). **44 Dorling Kindersley:** Gerard Brown / Bicycle Museum Of America (cra/X2, bl/x2); Gary Ombler, Courtesy of Deutsches Fahrradmuseum, German (cla/X2); Gary Ombler, Courtesy of Deutsches Fahrradmuseum, Germany (ca/X2); Gary Ombler / Jonathan Sneath (clb/X2); Gerard Brown / Llandrindod Wells National Cycle Museum Wales (br). **45 Dorling Kindersley:** Gerard Brown / Bicycle Museum Of America (tr); Gary Ombler / Stuart's Bikes (cla); Gary Ombler / London Green Cycles (crb); Gary Ombler / J.D Tandems (bl). **Karbon Kinetics Ltd.:** (br/x3). **46 Dorling Kindersley:** Dave King / National Motor Museum, Beaulieu (br); Dave King / Motorcycle Heritage Museum, Westerville, Ohio (fcrb). **Dreamstime.com:** Aramen4 (tr/x2). **47 Alamy Stock Photo:** David Davis Photoproductions (cl). **Dorling Kindersley:** Dave King / Motorcycle Heritage Museum, Westerville, Ohio (tr/x2, bc/x2); James Mann / Micheal Penn (cla). **Dreamstime.com:** Mariusz Burcz (cr/x2). **Shutterstock.com:** FernandoV (c/x2). **48-49 TurboSquid:** nitacawo (c). **48 Alamy Stock Photo:** ZarkePix (bl). **50 Alamy Stock Photo:** Paul Boyes (cr/x2); imageBROKER.com GmbH & Co. KG (bl/x2); Dave Ellison (br/x2). **Dreamstime.com:** Falun1 (cl/x2). **51 Alamy Stock Photo:** Bailey-Cooper Photography (bc); ZarkePix (tr/x2). **Dreamstime.com:** Artzzz (tl/x2); Walter Eric Sy (tc/x2); Anthony Baggett (cl/x2); Fabioma9001 (c/x2); Taina Sohlman (cr/x2). **52 Alamy Stock Photo:** D. Callcut (ca); Image Professionals GmbH (clb). **Dreamstime.com:** Welcomia (tr). **TurboSquid:** 3d_molier International (tl). **54 Alamy Stock Photo:** Alex Thomson (crb). **Dorling Kindersley:** Gary Ombler / Doubleday Swineshead Depot (c/x2); Gary Ombler / Daniel Ward (tl/x2); Gary Ombler / Paul Rackham (tr/x2); Gary Ombler / Chandlers Ltd (bl/x2); Gary Ombler / David Wakefield (br). **55 Dorling Kindersley:** Gary Ombler / Doubleday Swineshead Depot (tl/x2, ca, c); Gary Ombler / Chandlers Ltd (cl/x2, bl/x2). **Dreamstime.com:** Flavijus (br/x2). **56-57 TurboSquid:** iljujjkin (c). **57 Alamy Stock Photo:** Ed Buziak (crb); MediaWorldImages (br). **58 Alamy Stock Photo:** Washington Imaging (b/x2); Alberto Rigamonti (c); Vladimir Zaplakhov (clb). **59 Alamy Stock Photo:** Carlsson (tc). **Dorling Kindersley:** Gary Ombler / James River (cl/x2, cr/x2, bl). **Dreamstime.com:** Stefan11 (clb); Supertrooper (tr); Jonathan Weiss (br/x2). **Thwaites Limited, United Kingdom:** (tl/x2). **60 Alamy Stock Photo:** dpa picture alliance (bl/x2); Werner Otto (tc/x2); FirePhoto (cl/x2); Elizabeth Nunn (cr/x2). **Depositphotos Inc:** cobalt-70 (c/x2). **61 123RF.com:** Artem Konovalov / artzzz (bl/x2). **Alamy Stock Photo:** Amel Emric / Associated Press (ca/x2); Bob Graham (cr). **Dorling Kindersley:** Richard Leeney / Cockermouth Mountain Rescue Team, England (br/x2). **62-63 CGTrader:** cgjoe (c). **63 Alamy Stock Photo:** Dave Donaldson (tl). **Getty Images:** China News Service (br). **64 Dreamstime.com:** Jarnogz (br). **65 Alamy Stock Photo:** B Christopher (bc); Stephen Frost (ftr); ZUMA Press, Inc. (tr); Sipa US (fcr); Bob Hurley (c). **Getty Images:** Kyodo News (br). **Getty Images / iStock:** stockstudioX (bl). **66 Dorling Kindersley:** Alan Keohane (bc). **Dreamstime.com:** Ricky Corey (tc/x2); Peter Moulton (cl/x2). **Shutterstock.com:** Art Konovalov (cr). **67 123RF.com:** tritooth (cr). **Alamy Stock Photo:** Boaz Rottem (ca). **Dorling Kindersley:** Barnabas Kindersley (cra). **Dreamstime.com:** Ifeelstock (br); Photodynamx (tr); Bjorn Wylezich (clb). **Shutterstock.com:** saisnaps (tl). **Pulsar Expo s.r.o. - Torsus:** (cl/x2). **68-69 TurboSquid:** natman100. **68 Dreamstime.com:** Steve Woods / Woodsy007 (bl). **69 Alamy Stock Photo:** David Gee (tr). **70 Alamy Stock Photo:** Historical Views / agefotostock (cl); Jim West (c); B. David Cathell (crb). **71 Alamy Stock Photo:** ADB Travel (br). **Dorling Kindersley:** Mike Dunning / London Transport Museum (clb); Mike Dunning / National Railway Museum, York (bl). **72-73 Keith Fender:** (c/x2). **72 Science Museum Group Collection © The Board of Trustees of the Science Museum:** David Shepherd (br/x2). **Photo Georg Trb:** (tc/x2). **73 Alamy Stock Photo:** Chris Craggs (br/x2); Bilderbox / INSADCO GmbH (tc). **Dreamstime.com:** Jiawangkun (bl/x2). **Keith Fender:** (bc/x2). **75 Alamy Stock Photo:** Alan Mather (cb). **Getty Images:** Hulton Archive / Construction Photography / Avalon (tr). **76 Alamy Stock Photo:** Radharc Images (br/x2); Marc Tielemans (tr). **Dorling Kindersley:** Gary Ombler / Railroad Museum of Pennsylvania (bc/x2). **Dreamstime.com:** Vinayak Jagtap (bl/x2); Lightwavemodels (c). **Getty Images / iStock:** AFransen (ca). **77 Alamy Stock Photo:** dpa picture alliance (cr/x2); rebaixfotografie (tr); MARK HICKEN (br/x2). **Dreamstime.com:** Boarding1now (cl/x2); Tupungato (fcra); Yinglina (bl/x2). **78-79 TurboSquid:** Treapl (c). **78 Alamy Stock Photo:** MoiraM (tl). **79 Dreamstime.com:** Thinglass (crb). **82-83 Alamy Stock Photo:** Doug Houghton. **84 Dorling Kindersley:** The Tank Museum, Bovington (cla, c/x2, br/x2); Gary Ombler / Tank Museum (tl); Gary Ombler / The Tank Museum, Bovington (tc); Gary Ombler / Chris Till (cr/x2); Roger Dixon / Second Guards Rifles Division (bl). **Dreamstime.com:** Sergey Zavyalov / Saz1977 (fclb). **85 123RF.com:** Jordan Tan / prestonia (cl/x2). **Dorling Kindersley:** Andrew Baker, Tanks, Trucks and Firepower Show (cr/x2); War and Peace Show (tr/x2); Gary Ombler / The Tank Museum, Bovington (crb); The Tank Museum, Bovington (bc). **86-87 TurboSquid:** Alpen wolf. **86 Dreamstime.com:** Nico Kelder (bl). **Getty Images:** Scott Nelson / Stringer (tl). **TurboSquid:** Katie Byrne (tc). **87 Alamy Stock Photo:** dpa picture alliance (br). **Shutterstock.com:** Karlis Dambrans (tl). **88 123RF.com:** Rostislav Ageev / rostislavv (cr). **Dreamstime.com:** Guangliang Huo (c). **Getty Images:** Justin Sullivan / Stringer (cr/Aviator). **TurboSquid:** ACE_POLY (cl). **90 Alamy Stock Photo:** Design Pics Inc (tr); Newscom (c); Larry Morgan (bc); Mylam (br). **Dorling Kindersley:** Gary Ombler / Fleet Air Arm (cla). **Division of Work and Industry, National Museum of American History, Smithsonian Institution:** (ca). **Science Photo Library:** MIKKEL JUUL JENSEN (tl). **90-91 Alamy Stock Photo:** Jon Lord (tc). **91 Alamy Stock Photo:** ZUMA Press, Inc. (c). **Dorling Kindersley:** The Science Museum / Dave King (bl). **Dreamstime.com:** James Group Studios, Inc. (ca). **Getty Images:** Science & Society Picture Library (cla). **National Museum of New Zealand Te Papa Tongarewa:** (crb). **93 Alamy Stock Photo:** Wolfgang Diederich (c). **Dreamstime.com:** Solarisys13 (cl). **Getty Images:** Songphol Thesakit (cr). **94 Alamy Stock Photo:** Associated Press (cl/x2); megapress images (cra); Zev Radovan / BibleLandPictures (c/X2). **Dorling Kindersley:** Frank Greenaway / Town Docks Museum, Hull (tc); Richard Leeney / Maidstone Museum and Bentliff Art Gallery (bl). **Dreamstime.com:** Per Bjorkdahl (cr). **National Museum of New Zealand Te Papa Tongarewa:** (bc). **Shutterstock.com:** Jyoti Singh (tl/x2). **95 Alamy Stock Photo:** mauritius images GmbH (bl); Newscom (cr/x2); ZUMA Press, Inc. (br/x2). **Dreamstime.com:** Mr1805 (tl/x2); Photo25th (cla). **Getty Images:** Science & Society Picture Library (clb). **96-97 TurboSquid:** IMPERIUM Design (c/spartans); NKCGArtist (c). **96 Alamy Stock Photo:** Album (bc). **98-99 Dreamstime.com:** Guangliang Huo. **101 Alamy Stock Photo:** IanDagnall Computing (cra). **102 Alamy Stock Photo:** Glyn Genin (tl). **103 Alamy Stock Photo:** Glyn Genin (crb). **104 Getty Images:** Science & Society Picture Library (fclb, c, cr/x2, bc/x2, br/x2). **105 Alamy Stock Photo:** Siegfried Grassegger / imageBROKER.com GmbH & Co. KG (c/x2); Lanmas (tr/x2); Michael Lidski (cl/x2); Hemis (bc/x2). **Dreamstime.com:** Enrico Powell (cr/2). **Division of Work and Industry, National Museum of American History, Smithsonian Institution:** (tl/x2). **106 Alamy Stock Photo:** Matthew Richardson (tc). **Dreamstime.com:** Ypkim (br). **107 Alamy Stock Photo:** Chronicle (crb). **Getty Images:** Hulton Archive / Stringer / Hudson (bc). **108 Alamy Stock Photo:** Design Pics Inc (crb); Gilberto Mesquita (cl/x2); PA Images (bl). **Depositphotos Inc:** naiyyer (clb). **109 Alamy Stock Photo:** Duncan Astbury (tr/x2); Peter Titmuss (tl/x2); Steve Hawkins Photography (bl/x2). **Dreamstime.com:** Svetlin Yosifov (br). **Shutterstock.com:** Alexandre Arocas (c/x2). **110-111 TurboSquid:** LCYdesign (cr). **110 Alamy Stock Photo:** Les Breault (bc); Rob Taggart (tl); North Wind Picture Archives (bl). **Getty Images:** AFP / Jose Jordan (bc/Multihulls). **112 Alamy Stock Photo:** Chris Laurens (tl/x2); Jon Lord (c/x2). **Getty Images / iStock:** Nerthuz (bl/x2). **Shutterstock.com:** MartinLueke (tr/x2). **113 Alamy Stock Photo:** Adwo (tl/x2); Jouni Niskakoski (tr/x2); Associated Press (cr/x2); Jennifer Wright (br/x2). **Shutterstock.com:** MartinLueke (cl/x2). **114 Alamy Stock Photo:** JLBvdWOLF (cla); Cynthia Lee (tl). **115 Alamy Stock Photo:** Sipa US / Costfoto (cra). **116 Alamy Stock Photo:** DB Pictures (fbr). **Dreamstime.com:** Rolandm (br). **Getty Images:** Eric CHRETIEN (bc). **116-117 Tobias Bender / Marigraph GmbH:** (c). **117 Alamy Stock Photo:** Sina Schuldt / dpa picture alliance (br). **Dreamstime.com:** Oleksandr Kalinichenko (bc). **118 Alamy Stock Photo:** Robert Evans (cra/X2); Jochen Tack (b). **Dreamstime.com:** Franco Nadalin (c/X2); TasFoto (cla/X2). **119 Alamy Stock Photo:** Graham Hardy (cla/X2); Martin Lke (cra/X2); mvlampila (c/X2); David Maddock (bl/X2); Shipspotter ME (br/X2). **120 Science Photo Library:** British Antarctic Survey (tl). **121 Alamy Stock Photo:** DPA Picture Alliance (br); Rob Powell (tr). **Science Photo Library:** British Antarctic Survey (bc). **122-123 Alamy Stock Photo:** Jens Bttner / dpa picture alliance. **124 Alamy Stock Photo:** Kostiantyn Ablazov (bc). **Dorling Kindersley:** Gary Ombler / Fleet Air Arm (bl/x2). **Dreamstime.com:** Adam Fleks (br/x2). **Shutterstock.com:** Meng Luen (cb). **125 Alamy Stock Photo:** APFootage (bc); Christian Valverde via Planetpix / NATO (tc/x2); Florian Edus / US Navy Photo (tr/x2); ZapperSiR (c/x2). **Dreamstime.com:** Burnstuff2003 (cl/x2); Dereksmith09 (tl/x2); ZapperSiR (fcrb). **Shutterstock.com:** yanchi1984 (cr). **126-127 TurboSquid:** Lennon3DModelStudio (cb). **126 Alamy Stock Photo:** Chronicle (tl). **127 Getty Images:** Universal History Archive / Universal Images Group (cra). **128 Alamy Stock Photo:** John Cairns (cla/x2). **Dorling Kindersley:** Andy Crawford / The Royal Navy Submarine Museum (tr); Gary Ombler / Scale Model World, Allan Toyne (br/x2). **Shutterstock.com:** Olga Oggi (tr/Turtle). **129 Dorling Kindersley:** Tina Chambers / IFREMER, Paris (clb/x2); James Stevenson / Science Museum, London (tl/x2). **Getty Images:** Justin Sullivan / Stringer (bc/x2). **Science Photo Library:** MIKKEL JUUL JENSEN (tr/x2). **130 Alamy Stock Photo:** Renaud Rebardy (tl). **131 Alamy Stock Photo:** Ludovic MARIN / Pool / Abaca Press (tl, clb). **132-133 TurboSquid:** ptstudio (c). **132 Alamy Stock Photo:** Aclosund Historic (bc). **133 Alamy Stock Photo:** MC2 Justin McTaggart / US Navy Photo (tl). **134 Alamy Stock Photo:** Operation 2022 (bl); PJF Military Collection (bc). **135 Alamy Stock Photo:** APFootage (cla). **136-137 Alamy Stock Photo:** Planetpix. **138 Dreamstime.com:** Byheaven87 / Marina Pissarova (cl). **Getty Images / iStock:** 3DSculptor (cr). **140 Alamy Stock Photo:** Aviation History Collection (cb); Gordon Zammit (tr); David Gee (cr); Pictures Now (crb). **Dorling Kindersley:** Andy Crawford / Bob Gathany (cra). **Science Photo Library:** Carlos Clarivan (tl). **TurboSquid:** 3d_molier International (ca). **141 Alamy Stock Photo:** Stuart Hough (tc); Peter Lane (c); Steve Speller (cb). **Shutterstock.com:** Flying Camera (crb). **142 123RF.com:** Brian Kinney (cr). **Alamy Stock Photo:** Holden Wildlife (tc). **Dreamstime.com:** Elenatur (br). **Getty Images:** Science & Society Picture Library (tr). **143 Alamy Stock Photo:** Sean Bolton (crb). **144 Bridgeman Images:** Leonard de Selva (bl). **Dorling Kindersley:** Musee Air & Space Paris, La Bourget / Gary Ombler (ca/X2); Real Aeroplane Company / Gary Ombler (br/X2). **Getty Images:** Science & Society Picture Library (crb/X2). **145 Alamy Stock Photo:** Roland Bouvier (br/X2); DPA Picture Alliance Archive (bl). **Dorling Kindersley:** Flugausstellung / Gary Ombler (crb/X2); Shuttleworth Collection / Gary Ombler (tl/X2, clb); Musee Air & Space Paris / Gary Ombler (cra/X2). **146-147 Alamy Stock Photo:** Sueddeutsche Zeitung Photo. **149 Alamy Stock Photo:** Sddeutsche Zeitung Photo / Scherl (crb); Vintage_Space (br). **150 Alamy Stock Photo:** Angus McComiskey (b/X2). **Dorling Kindersley:** Fleet Air Arm Museum / Gary Ombler (crb/X2); Royal Airforce Museum, London / Gary Ombler (cla/X2). **Philip Stevens:** (tc/x2). **151 Alamy Stock Photo:** Ivan Batinic (br). **Dorling Kindersley:** Fleet Air Arm Museum / Gary Ombler (crb/X2); Musee Air & Space Paris, La Bourget / Gary Ombler (ca/X2); Shuttleworth Collection / Gary Ombler (clb/X2). **Shutterstock.com:** Flying Camera (c/X2). **152-153 TurboSquid:** kosta58 (c). **152 Dorling Kindersley:** Gary Ombler / Fleet Air Arm Museum (bc). **153 Dorling Kindersley:** Gary Ombler / Paul Ford (cra). **TurboSquid:** Vlada018 (c). **154 Alamy Stock Photo:** Chronicle (crb/Boeing); Interfoto / History (crb); Granger - Historical Picture Archive (br/BOEING 707); Antony Nettle (br/Concorde). **Getty Images:** Bettmann (bc); Stringer / Michael Ochs Archives (br). **National Air and Space Museum, Smithsonian Institution:** (tl). **155 Alamy Stock Photo:** Aviation History Collection (cr); Chronicle (tc); Em Campos (tr); Dappled History (cl); JSM Historical (fcr). **156 Alamy Stock Photo:** Heritage Image Partnership Ltd / Curt Teich Postcard Archives (bc). **Getty Images:** Ullstein Bild (tc). **156-157 Dreamstime.com:** Allexxandar (b). **157 Alamy Stock Photo:** Pictures Now (tr, br); Stephen Power (tl). **158 Alamy Stock Photo:** Bill Crump (fbl); Antony Nettle (bl); Andrew Harker (c). **Dorling Kindersley:** Model Exhibition, Telford / Gary Ombler (clb); Gary Ombler (crb); Planes of Fame Air Museum, Chino, California / Peter Cook (tr). **Shutterstock.com:** BlueBarronPhoto (br); Angel DiBilio (cra). **159 Alamy Stock Photo:** Album (br); Flight / Anthony Kay (tc). **Dorling Kindersley:** Gatwick Aviation Museum / Gary Ombler (clb, crb/Junkers, bl/X2); Avro Lancaster (tr); Scale Model World, Steve Abbey / Gary Ombler (ca). **Dreamstime.com:** Kevin M. Mccarthy (cb). **Shutterstock.com:** BlueBarronPhoto (crb/X2); Arkady Zakharov (clb/X2); Chris Smithe (cra). **160-161 TurboSquid:** file404 (c). **160 Alamy Stock Photo:** Lordprice Collection (cr). **161 Alamy Stock Photo:** Avpics (br). **Getty Images:** SSPL / Daily Herald Archive (cra). **162 Alamy Stock Photo:** Haiyin Wang (cb). **163 Alamy Stock Photo:** Sueddeutsche Zeitung Photo (cl). **Getty Images:** Corbis Historical / Hulton Deutsch (cla); John Keeble (bl). **Shutterstock.com:** Sam-Whitfield1 (crb). **164-165 Alamy Stock Photo:** Abaca Press. **166 Alamy Stock Photo:** Gerard van Bree (cra); Niels Quist (crb); Gary Moseley (bl). **Dreamstime.com:** Robert Buchel (tc); Simon Greig (crb/Airbus). **167 Alamy Stock Photo:** Chronicle (br/X2); Zoonar / Markus Mainka (bl). **Dreamstime.com:** Robert Buchel (cra/X2); Ventura69 (tc/X2); Michael Clarke (c/X2). **168-169 TurboSquid:** mach 3 graphics (c). **170-171 Alamy Stock Photo:** Alex Federowicz. **172 Alamy Stock Photo:** Cultura Creative RF / Carol Kohen (bl). **174 Alamy Stock Photo:** SuperStock / Associated Press Photo / Sydney Morning Herald (cra). **175 Alamy Stock Photo:** Vladimir Glinskii (cl); Peter Lane (clb); Peter Michael Rhodes (bl). **176 Alamy Stock Photo:** David Gee (cra/X2); Stocktrek Images / Artem Alexandrovich (br/X2). **Dreamstime.com:** Ryan Fletcher (bl/X2). **Shutterstock.com:** Simon_g (tl/X2). **177 Alamy Stock Photo:** Blakeley (cla/US Huey helicopter); Stocktrek Images / Andrew Chittock (tr); PhotoStock-Israel / Amos Dor (cl/X2); Tami Freed (bl/X2). **Dorling Kindersley:** Gatwick Aviation Museum / Gary Ombler (cla). **178-179 TurboSquid:** CGShape (bl); ES3DStudios (cl); rfarencibia (cl/helmet). **178 Alamy Stock Photo:** PJF Military Collection (tl). **179 Alamy Stock Photo:** Stocktrek Images, Inc. / Terry Moore (cla); Nir Ben-Yosef (cr). **Getty Images:** 500Px Unreleased Plus / Mike Potter (bc). **180-181 Getty Images:** Stocktrek Images. **180 Alamy Stock Photo:** AV8Photo (bl). **Dorling Kindersley:** Royal International Air Tattoo 2011 / Gary Ombler (crb). **Dreamstime.com:** Serge Goujon (clb). **Shutterstock.com:** Eliyahu Yosef Parypa (br). **181 Alamy Stock Photo:** PJF Military Collection (b). **Dorling Kindersley:** Royal International Air Tattoo 2011 / Gary Ombler (cra). **Dreamstime.com:** Andrewharker402 (ca); Brett Critchley (crb). **Getty Images:** Education Images (fcrb). **Shutterstock.com:** VanderWolf Images (tr, clb). **182-183 TurboSquid:** f_san_wu (c/x3). **182 Alamy Stock Photo:** Photo12 / Collection Bernard Crochet (cla). **Dorling Kindersley:** Royal Airforce Museum, London / Gary Ombler (cb). **184 Alamy Stock Photo:** CPC Collection (br/X2). **185 Alamy Stock Photo:** Associated Press / Rich Pedroncelli (bl/X2); Everett Collection Inc (cla); DPA Picture Alliance (br/X2). **Getty Images:** Bloomberg (c/X2). **186 Dreamstime.com:** (cl). **187 NASA:** (cb). **Science Photo Library:** Carlos Clarivan (cr). **Shutterstock.com:** Marc Ward (br). **188 Alamy Stock Photo:** RGB Ventures / SuperStock / NASA (c/X2); Bill Waterson (br). **Anatoly Zak:** Anatoly Zak / RussianSpaceWeb.com (bl/X2). **Dorling Kindersley:** RAF Museum, Cosford / Gary Ombler (tr/X2). **Dreamstime.com:** Konstantin Shaklein (bc/X2). **Getty Images / iStock:** Nerthuz (cra). **189 NASA:** Joel Kowsky (r/X2). **190-191 NASA:** Danny Nowlin. **192 Alamy Stock Photo:** Danita Delimont / Kevin Oke (cl/X2); Sueddeutsche Zeitung Photo / Weltbild Lppert (bc/X2). **Dreamstime.com:** Mechanik (bl/X2); Konstantin Shaklein (br/X2). **Shutterstock.com:** Mechanik (t/X2). **193 NASA:** (bc/x2, br/x2, c/x2). **Science Photo Library:** Carlos Clarivan (tc/X2); Detlev Van Ravenswaay (cl/X2). **194-195 TurboSquid:** 3d_molier International (c/Telescope model); AlbinMERLE (c). **194 Alamy Stock Photo:** SuperStock / RGB Ventures / Ben Cooper (br). **195 NASA:** (br). **196 Dreamstime.com:** Aleks49 (br, crb). **Getty Images:** Science & Society Picture Library (crb/soft-lander, cra/soft-lander). **Science Photo Library:** Giphotostock (c, cb); Millard H. Sharp (cra/X2). **197 ESA:** Solar Orbiter: ESA / ATG medialab; Parker Solar Probe: NASA / Johns Hopkins APL (crb/x2). **Getty Images:** Stocktrek Images (tr/X2). **NASA:** JPL-Caltech (br/x2); JPL (tc/x2, clb/X2); McREL (c/x2); JPL-Caltech / MSSS (bc/x2). **Science Photo Library:** Detlev Van Ravenswaay (tl, cla). **198-199 CGTrader:** albin (c, c/astronauts); gsanimation (c/interior); brianzero (c/Cargo Capsule); vfxcgartist (c/background). **Getty Images / iStock:** AFransen (b). **198 Alamy Stock Photo:** Geopix (clb); UPI (cb). **199 Alamy Stock Photo:** Geopix (br). **200 Dreamstime.com:** Lightwavemodels (bl). **Getty Images / iStock:** AFransen (cb). **201 Dreamstime.com:** Ryan Fletcher (t). **203 Alamy Stock Photo:** Bill Waterson (r)
All other images © Dorling Kindersley

WHAT WILL YOU DISCOVER NEXT?

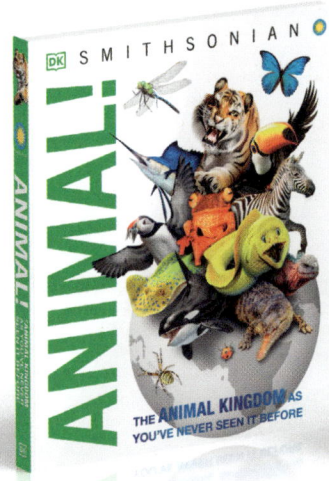

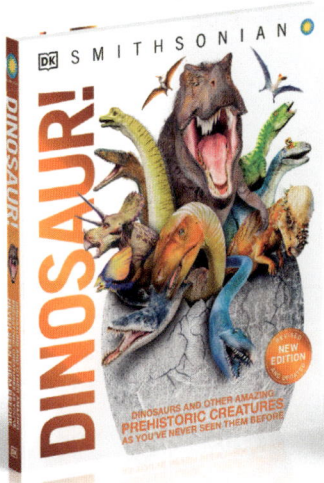

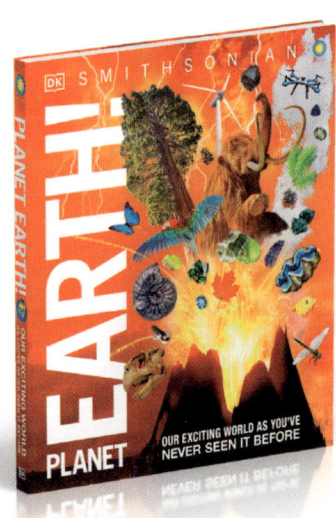

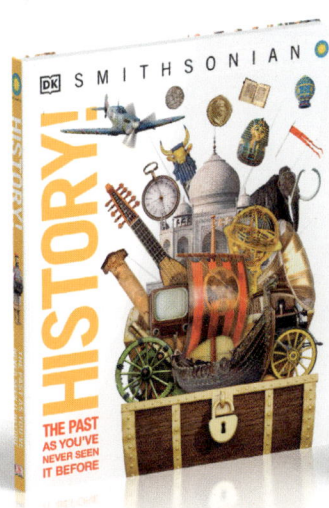

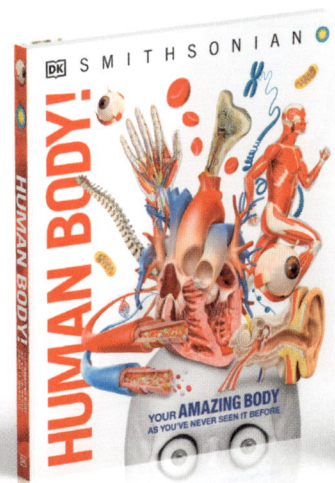

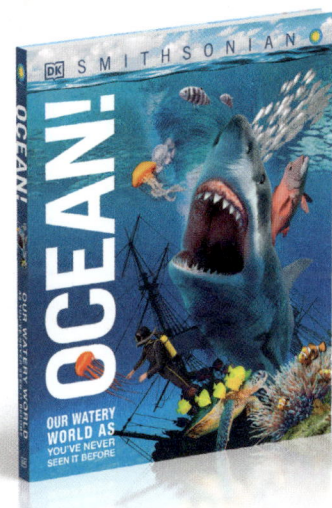

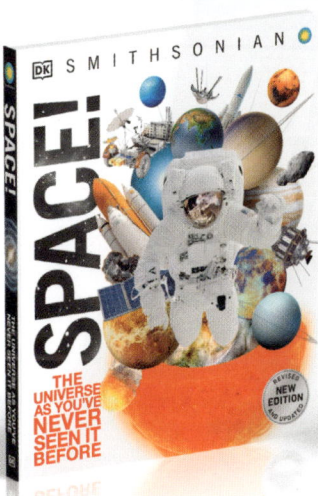

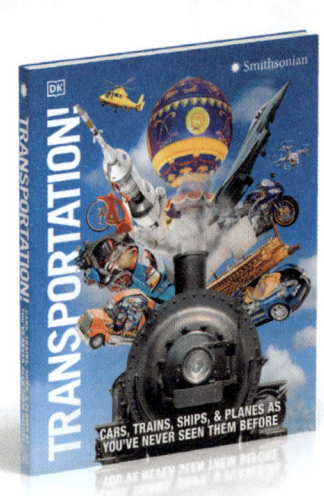